# Drug Targeting
# Technology

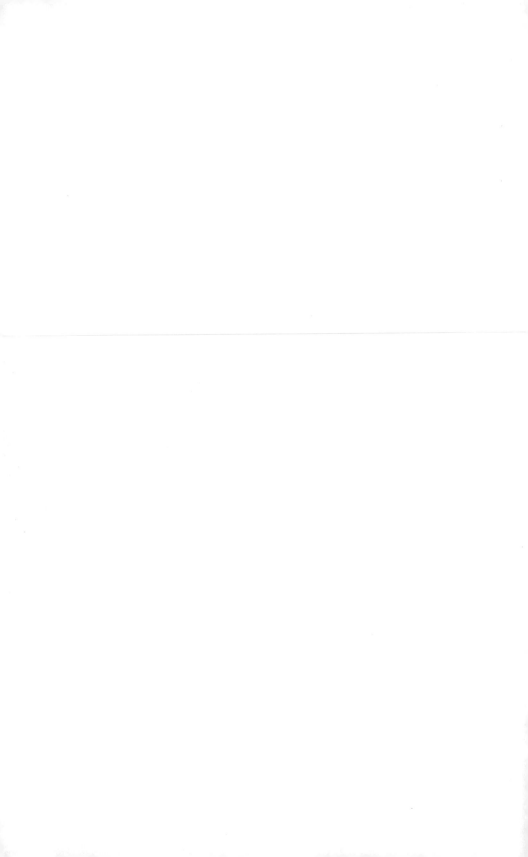

# DRUGS AND THE PHARMACEUTICAL SCIENCES

## A Series of Textbooks and Monographs

# Drug Targeting Technology

## Physical · Chemical · Biological Methods

edited by

## Hans Schreier
*H. Schreier Consulting*
*Langley, Washington*

MARCEL DEKKER, INC.

NEW YORK · BASEL

ISBN: 0-8247-0580-7

*Cover photo:* Liposomes attached to the surface of an alveolar macrophage, some in the process of being engulfed by the macrophage. (Scanning electron micrograph by G. W. Erdos, Electron Microscopy Core Lab, University of Florida, Gainesville, FL. Reprinted from H. Schreier, R. J. Gonzalez-Rothi, A. A. Stecenko, Pulmonary delivery of liposomes, *Journal of Controlled Release* 24:209–223 (1993), with permission from Elsevier Science.)

This book is printed on acid-free paper.

**Headquarters**
Marcel Dekker, Inc.
270 Madison Avenue, New York, NY 10016
tel: 212-696-9000; fax: 212-685-4540

**Eastern Hemisphere Distribution**
Marcel Dekker AG
Hutgasse 4, Postfach 812, CH-4001 Basel, Switzerland
tel: 41-61-261-8482; fax: 41-61-261-8896

**World Wide Web**
http://www.dekker.com

The publisher offers discounts on this book when ordered in bulk quantities. For more information, write to Special Sales/Professional Marketing at the headquarters address above.

Current printing (last digit):
10 9 8 7 6 5 4 3 2 1

**PRINTED IN THE UNITED STATES OF AMERICA**

# Preface

A book on the subject of drug targeting could not possibly omit Paul Ehrlich's theory of the magic bullet. It was on the occasion of attending *Der Freischütz*, a romatic opera by the German composer Carl Maria von Weber, that he coined his famous phrase. As the story goes, the young hunter apprentice (to ensure success in the imminent shooting competition that would award him the degree of master shooter) draws up a pact with the Devil, who promises three "free shots" (whence *Der Freischütz*)—shots that will unfailingly hit the target—in exchange for the apprentice's soul, naturally. The young hunter's bullet hits a white dove flying above. Alas, the mortally wounded dove is his beloved Agatha in disguise, whom he was to marry the same afternoon. Since this is a romantic opera, all is well in the end. In the real world, however, Paul Ehrlich's goal of finding the magic bullet to cure disease has been rather elusive indeed.

Gauging success of drug targeting depends to a great degree on how one defines "targeting": 1) in the sense of manipulating the release and uptake of a drug in specific body compartments in order to accomplish prolonged action, or protection of a drug or body compartments from unwanted side effects; 2) in the sense of selective delivery of a drug to a specific tissue, organ, or cell population; or 3) just as challenging, in the sense of avoiding a specific tissue or organ that would be particularly affected by cytotoxic side effects.

Success in selective targeting, to or away from organs or cells, has been scarce to date, although extensive and ingenious efforts are being made, as discussed further in the text that follows. However, we have been very successful in targeting oral tablets, capsules, granules, and other drug forms to particular segments of the intestinal tract by coating them with pH-sensitive polymers that dissolve only within a specific pH range. Enteric coating is now a classic means of protecting sensitive agents during gastric transit, as well as a guard for stomach and intestinal lining against irritating effects of the agent delivered. In addition to the gastrointestinal tract, the colon, with its low concentration of peptidases, has become a focus of particular interest for targeted delivery of peptides and proteins. The first two chapters (Chapter 1 by Brögmann and Beckert and Chapter 2 by Bauer) bring the reader up to date on recent developments in oral and colonic targeting technology using pH- and enzyme-sensitive coating materials and novel polymer systems, respectively.

Similarly, topical targeting techniques (including advanced aerosol systems and dermal or transdermal formulations) have been introduced or are about to be introduced that provide prolonged localized delivery or alternative means of systemic delivery. Both pose problems of a different kind (and perhaps magnitude), depending on whether the lungs or the skin, respectively, are the site of action or the portal of entry for systemic delivery.

Pulmonary drug delivery has experienced a renaissance recently with much improved mechanical delivery equipment becoming available. At the same time, it has been shown unequivocally that the lung can serve as portal of entry for a great variety of drugs that, in many instances, may eventually replace parenteral forms of administration. Yet manipulating pulmonary pharmacokinetics has been challenging and not always as straightforward as we may have liked to believe. For instance, it was thought intuitively that liposomes would be an excellent matrix to prolong pulmonary activity of corticosteroids when it was found that in practice the liposomal membrane was not rate-limiting, hence did not prevent rapid systemic distribution, and thus failed to be a suitable lung-targeted carrier for lipophilic corticosteroids. The concept is valid, however, for water-soluble derivatives or other water-soluble compounds that are encapsulated within the interior aqueous compartment of the carrier. Mobley and Hochhaus (Chapter 3) discuss the pharmacokinetics of drugs and how they determine their usefulness of pulmonary delivery and also show which drugs benefit from delivery as a liposomal delivery system and which ones do not.

A number of transdermal patches, such as nitroglycerin, nicotine, and estrogen patches, have to a degree replaced parenteral injections. However, it is again rather challenging to target drugs to selective skin layers without instantaneous removal by the bloodstream. Flexible polymer and liposome systems are now being designed that address this problem, although no such product has been introduced into clinical practice yet. Bouwstra and colleagues (Chapter 5) outline the basic parameters required to formulate successful dermal and transdermal delivery systems.

In the wake of the commercial success of several liposome drug products, lipid-based carriers are on the verge of becoming an accepted and versatile tool in drug delivery. Porter and Charman (Chapter 4) add an entirely new dimension to the usefulness of lipid-based carriers by showing that the systemic distribution of drugs can be modulated by carriers that use alternative routes of distribution, particularly lymph flow. This is a novel and truly innovative use of lipid systems to gain control over the distribution and accumulation of active agents in vivo and brings us closer to the ultimate goal of systemic organ targeting.

Chemical, and more recently biological, pathways to drug targeting and intelligent design approaches utilize physiological processes (e.g., intracellular enzymatic pathways or ligand-receptor interaction) or pathophysiological processes (such as the infectious pathway of viruses) to reach their site of action in a selective fashion. Chemical targeting has now matured to the point that products have passed clinical tests and are being introduced in the pharmaceutical market, whereas targeting using viral or viromimetic approaches is still in its infancy.

Chemical drug targeting is based on the premise that drugs can be chemically modified such that they accumulate in specific areas due to enzymatic metabolic processes, or are rapidly metabolized to inactive and nontoxic metabolites upon reaching the desired target, thus being removed efficiently from the body without further harm to other body compartments. The latter is the principal mode of action of soft drugs. This is described in exhaustive detail in the chapter on soft drug design (Chapter 6) by Bodor and Buchwald.

An example of a transitional state between chemical and biological drug targeting design is the work of Beljaars and coworkers (Chapter 7). They demonstrate how substitution of a variety of ligands, from carbohydrates to peptides, can render albumin a versatile targeting tool to selectively accumulate drug in the various types of cell populations in the liver, from hepatocytes to Kupffer cells to hepatic stellate cells. This is of particular interest because particulate carriers have consistently failed to target any hepatic cell population other than macrophages (Kupffer cells).

Clearly in its infancy, yet of enormous potential, are biological pathways that employ biotechnologically modified viruses or, alternatively, synthetic viruses (constructs that mimic the mode of operation of viruses but are composed of synthetic chemicals). Drug delivery system designers have enthusiastically embraced biotechnology and molecular biology as a means to devise ''intelligent'' drug delivery devices. A logical means to exploit biological pathways is to harness viruses with therapeutic genes that the virus will dutifully express once it has arrived at the desired site of action.

Synthetic pathways that supposedly avoid the pitfalls of viral delivery systems (see below) have been explored. Both Kaneda and coworkers (Chapter 9) and Sorgi and coworkers (Chapter 8) have been successful in designing viruslike liposomal delivery systems that provide some of the advantages of viral carriers—in other words, cell surface recognition and fusion with target cells (or intracellular compartments, i.e., endosomes, respectively)—without the detrimental immune response that viral systems generate. The efficiency of these systems is still orders of magnitude less than that of viral carriers; however, cytotoxic side effects typical for viruses, particularly fulminant inflammatory responses, have not been observed. Since this has not been shown in the clinical setting, the clinical and commercial viability of such systems remains to be seen.

Viral targeting may well be the ultimate approach to harnessing biological processes for the delivery or expression of pharmacologically active agents. This has been shown with a variety of viruses and has, overall, been extremely successful because viruses naturally exhibit a high expression efficiency, leading to the expression of (potentially) therapeutic levels of proteins or hormones, or the expression of immune response–stimulating proteins or antigenic moieties at cell surfaces. Nettelbeck and Müller (Chapter 10) provide an example of how viral delivery systems can be employed to target cancer sites. However, a major obstacle in the exploitation of viruses as targeted drug carriers has been immune and inflammatory responses that have made this approach, as ultimately intelligent as it may appear, less likely to succeed as a ubiquitously applicable mode of delivering genetic information to the human body in a targeted fashion.

It is not the aim of the book to trace the history of drug targeting, although its three main topics—physical, chemical, and biological targeting—correspond roughly to the historic development of the field. Nor is it intended to be an exhaustive anthology of drug targeting techniques and strategies. I envisioned it rather as a "reader" for interested scientists, experts, and students who are open for lateral views beyond the boundaries of their own field of interest. Its intention is to highlight the diversity of approaches and invite the reader to match—be it as an intellectual matching game or a very real novel design strategy—existing delivery technologies with novel ideas and strategies of drug targeting. Although questions are frequently raised as to the feasibility of combining macroscopic carrier systems (for instance, tablets or capsules) or aerosol devices with targeted microscopic carriers such as liposomes or derivatized albumins, no serious developments beyond the boundaries of macrosystems on the one hand and microsystems on the other have materialized to date. It is the book's mission to make these boundaries more fluid and transparent.

Ehrlich's inspiration continues to be a major driving force behind the design and development of modern drug delivery systems. The precise targeting of a tissue, organ, or cell population will eventually create distinctly superior delivery systems that will result in better treatment outcome, better patient compliance, and greatly improved economics of drug treatment.

*Hans Schreier*
*hansschreier@attglobal.net*

# Contents

# Contributors

**Kurt H. Bauer**  Pharmaceutical Technology Department, Freiburg Materials Research Center, Freiburg, Germany

**Thomas E. Beckert**  Technical Customer Service Pharma Polymers, Röhm GmbH, Darmstadt, Germany

**Leonie Beljaars**  Pharmacokinetics and Drug Delivery, Groningen University Institute for Drug Exploration, Groningen, The Netherlands

**Nicholas Bodor**  Center for Drug Discovery, University of Florida, Gainesville, Florida

**Joke A. Bouwstra**  Leiden/Amsterdam Center for Drug Research, Leiden University, Leiden, The Netherlands

**Bianca Brögmann**  Pharmaceutical Development, Mundipharma GmbH, Limburg, Germany

**Peter Buchwald**  Center for Drug Discovery, University of Florida, Gainesville, Florida

**William N. Charman**  Victorian College of Pharmacy, Monash University, Parkville, Victoria, Australia

**Lucie Gagné**  School of Pharmacy, University of California, San Francisco, California

**Günther Hochhaus**  College of Pharmacy, University of Florida, Gainesville, Florida

**Yasufumi Kaneda**  Division of Gene Therapy Science, Graduate School of Medicine, Osaka University, Suita, Osaka, Japan

**Dirk K. F. Meijer**  Pharmacokinetics and Drug Delivery, Groningen University Institute for Drug Exploration, Groningen, The Netherlands

**Barbro N. Melgert**  Pharmacokinetics and Drug Delivery, Groningen University Institute for Drug Exploration, Groningen, The Netherlands

**Cary Mobley**  College of Pharmacy, Nova Southeastern University, Fort Lauderdale, Florida

**Grietje Molema**  Pharmacokinetics and Drug Delivery, Groningen University Institute for Drug Exploration, Groningen, The Netherlands

**Ryuichi Morishita**  Division of Gene Therapy Science, Graduate School of Medicine, Osaka University, Suita, Osaka, Japan

**Rolf Müller**  Institute of Molecular Biology and Tumor Research, Philipps University, Marburg, Germany

**Dirk M. Nettelbeck**  Institute of Molecular Biology and Tumor Research, Philipps University, Marburg, Germany

**Klaas Poelstra**  Pharmacokinetics and Drug Delivery, Groningen University Institute for Drug Exploration, Groningen, The Netherlands

**Christopher J. H. Porter**  Victorian College of Pharmacy, Monash University, Parkville, Victoria, Australia

**Yoshinaga Saeki**  Division of Gene Therapy Science, Graduate School of Medicine, Osaka University, Suita, Osaka, Japan

**Hans Schreier**  H. Schreier Consulting, Langley, Washington

**Frank L. Sorgi**  Research and Development, OPTIME Therapeutics, Petaluma, California

**M. Suhonen**  Leiden/Amsterdam Center for Drug Research, Leiden University, Leiden, The Netherlands

**B. A. I. van den Bergh**  Leiden/Amsterdam Center for Drug Research, Leiden University, Leiden, The Netherlands

# 1
# Enteric Targeting Through Enteric Coating

**Bianca Brögmann**
*Mundipharma GmbH, Limburg, Germany*

**Thomas E. Beckert**
*Röhm GmbH, Darmstadt, Germany*

## I. INTRODUCTION

The aim of this review is to provide an up-to-date technological and pharmacological assessment of an "old" but still widely used dosage form, the enteric-coated tablet, capsule, or granule.

Enteric-coated formulations are suitable vehicles to modify the release of active substances such that release at specific target areas within the gastrointestinal (GI) tract can be affected, although the effectiveness of this methodology has long been a point of discussion. Krämer et al. [1] investigated the use of enteric coatings of 261 pharmaceutical products (Figure 1). The intended use included taste (9.6%) and odor (1%) masking, drug stabilization (31%), protection against local irritation (38%), and release directed to defined segments in the digestive tract (51%).

A major aim of enteric coating is protection of drugs that are sensitive or unstable at acidic pH. This is particularly important for drugs such as enzymes and proteins, because these macromolecules are rapidly hydrolyzed and inactivated in acidic medium. Antibiotics, especially macrolide antibiotics like erythromycin, are also rapidly degraded by gastric juices. Others, such as acidic drugs like NSAID's (e.g., diclofenac, valproic acid, or acetylsalicylic acid) need to be enteric coated to prevent local irritation of the stomach mucosa.

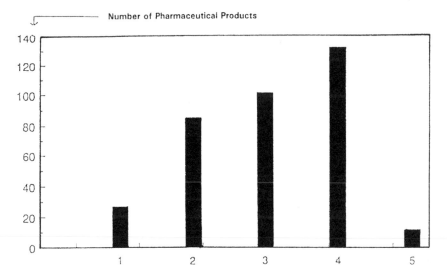

**Figure 1**  Functions of enteric coatings according to the statements of the pharmaceutical manufacturer. 1, Taste masking; 2, stability; 3, protection against local irritation; 4, drug release in specific parts; 5, odor masking.

Another purpose of enteric coating is drug targeting, as in the case of 5-aminosalicylic acid or the prodrugs mesalazine and sulfasalizine. In these cases, enteric coating is applied such that the drug concentration is increased in the lower parts of the GI tract.

Although the use of enteric coating to achieve modified release has been known for a long time, it has always been criticized as to its true value of providing protection and targeted release of the coated active agents [2]. The conclusion of this review is that, from a technical point of view, progress in film-forming polymers, together with advances in excipient technology and modern coating equipment design, has greatly facilitated the design of enteric-coated formulations that fulfill the requirements for controlled and targeted release.

## II. PHYSIOLOGICAL CONSIDERATIONS

### A. pH

By far the most important physicochemical and physiological parameter determining the functionality of enteric-coated drug delivery systems is the pH of the GI tract. Figure 2 [3] illustrates the variability of pH in the gastrointestinal

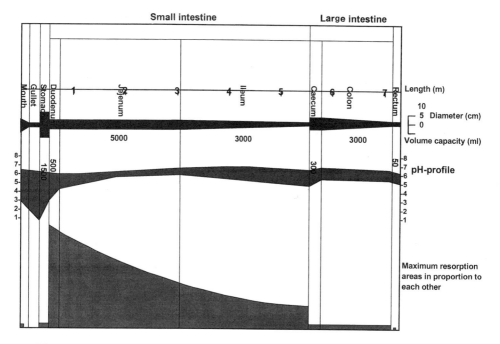

**Figure 2**   Schematic model of the gastrointestinal tract.

(GI) tract, also with reference to the fed and fasted state. The lowest pH is generally encountered in the stomach, whereas the highest pH is usually observed shortly before the beginning of the colon. The principle of enteric coating is based on this variance in pH in the GI tract: film coatings are composed of one or a mixture of polymers in such a fashion that the coat dissolves at a predetermined pH, hence the drug is released in a predetermined segment of the GI tract.

## B. Ionic State

Because most of the polymers that are used for enteric coatings are of anionic character, their dissolution is in some way dependent on the ionic character of the medium surrounding the dosage form. Frodtran and Locklear [4] were among the earliest to investigate the ionic concentrations in the GI tract of man in the fed and nonfed status. They found that the sodium content rises from the stomach toward the small intestine, starting from 0.04 mol/L and reaching blood level at 0.14 mol/L. For standard applications this will be of minor importance, because the dosage form is commonly assumed to release the drug in the small intestine, where the concentration of cations is high and supports the dissolution of the polymers. However, for specific applications (e.g., drug delivery within lower parts of the small intestine) this might be an important point to consider.

Spitael and Kinget [5] investigated the influence of ionic states on the dissolution of different enteric film coatings. They pointed out that buffers have to be carefully selected for dissolution studies and *in vitro–in vivo* correlation analysis. Ideally, their ionic state should be as close as possible to the *in vivo* conditions. As a minimum condition, the nature and concentration of cations should be matched as closely as possible.

## C. Food Effects

The intake of food or beverages leads to an increase in gastric pH. In addition, depending on the composition of the food, the hydrophilicity of the gastric medium may differ from that under fasted conditions. Food intake may also alter, mostly prolong, the GI transit time of dosage forms [6, 7]. However, other physiological parameters are also influenced. For instance, blood flow in stomach and liver is increased, protein binding may be increased, and ionic concentrations may be altered [8]. For some aqueous ethyl cellulose dosage forms, it was found that the film coating detached from the bead surface after simulation of a fatty meal [9], which was attributed to leaching of the plasticizer from the film into the lipophilic bulk environment.

It is, therefore, usually recommended that enteric-coated dosage forms be administered before the intake of a meal, although this may be in conflict with dosing based on pharmacodynamic reasons in the clinical situation. Also, owing to safety reasons, it does not seem feasible to select polymers that dissolve in an environment lower than pH 5.0, because the pH may rise above 5.0, and the intended protection of the gastric mucosa may no longer be guaranteed. Therefore, for some enteric-coated dosage forms it is recommended that they be taken after meals, when the gastric pH is within the operational range of the polymer coating.

## D. Gastric Emptying

Gastric emptying is an important parameter that needs to be taken into account in the design of enteric film–coated dosage forms. Standard enteric-coated tablets will usually maintain their integrity during passage through the stomach. These "large" particles will empty from the stomach with housekeeper waves only, whereas the gastric emptying of pellets with a size less than 1.4 mm and a density less than 2.4–2.6 g/cm$^3$ is almost as fast as liquid emptying, occurring within minutes of intake [10, 11] and independent of food. The transit of heavier and larger particles appears to be delayed, which can be attributed to the influence of peristalsis on gastric emptying and the GI [12]. Gamma scintigraphic studies have been used to investigate the GI transit of pellets. It was found, that densities up to 2.4 g/cm$^3$ did not alter GI transit of pellets of a size ranging from 1.2–1.4 mm compared with a control with 1.5 g/cm$^3$. However, densities of 2.8 g/cm$^3$ showed delayed transit. Thus, the critical density lies between 2.4 and 2.8 g/cm$^3$ [13, 14]. Additional investigations on pellet size (0.5 mm and 4.75 mm) showed only little effect on GI transit but a significantly prolonged transit in the small intestine [14]. Food, however, did not affect the emptying of the pellet formulation from the stomach [15]. Gastric emptying becomes of particular importance in patients in whom this function is impaired (e.g., in the elderly population or in persons restrained in resting position for a prolonged time).

The viscosity of the stomach contents should have an influence on gastric emptying. However, despite extensive studies, no correlation between viscosity and gastric motility could be established [16]. Although, it could be demonstrated that a high-viscosity hydroxypropylmethylcellulose (HPMC) reduced postprandial glucose concentrations by reducing the uptake rate of glucose [17]. This could also be an important factor for the time-to-action in the enteric targeting of drugs.

### E. Enzymatic Status, Proteolysis

Proteolysis of peptides and proteins commences in the stomach when pepsin is present. As a result, protein or peptide drugs will be hydrolyzed into smaller fragments like amino acids or oligopeptides, which are absorbed through the mucosa either by diffusion or by a carrier-mediated transport [18]. In an average individual, 94–98% of the total protein is completely digested and absorbed [19]. Proteolysis continues in the intestine with pancreatic enzymes like trypsine and brush-border enzymes.

Prevention of proteolysis will likely be one of the most important reasons for enteric coatings in the future, because more and more drugs are developed that are based on proteins (e.g., hormones, enzymes, antigens). Release of these drugs at the desired site of absorption in the intestine can be controlled by varying the thickness of the film coating. Also, combinations of enteric-coated protein drugs and enteric-coated protease-inhibiting molecules may be used to increase the absorption of protein drugs.

The use of enteric coatings will increase as the availability of protein drugs increases and these agents become less expensive. The interest in the oral administration of these proteins will increase despite the lingering problem of very low bioavailability (e.g., insulin).

### F. Age

In general, no significant differences were found between elderly and younger people in the adjustment of pH under fasted and fed conditions [20]. In about 10% of the elderly (aged 71 ± 5 years) gastric pH was found to be markedly elevated in the fasted state, reaching pH $>$ 5.0. Accordingly, the application of polymers with higher dissolution pH is recommended for dosage forms intended for the elderly.

### G.  Pathophysiological Changes

Inflammatory diseases of the GI tract like Morbus Crohn and ulcerative colitis greatly impair drug absorption. Depending on the condition, ulceration can be both in the upper or the lower intestine as well as in the colon. Changes in pH and ionic state may accompany the pathological condition, which further complicates dissolution of coating polymers and jeopardizes release or targeting, respectively, of the drug at the precise point of action. Few literature references exist describing the preceding with regard to the small intestine.

### III.  TECHNOLOGY: TYPES OF POLYMERS, PROCESSING, AND DISSOLUTION CHARACTERISTICS

### A.  Types of Polymers

Shellac is the oldest known material that has been used as enteric coating material. However, as a natural material, it lacks a crucial quality criterion of more modern polymers (i.e., batch-to-batch reproducibility). Hence, the most commonly used polymers today are the synthetic methacrylate copolymers or semisynthetic derivatives of cellulose. The main structural element of these polymers is an acidic function (either phthalate or methacrylic acid), which is responsible for the pH-dependent dissolution.

   A survey of the German market showed that more than 50% of enteric formulations were coated with methacrylate copolymers, about 40% with cellulose derivatives, 5% with shellac, and 3% with other materials [1]. Enteric coating materials (Table 1) are described in various publications [21, 22]. In addition to polymers mentioned in Table 1, others are being studied (e.g., to obtain release at lower pH) [23]. Polymers with a dissolution at lower pH are intended for the protection of drugs in acidic medium and not for the protection of the gastric mucosa.

### B.  Stability of Polymers

Stability of the polymers at usual storage conditions depends on their molecular structure. Several studies have documented that methacrylic acid copolymers are the most stable polymers as bulk and finished dosage form (Figure 3) [24]. The limiting factor of stability of the polymers is their ester function, which tends to hydrolyze. This happens more easily with cellulose phthalates than with methacrylic acid copolymers.

All polymers tend to lose carboxylic functional groups when they are subjected to ionizing or energetic radiation for sterilization [25].


## C.  Coating Processes

### 1.  Single/multiple Layer Coating

Enteric film coatings are usually applied as a simple single-layer film coat, although in some instances it may be necessary to apply multiple-layer film coatings.

Coating formulations usually contain the enteric polymer, a plasticizer if necessary, pigments, and sometimes also a glidant to prevent sticking of the film coatings during the manufacturing process. Colorants might be added to the enteric film coating formulations directly or be applied in a separate coating layer. By use of methacrylates for colored film coatings, a one-layer system can be easily applied because of the high pigment-binding capacity of the polymethacrylates.

The polymers are usually applied either from aqueous or from organic solvent. More recently, aqueous dispersions or redispersible powders of the polymers have become available that ensure economical, fast, and environmentally safe processing of the film coatings. In some cases it might be necessary to prevent an interaction between the acidic functional groups and the drug. In these cases a subcoat (e.g., of hydroxypropylmethylcellulose [HPMC]) is recommended.

With acid-sensitive drugs (e.g., proteins, bacteria, enzymes) the use of water instead of organic solvent is preferred. Partially neutralized aqueous dispersions are applied to form a spray-dried "subcoat." An important point in this process is to prevent the contact of the substrate or drug with the solvent (e.g., for the aqueous EUDRAGIT L 30 D-55 dispersion), originally with a pH around 2.6, it is possible to increase the pH to 5 by adding sodium hydroxide. With multiple unit dosage forms, it becomes possible to prepare a mixture of pellets coated with polymers of different pH-dependent release profiles (i.e., EUDRAGIT L 30 D-55 dissolves at pH 5.5, EUDRAGIT L 100 at pH 6, and EUDRAGIT S 100 at pH 7). Mixing pellets, which are coated with these three different polymers, an enteric dosage form with a specific sustained-release profile is achieved. Drug release will occur at different locations in the small intestine as a function of both pH and the thickness of the polymer coat. It has been reported that the application of 5 mg/cm$^2$ of a methacrylate copolymer, which dissolves at pH 5.5, prevented drug release in fasted male subjects until the formulation had reached the colon [26].

**Table 1**  Properties and Applications of Enteric Coating Materials (From Literature 1)

| Chemical name abbreviation | Functional groups | Soluble above pH | Trade name (company) | Application form | Remarks |
|---|---|---|---|---|---|
| Cellulose acetate phthalate CAP USP 23/NF 18 | Acetyl, phthalyl | 6 | CAP (Eastman Comp.) Aquateric (Lehmann & Voss) | Organic solution Aqueous dispersion (pseudolatices) | Sensitive to hydrolysis, 5–30% plasticizer required Micronized powder (0.05–3 μm) |
| Hydroxypropyl methyl cellulose phthalate HPMCP USP 23/NF 18 | Type 200731 Methoxy, hydroxypropoxy, phthalyl Type 220824 Methoxy, hydroxypropoxy, phthalate | 5 | HP 50, HP 55 (Syntapharm) HP 50 F, HP 55 F (Syntapharm) | Organic solution Aqueous dispersion (pseudolatices) | Less sensitive to hydrolysis, plasticizer not essential Powder <20 μm, redispersible in water |
| Hydroxypropyl methylcellulose acetate succinate HPMCAS | Methoxy, hydroxypropoxy, acetyl, succinyl | 5 | HPMCAS-L HPMCAS-M HMPCAS-H (Syntapharm) | Aqueous dispersion | Powder <5 μm Elastic properties, plasticizer not essential Slightly hygroscopic |
| Carboxymethyl ethylcellulose CMEC (Standard of Pharmaceutical Ingredients, Japan) | Carboxymethyl, ethoxy | 5 | Duodcell OQ Duodcell OQ (Lehmann & Voss) | Organic solution, aqueous dispersion | Not micronized Micronized Stable, not sensitive to moisture |

| Polymer | Composition | pH | Trade name | Physical form | Comment |
|---|---|---|---|---|---|
| Methacrylic acid-methyl methacrylate copolymers USP 23/NF 18 | Type A: methacrylic acid | 6 | EUDRAGIT L 100 | Organic solution, dispersion in water | Solvent isopropanol with partial neutralization |
|  | Type B: methacrylic acid | 7 | EUDRAGIT S 100 | Organic solution, dispersion in water | Solution isopropanol with partial neutralization |
| Methacrylic acid ethyl acrylate copolymer USP 23/NF 18 | Type C: methacrylic acid | 5.5 | EUDRAGIT L 30 D-55 Kollicoat MAE 30 DP | Aqueous dispersion | 30% dry polymer with partial neutralization |
|  |  |  | EUDRAGIT L 100-55 | Dispersion in water, organic solution | Solvent isopropanol |
| Polyvinylacetate phthalate PVAP USP 23/NF 18 | Acetyl, phthalate | 5 | Opadry enteric Opadry aqueous Enteric (Colorcon) Coating CE 5142 (BASF AG) | Organic solution Aqueous dispersion | Plasticizer required |
| Crotonic acid vinyl acetate copolymers | Vinylacetate: crotonic acid ratio 90:10 |  |  | Aqueous solution of the salt | Aqueous neutralized solutions are recommended |
| Methacrylic acid-methylacrylate-methylmethacrylate Copolymer | Methacrylic acid | 7 | EUDRAGIT FS 30 D | Aqueous dispersion | Requires only low amounts of plasticizer; flexible film former |

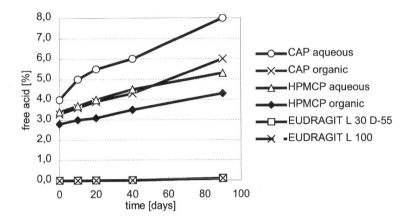

**Figure 3** Hydrolysis of different film coatings after application or storage, indicating the stability of these polymers. These data were obtained under storage at 20°C/100% r.h. (Source: Ref. 24 and Ref. 72).

An interesting example for multiple component coating is the following formulation. To achieve stable enteric lansoprazole granules, the enteric coating polymer EUDRAGIT L 30 D-55 and magnesium carbonate as alkaline stabilizer were added to lansoprazole [27]. To reduce friability of the granules, hydroxypropyl cellulose (HPC) was added, which did not affect the dissolution profile. Granules were manufactured by an extrusion method and a centrifugal fluid-bed granulator, the latter of which generated more stable enteric granules.

## 2. Organic and Aqueous Coating

Organic coating was the first choice in earlier times. Film formation takes place when the solvent evaporates. However, because of increasing environmental concerns and increasing regulatory restrictions of residual solvent content in the final formulation, coating material suppliers have been highly motivated to develop aqueous systems.

Aqueous systems are either aqueous latex dispersions or pseudolattices. The viscosity of organic polymer solutions is generally much higher than that of aqueous dispersions. Therefore, the concentrations of organic polymer solutions for processing are usually about 10%, whereas the concentration of an aqueous dispersion for processing is about 20%. Aqueous formulations, therefore, can provide better and faster process conditions, thus reducing the coating time and production cost. Aqueous systems of the most common suppliers have been reviewed by McGinity [28]. Comparisons of coats generated from organic vs. aqueous solutions indicate that the functionality of the polymer is retained, regardless of the medium [29, 30].

## 3. Dry Coating

For hydroxypropyl methylcellulose acetate succinate, a novel enteric "dry coating" method has been developed [31]. The unique feature of this method is that the enteric polymer is added in powder form (e.g., mixed with talcum directly to tablets or pellets) whereas a plasticizer diluted with paraffin is sprayed separately. The tablet core temperature is around 40°C, and the film is cured for a short time. To achieve a homogenous film, the rates of powder feeding and plasticizer spray have to be adjusted such that the two processes start and end simultaneously.

Modification of the coating process will also influence drug release, because the porosity of the film coating may be influenced by the spray rate or the inlet air temperature. This dry coating process provides short processing times and is thus more economical.

## D. Polymer Dissolution

The dissolution of polymers, regardless of whether they are cellulose based, methacrylates, or other, depends on a variety of factors that may also influence the release of the drug. These are discussed in detail following. Some of these factors are important *in vivo*, whereas others play a role *in vitro*.

## 1. Thickness

To achieve enteric protection of the core, at least 3–4 mg/cm$^2$ of polymer have to be applied (Figure 4). The precise amount of film coating depends on the type of polymer that is applied. Cellulose derivatives usually require higher amounts of polymer to obtain the same protection as methacrylic acid copolymers. A thin layer of 4 mg/cm$^2$ of methacrylate copolymer will dissolve within approximately 10 minutes. However, if increasing layers of polymer are applied, the dissolution time will be prolonged, which can be used to delay the dissolution of the drug in the small intestine (Figure 5).

The salt form of the polymer may also play a role in determining the performance of the formulation. Kané et al. [32] found that cellulose acetate phthalate was more effective than cellulose acetate trimellitate in controlling the dissolution of sulfothiazole-sodium tablets with cellulose acetate. The enteric properties of hydroxypropylmethylcellulose phthalate (HPMCP) were found to depend on the solubility of the drug that was coated.

00002142                              ——————— 100 µm

**Figure 4**   SEM of an aspirine crystal film coated with EUDRAGIT L 30 D-55.

## 2.  pH

Dissolution of polymers intended for enteric targeting depends on the pH of the dissolution medium [33, 34]. This is mainly influenced by the composition of the polymer, the monomers, or the type and degree of substitution. pH dissolution profiles can also be modified by the addition of other polymers, as demonstrated for EUDRAGIT L 100 and EUDRAGIT S 100 [35] (Figure 6). Such mixtures provide a variety of different pH dissolution profiles, which allows for specific targeting anywhere between the pylorus and the colon. This is also illustrated in Figure 2.

## 3.  Other Excipients

Other excipients used in film-coating may influence the dissolution of the polymers [36]. For instance, plasticizers may increase or decrease dissolution rate, depending on whether a lipophilic or a hydrophilic plasticizer was used. Using this effect, the time-to-action of a drug may be improved (e.g., by using a hydrophilic plasticizer like triethyl citrate). Usually these effects are not detectable, if

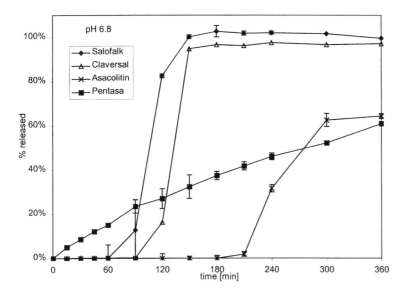

**Figure 5** *In vitro* dissolution profile of different dosage forms at pH 6.8, simulating the small intestine. Dosage forms that are film coated with an enteric polymer, EUDRAGIT L, (Salofalk, Claversal), show a delay in drug release at the dissolution pH of the polymer, which is probably due to the higher amounts of polymer that are applied (Rudolph et al., in press). Other film coatings: Pentasa (Ethylcellulose), Ascacolitin (EUDRAGIT S 100).

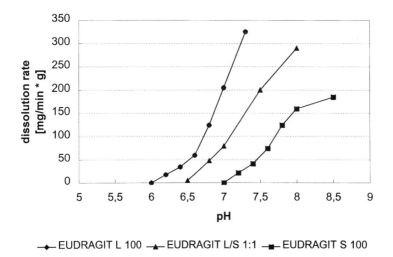

**Figure 6** Different dissolution rates from different methacrylate copolymers and mixtures of these polymers from their organic solutions (Röhm GmbH, technical information).

the USP 23 dissolution test [37] is used. However, *in vivo*, the dissolution times may vary between 10 minutes and 40 minutes, which might cause a considerable delay in the action of the drug. As a caveat, it is worth mentioning here that many of the pharmacopoeial dissolution tests for dosage form release do not take into consideration the *in vivo* conditions in some of the important aspects like the composition.

## 4. Ionic State

It could be shown [5] that the dissolution of polymers depends on the type of ions present in the dissolution medium. Dissolution is base catalyzed and can be described by the Brönstedt dissolution law [38]. At a given pH, a linear relationship exists between the logarithm of the dissolution rate and the $pK_a$ of the acidic component of the salt present in the dissolution medium. Cellulose acetate phosphate, especially, showed a strong dependency of the dissolution rate on the type of ions added. Sodium chloride prevented the dissolution of some polymers, because the base catalysis was at a minimum level.

## E. Dosage Forms

In general, film-coated dosage forms can be divided into multiple-unit and single-unit dosage forms. Single units comprise tablets and film-coated capsules or other forms, usually monolithic structures. Multiple-unit dosage forms can be packages containing granules, capsules containing pellets, or compressed film-coated particles. In the latter situation, total dosage is divided into multiple units that are dispersed in the GI tract, which often results in safer and usually faster action of the drug. Recently, it has also been reported that aqueous dispersions or suspensions can be produced, in which the drug is present in enteric-coated form.

    The enteric-coated Time Clock System consists of a tablet core coated with a mixture of hydrophobic material and surfactant, which is applied as an aqueous dispersion [39]. The drug release from the core of the Time Clock system occurs after a predetermined lag time. This lag time mainly depends on the thickness of the hydrophobic layer and thus is independent of GI pH. Investigations that used scintigraphic studies demonstrated that the method for *in vitro* testing was a good predictor of *in vivo* release. A greater targeting specificity can be achieved when an enteric coat is additionally applied to this system to avoid problems caused by longer gastric resistance time.

## 1. Tablets

Tablets can easily be enteric coated, and a variety of products are available on the market, including drugs like acetylsalicylic acid, diclofenac, naproxen [40], omeprazole, lansoprazole [27], sodium valproate, and many others. Generally, increased bioavailability, improved patient acceptance, and formulation stability result from the coating process.

Lehmann investigated the increased stability of acetylsalicylic acid tablets when using enteric film coatings [41]. Reduction of side effects and increase in patient compliance of enteric-coated acetylsalicylic acid tablets has also been shown in various clinical studies [42, 43]. In another study [44] different enteric film coatings on pancreatic enzymes were compared. It was found that products containing HPMCP adhered to the gastric mucosa, causing unwanted side effects, including irritation and inflammation of the gastric wall, whereas methacrylic acid copolymers and CAP encountered no such problems. The residence time of the tablets in fed dogs was found to be 6–8 h, which is undesirably long and requires a revised dosage regimen of the tablet (fasted or preprandial).

## 2. Capsules

Capsule coating often requires extra precautions (e.g., increased plasticizer content or sometimes an insulating layer), otherwise film coatings or capsule shells may become brittle during storage. Usually the thickness of the film coating layer has to be increased to ensure proper coating of the capsule closure. Vilivalam and Baugher [45] demonstrated enteric film coatings with methacrylic copolymers on starch capsules filled with 5-ASA resulted in good storage stability. Good stability was also reported for the enteric coating of hard gelatin capsules containing acetaminophen [46]. Cellulose acetate phthalate was used for an enteric coating on hard gelatin capsules filled with aspirin crystals [47]. Water uptake into the capsule was found to be unacceptably high, which was attributed to high water vapor permeabilities of cellulose film coatings compared with the more "dense" methacrylate copolymers (Table 2). Soft gelatin capsules were also coated with transparent film coatings, and good stability on storage was observed [48].

## 3. Multiple Units

A widely used method to produce multiple-unit dosage forms has been the production of sachets that contain film-coated granules. More common is the use of capsules in which enteric-coated particles are filled. A study that used radioactive tracers revealed that enteric-coated erythromycin pellets in capsules were superior to enteric-coated tablets with respect to faster action of the drug caused by a shorter passage time of the coated granules in the stomach [49–51].

In 1998 the first tablet containing enteric-coated particles was marketed (Losec Mups, Omeprazole-Magnesium by ASTRA, Sweden). This is a new prin-

**Table 2** Overview About Different Permeabilities of Different Film Coatings (Water Vapor Permeability) [73]

| Film types | Permeability coefficient $P(H_2O)$ $10^{-4} \times$ [g cm$^{-1}$d$^{-1}$bar$^{-1}$] |
|---|---|
| Pharmacoat 603 from aqueous solution | 83.5 |
| Pharmacoat 603 from organic solution | 82.4 |
| CAP + 10% triacetin | 19.5 |
| HP-50 | 25.9 |
| HP-55 | 14.5 |
| EUDRAGIT L 100 + 10% triacetin | 12.3 |
| EUDRAGIT S 100 + 10% triacetin | 11.7 |
| EUDRAGIT L 30 D-55 + 10% triacetin | 9.1 |

ciple and may serve as a paradigm of how enteric dosage forms may be designed in the future. However, flexible polymers are required for this purpose, and a variety of other factors have to be considered [52, 53] (Table 3). In addition to flexibility of the film coating, suitable larger sized filler-binders and stable and strong pellet cores also have to be taken into account. Only the methacrylic acid copolymers seem to have suitable properties necessary to produce these dosage forms (Figure 7).

As another example, small microcapsules of ibuprofen were film coated with cellulose acetate phthalate and dispersed in water before administration [54]. Plasma levels were as expected and did not differ from those of a conventional enteric-coated tablet.

**Table 3** Elongation at Break of Different Film Coatings, Measured According to ISO 527. High Values of the Elongation at Break Are Indicators for Flexible Film Coatings. Flexible Enteric Films are Required for the Compression of Enteric-coated Particles

| Filmcoating type | Plasticizer (amount on dry polymer) | Elongation at break |
|---|---|---|
| EUDRAGIT L 30 D-55 | Triethyl citrate, 10% | <5% |
| EUDRAGIT L 30 D-55 | Propylene glycol, 10% | <5% |
| EUDRAGIT L 30 D-55 | Propylene glycol, 20% | 180% |
| EUDRAGIT L 30 D-55 | Triethyl citrate, 10% | 112% |
| EUDRAGIT NE 30 D (1:1) | | |
| EUDRAGIT NE 30 D | | 600% |

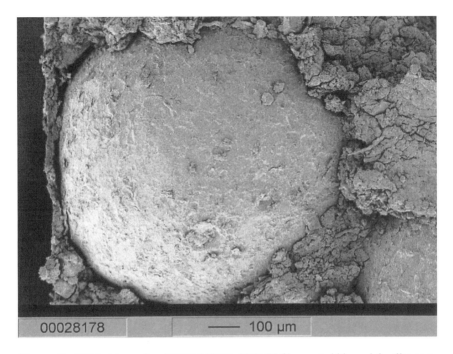

**Figure 7**  SEM picture of an EUDRAGIT L 30 D-55 film-coated bisacodyl pellet compressed into a tablet. The picture has been taken at the inside of a random broken tablet.

## IV.  *IN VIVO* EVALUATION

### A.  *In Vitro–In Vivo* Correlation

Conventionally, formulation development is based on *in vitro* test methods to prove the functionality of the enteric coating properties. According to the pharmacopeias' disintegration test methods, hydrochloric acid of different molarities or other dissolution methods are required. *In vitro–in vivo* correlation has been investigated for various drugs like *p*-aminosalycilic acid, riboflavin, thiamin, and niacinamide [55]. It was shown that in some cases physiological availability and other parameters derived from *in vivo* studies in man correlate well with *in vitro* disintegration time. However, when dissolution from agglomerates or granules is rate-limiting after disintegration, a better correlation between *in vivo* data and the *in vitro* rate of dissolution is to be expected.

The enteric protection properties of a methacrylate polymer were investigated with the use of naproxen tablets [56] enteric coated with an aqueous dispersion of methacrylic acid copolymer (EUDRAGIT L 30 D-55, USP NF type C).

The amount of polymer was 6 mg/cm² calculated to tablet surface. *In vitro* dissolution testing showed a drug release of 81% within 15 minutes in phosphate buffer (pH 6.8). *In vivo* tablet disintegration started after 38 min ± 15 min after leaving the stomach and was completed 10 ± 6 min later. The onset of drug release was observed by scintigraphy and correlated with the absorption of naproxen as assessed by simultaneous pharmacokinetic evaluation. It was confirmed that the enteric coating provided excellent gastric resistance and rapid drug release in the small intestine after gastric emptying. Increased layer thickness of the methacrylic acid copolymer mentioned earlier was evaluated for a colonic delivery system of nisin, using samarium oxide for gamma scintigraphy [57]. The proof for a colonic target was achieved by the scintigraphic studies, whereas *in vitro* drug dissolution according to USP/NF did not correlate to the *in vivo* results.

Because it has been shown in multiple situations that these methods do not entirely reflect *in vivo* conditions during passage of dosage form through the GI tract, various *in vivo* methods have been described to evaluate enteric dosage forms.

## B.  In Vivo Testing

A radiotelemetric method has been described to test the disintegration property of a tablet [58]. A tablet containing a citrate buffer was coated with hydroxypropyl methylcellulose phthalate and then placed in a Heidelberg capsule that also contained a pH-electrode (Figure 8). An antenna is strapped around the midriff of the dog to detect the output from the Heidelberg capsule.

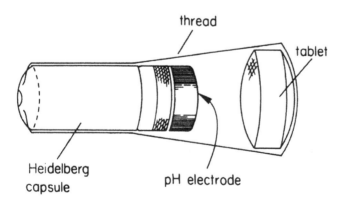

**Figure 8**  Schematic for attachment of the enteric-coated tablet to the Heidelberg pH detector capsule (From Ref. 58).

Thus, important parameters such as the baseline gastric pH, gastric empty-ing time, intestinal pH subsequent to gastric emptying, and time of onset of core disintegration can be determined by analyzing the pH time profile. If enteric properties are achieved, the tablet disintegration will start after the gastric empty-ing time, which is indicated by an increase in pH up to 6 or 7. After tablet disinte-gration, the pH decreases because of the buffer released from the tablet to around 4. This method requires less experimental expertise and less equipment than other techniques; however, up to now it is limited to the development of monolithic forms, and complete disintegration cannot be detected.

The relationship between gastric residence time (GRT) and the variability in aspirin absorption from enteric-coated tablets has also been evaluated with the Heidelberg capsule [59]. The study design was a single dose crossover study with men and women, fasted and fed. In addition to pH monitoring, plasma con-centrations of salicylic acid and salicyluric acid in plasma and urine were mea-sured. GRT was significantly delayed in the fed state, and women had longer GRT compared with men.

A new noninvasive method for assessing regional drug absorption from the GI tract has been described recently [60]. The InteliSite capsule (Figure 9) is available in two versions, for delivering either solutions and suspensions or pow-der formulations. The capsule consists of an outer sleeve composed of a chloro-fluorethene homopolymer and an inner sleeve made from polysulfone. Both contain drug release openings that expose the reservoir content when the inner sleeve is rotated through an electronic impulse and mechanical assembly such that the slots of the inner sleeve are aligned with the slots of the outer sleeve. A gamma-ray–emitting radio tracer is incorporated into the formulation and then recorded by a gamma camera showing the site of exposure in the body. The proof of concept for this system has been demonstrated with furosemide and theophylline.

In addition to the evaluation of enteric dosage forms with special systems like the Time Clock or others mentioned earlier, conventional pharmacological studies are commonly done with the final formulation to assess the desired activ-ity profile.

As previously discussed, food effects are an important parameter for enteric-coated systems, especially for drugs, that are sensitive to food. Pancreatic enzyme-containing products fail when they come in contact too early with lipids, proteins, and carbohydrates present in food. The clinical efficacy of pancreatic enzymes formulated as enteric-coated tablets was investigated in man and dog [44]. The enteric materials examined were hydroxypropyl methylcellulose phthal-ate (HPMCP), cellulose acetate phthalate (CAP), and the methacrylic acid copol-ymer USP/NF Type C. *In vivo* behavior monitored by x-ray scintigraphy showed clear differences between the three coating formulations. HPMCP-coated prod-ucts adhered to the gastric mucosa, whereas CAP and methacrylate copolymer

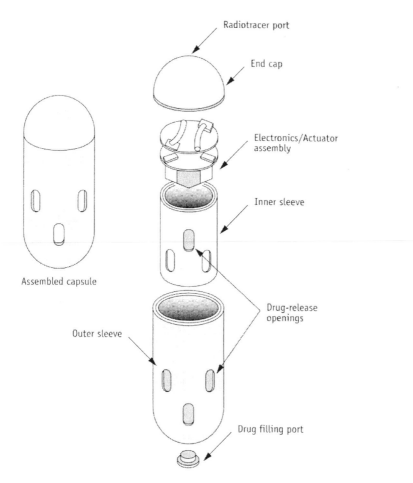

**Figure 9**   The components of the InteliSite capsule (From Ref. 60).

showed less adhesion. Marvola et al. [61] also reported that enteric film coatings made of CAP or methacrylate copolymer hindered the adhesion of tablets to the esophagus compared with HPMC film coatings. HPMCP, which dissolves above pH 5, already shows tablet disintegration in the stomach. In addition, it was found that there was a failure of tablets to pass through the pylorus in conjunction with food. The housekeeper waves transferred the enteric tablets to the duodenum in the fasting mode only. If formulations are to be used postprandially during this digestive mode, the size of nondisintegrating particles must be less than 2 mm.

In contrast to this investigation, in which pharmacological results from humans and dogs were comparable, a coating formulation with CAP showed no comparable results in humans, dogs, and pigs [62]. The results *in vitro*, however, were comparable to the results found *in vivo* in humans.

The relative bioavailability of different enteric diclofenac products was investigated with normal and artificially decreased gastric acidity [63]. Only one generic product was found to be fully bioequivalent. Comparison with *in vitro* studies concluded that the general test on enteric-coated tablets according to Ph.Eur. did not detect any difference between the four products. A modified dissolution test without mechanical stress gave an indication on differences in the lag time of the different products.

Naproxen sodium tablets were used in an *in vivo–in vitro* evaluation with four different polymeric dispersions, cellulose acetate phthalate (CAP), cellulose acetate trimellitate (CAT), 50:50 CAP/CAT, and methacrylic acid copolymer [64]. The study indicated that coating material that dissolves at a more acidic pH *in vitro* (such as CAT at pH 4.5) will also dissolve at a more acidic pH *in vivo* (i.e., the coating dissolves higher up in the GI tract). In addition, it was found that aging did not markedly affect dissolution characteristics of CAT or methacrylic acid copolymer–coated tablets.

The influence of a single dose of omeprazole on the pharmacokinetics of enteric-coated ketoprofen tablets was tested [65]. There was no significant difference with or without single-dose omeprazole administration for the systemic bioavailability of the ketoprofen products. A trend in higher plasma concentrations with omeprazole indicates a possibility of drug release from enteric-coated products at potentially elevated stomach pH values.

The pharmacological effects of oral sulfadoxine, administered as enteric-coated tablets, were investigated applying 100 mg and 200 mg/day as a single and repeated dosage form and comparing the activity on fibrinolytic and coagulation parameters [66]. Compared with the placebo, sulfadoxine in single and repeated oral administration of a gastroresistant formulation is well tolerated, devoid of anticoagulant activity and exhibits a profibrinolytic effect, whose trend is in line with the pharmacokinetic profile of the drug. In addition, the potency can be increased by repeated administration.

## V. APPLICATIONS

Aspirin, especially when used chronically in geriatric populations, is known for its side effects of GI discomfort and bleeding. Therefore, serum salicylate levels with regular and enteric-coated aspirin were compared in volunteers in a chronic disease hospital and residential home for the elderly [42]. It was found that there was no significant difference between the blood levels reached with conventional aspirin and enteric-coated aspirin. Because the side effects are mostly a direct effect of aspirin in the stomach on the gastric mucosa, enteric-coated aspirin is, therefore, favorable.

In another study, two different types of enteric-coated aspirin formulations, enteric-coated tablets, and granules were compared with conventional tablets [43]. Plasma levels and excretion of salicylic acid and some of its metabolites were investigated under steady-state conditions. It was demonstrated that the dosage form consisting of enteric-coated granules gives a more uniform plasma level during the studied 12-h time intervals and less inter- and intraindividual variations than enteric tablets. The urinary excretion of total salicylate showed no significant difference, the two enteric-coated formulations, however, provided significantly higher morning plasma concentrations than the conventional aspirin.

The pharmaceutical and biological availability of eight commercial furosemide preparations was compared including two products with modified release properties [67], an enteric-coated tablet and a sustained-release preparation, in the form of a capsule containing diffusion pellets [28]. A correlation between the rate of dissolution of different techniques and the area under the plasma concentration time curve was documented. The sustained-release preparation and the enteric-coated formulation clearly showed different pharmacokinetic behavior compared with conventional tablets. Although the literature mentions the maximal absorption at pH 5.5, the modified release formulations only showed a relative bioavailability of 80%.

Enteric-coated products with pancreatic enzymes have been marketed for a long time and are a well-known example for this application. In recent years, increasing numbers of peptide drugs are appearing on the market. Insulin is but one of the potential candidates for oral delivery of a peptide drug; however, this system would preferentially be targeted to the colon section, because the colon has been found to be a better absorption site for insulin. Ovalbumin and short ragweed antigens have recently been enteric-coated and applied successfully to desensitize allergic patients [68]. Without the film coating made of methacrylic acid copolymer, the antigens would be destroyed by hydrochloric acid in the stomach and not be available to trigger an immune reaction by way of Peyer's patches in the intestine. Targeting antigens to the small intestine will be an approach for the future to help allergic patients (Figures 10 and 11).

CGP 57813 is a peptidomimetic inhibitor of human HIV-1 protease. This lipophilic compound has been successfully entrapped in polylactic acid (PLA) and into pH-sensitive methacrylic acid copolymer particles (EUDRAGIT L 100-55) [69]. After the application of a film-coating, the plasma concentration was acceptable and reached similar levels as with injections of drug-loaded PLA carriers. To hinder the proteolytic degradation of a drug, two types of enteric-coated pellets were applied simultaneously. One contained the protease inhibitor coated

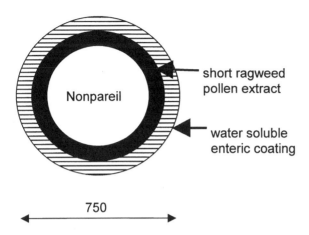

750

**Figure 10** Sample dosage form design for enteric-coated short ragweed pollen extract. (Source: Ref. 68.)

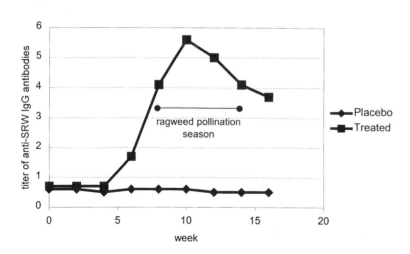

**Figure 11** Titer of anti-short ragweed IgG antibodies, treated and placebo. The amount of IgG was increased by administration of the enteric-coated dosage form. Without enteric film coating the antibodies would have been destroyed. (Source: Ref. 68.)

with methacrylic acid copolymer, which hindered the enzymatic degradation of the drug by inhibiting the responsible enzyme. The other pellet type contained the peptide drug, which was delivered successfully by way of Peyer's patches [21]. Summarizing these examples, two main fields of application can be recognized: (1) the protection of peptide drug against degradation (pancreatic enzymes, CGP 57813, antigens), and (2) products in (1) that partially contain a locally acting enzyme inhibitor or a penetration enhancer.

## VI.  OUTLOOK

On the way to multiple-unit dosage forms, a demand is seen for flexible polymers. These polymers can be summed up in a new class of polymers that have at least two main functions: They have the functionality for enteric targeting and they show physicochemical properties necessary to obtain the desired flexibility. A new market can be seen in this field for the manufacture of coating materials, because the functionality of these polymers should not be limited to enteric targeting but also be applicable to other fields of controlled release, like transdermal applications or oral sustained-release dosage forms.

The pH-dependent release was found to be the most advantageous option for the enteric targeting of drugs. The existing polymers cover most of the applications. However, some additional polymers may also be needed to improve the dissolution of a drug above a certain pH (e.g., pH 4.0) above which no more degradation of the drug takes place. Looking beyond enteric targeting, these polymers may also be of interest for the delivery of drugs in the vagina or for the delivery of drugs in the treatment of dermal inflammatory diseases. Furthermore, more specific applications for specific patient populations (e.g., elderly or pediatric use) will be developed.

To improve the safety of the dosage form and the mode of drug delivery, the time-to-action of a drug, new dosage forms will be developed. The development will lead to enteric multiunit dosage forms, as has been the case for many years with oral sustained-release dosage forms. Because the technological development to produce these multiple unit dosage forms has progressed over recent years, this has had an effect on the economic feasibility. As an additional application, enteric targeting or film coating will be used diagnostically for the determination of the absorption site of a drug [70]. This will allow effortless screening of sites of absorption.

Enteric targeting may also include other, new dosage forms, which remain located in the stomach and start releasing their drug from there. These gastric-retention systems consist of a swellable core, usually film coated with highly flexible polymers that disintegrate after a certain time of "floating" in the stomach. Deshpande et al. [71] have described such a system recently.

Enteric targeting will gain more importance as biotechnological products like proteins, hormones, and vaccines become available at reasonable cost and for a larger number of patients. Many of the proteins are sensitive to enzymes in the stomach but may be absorbed in the intestine. Thus, we foresee a merger of modern biotechnology with the traditional aspects of film coating in terms of drug targeting.

## REFERENCES

1. J. Krämer, H.-J. Pfefferle, H. Blume, Magensaftresistent überzogene Zubereitungen, *Encyclopedia of Pharmaceutical Technology*, Marcel Dekker, New York, *Volume 14*:355–384 (1996).
2. P. C. Schmidt, K. Teuber, Magensaftresistenz—gibt es die überhaupt?, *Pharmazie in unserer Zeit 20*:164–178 (1991).
3. K. Bauer, K. Fröming, C. Führer, *Lehrbuch der pharmazeutischen Technologie*, Verlag, 1999, p. 194.
4. J. Fordtran, T. Locklear, *Ionic Constituents and Osmolality of Gastric and Small-intestinal Fluids After Eating*, Harper and Row Publishers Inc., New York, *New Series:11(7)*: (1966).
5. J. Spitael, R. Kinget, Factors affecting the dissolution rate of enteric coatings, *Pharm Ind 39*:502 (1977).
6. S. S. Davis, J. G. Hardy, J. M. Taylor, D. R. Whalley, C. G. Wilson, The effect of food on the gastrointestinal transit of pellets and an osmotic device (Osmet), *Int J Pharm 21*:331–340 (1984).
7. J. N. Hung, D. F. Stubbs, The volume and energy content of meals as determinants of gastric emptying, *J Physiol 215*:209–225 (1975).
8. P. Mojaverian, M. L. Rocci, D. P. Conner, W. B. Abreuns, P. H. Vlasses, Effect of food on the absorption of enteric-coated aspirin, Correlation with gastric residence time, *Clin Pharmacol Ther 41(1)*:11–17 (1987).
9. H. H. Blume, B. S. Schug, H. Potthast, Influence of food on the bioavailability of controlled/modified release products, *Food-Drug Interactions* (J. Kuhlmann, T. R. Weihrauch, eds.), W. Zuckschwerdt Verlag, München, Germany, 1995, pp. 25–33.
10. R. O. Williams III, M. Sriwongjanya, J. Liu, An in vitro method to investigate food effects on drug release from film-coated beads, *Pharm Dev Technol 2*:1–9 (1997).
11. G. M. Clarke, J. M. Newton, Gastrointestinal transit of pellets of differing size and density, *Int J Pharm 100*:81–92 (1993).
12. J. H. Meyer, J. Elashoff, V. Porter-Fink, J. Dressman, G. L. Amidon, Human postprandial gastric emptying of 1–3 millimeter spheres, *Gastroenterology 94*:1315–1325 (1988).
13. M. Kamba, Y. Seta, N. Suzuki, A. Kusai, K. Nishimura, Evaluation of destructive force in human stomach and small intestine, *Proc Int Symp Control Rel Bioact Mater 23*:579–580 (1996).
14. G. M. Clarke, J. M. Newton, M. B. Short, Comparative gastrointestinal transit of pellet systems of varying density, *Int J Pharm 114*:1–11 (1995).

15. G. M. Clarke, J. M. Newton, M. B. Short, Gastrointestinal transit of pellets of differing size and density, *Int J Pharm* 100:81–92 (1993).

16. J. Coupe, S. S. Davis, D. F. Evans, I. R. Wilding, Do pellet formulations empty from stomach with food?, *Int J Pharm* 92:167–175 (1993).

17. P. Gupta, J. Robinson, Effect of meal viscosity on gastric pH and residual volume in the fasted dog, *Int J Pharm* 109:91–98 (1994).

18. N. Reppas, J. Dressman, Viscosity modulates blood glucose response to nutrient solutions in dogs, *Diabetes Research and Clinical Practice* 17:81–88 (1992).

19. T. L. Russell, R. R. Berardi, J. L. Barnett, L. C. Dermentzoglou, K. M. Jarvenpaa, S. P. Schmaltz, J. B. Dressmann, Upper gastrointestinal pH in seventy-nine healthy, elderly, North American men and women, *Pharm Res* 10:187 (1993).

20. K. Barnes, J. Ingram, A. J. Kenny, Proteins of the kidney microvillar membrane. Structural and immunochemical properties of rat endopeptidase-2 and its immunohistochemical location in tissues of rat and mouse, *Biochem J* 264:335–346 (1989).

21. P. Langguth, V. Bohner, J. Heizmann, H. P. Merkle, S. Wolffram, G. L. Amidon, S. Yamashita, The challenge of proteolytic enzymes in intestinal peptide delivery, *J Controlled Release* 46:39–57 (1997).

22. K. Lehmann, B. Brögmann, Tablet Coating, *Encyclopedia of Pharmaceutical Technology, Volume 14*:355–384 (1996).

23. H. Kokubo, S. Obara, K. Minemura, T. Tanaka, Development of cellulose derivatives as novel enteric coating agents soluble at pH 3.5–4.5 and higher, *Chem Pharm Bull* 45:1350–1353 (1997).

24. A. G. Eshra, Der Einfluß von sauer oder basisch reagierenden Modellsubstanzen auf die Stabilität von Filmumhüllungen beim Wechsel von organischen auf wäßrige Herstellungsmethoden, Dissertation, Freiburg 1982.

25. T. Waaler, S. A. Sande, B. W. Müller, G. Schüller-Lisether, The influence of thermal neutron irradiation on the in vitro characteristics of ASA oral dosage forms—Validation of neutron activation, *Eur J Pharm Biopharm* 43:159–164 (1997).

26. K. Steed, G. Hooper, P. Ventura, R. Musa, I. R. Wilding, The in vivo behaviour of a colonic delivery system: a pilot study in man, *Int J Pharm* 112:199–206 (1994).

27. T. Tabata, T. Makino, J. Kikuta, S. Hirai, N. Kitamori, Manufacturing method of stable enteric granules of a new antiulcer drug (lansoprazole), *Drug Dev Ind Pharm* 20(9):1661–1672 (1994).

28. J. McGinity, *Aqueous Polymeric Coatings for Pharmaceutical Dosage Forms*, Marcel Dekker, New York, 1996.

29. K. Lehmann, Acrylic latices from redispersible powders for peroral and transdermal drug formulations, *Drug Dev Ind Pharm* 12:265–287 (1986).

30. J. Plaizier-Vercammen, G. Suenens, Evaluation of Aquateric, a pseudolatex of cellulose acetate phthalate, for its enteric coating properties on tablets; *STP Pharma Sci* 5:307–312 (1991).

31. Shin-Etsu, *Technical Bulletin, No. A-3*, Synthapharm, Mühlheim-Ruhr, Germany, September 1996.

32. Y. Kané, J. Rambaud, H. Malliols, J. P. Laget, D. Gaudy, H. Delonca, Technological evaluation of three enteric coating polymers—II. With a soluble drug; *Drug Dev Ind Pharm* 20:1021–1034 (1994).

33. K. Lehmann, D. Dreher, Permeable acrylic resin coatings for the manufacture of depot preparations of drugs, *Drugs Made in Germany 11*:34 (1968).
34. P. Delporte, *Il Farmaco 30*:111 (1975).
35. K. Lehmann, Mixtures of aqueous polymethacrylate dispersions for drug coating, *Drugs Made in Germany 31*:101–102 (1988).
36. K. Bauer, K. Lehmann, H. Osterwald, G. Rothgang, *Coated Pharmaceutical Dosage Forms*, Medpharm Scientific Publishers, Stuttgart, 1998.
37. *USP 24*, Delayed Release (Enteric-Coated) Articles—General Drug Release Standard ⟨724⟩, U.S. Pharmacopoeial Convention Inc., Rockwell, MD, 1999.
38. M. L. Bender, L. J. Burbacher, *Catalysis and Enzyme Action*, McGraw Hill, New York, 1973, p. 52.
39. R. Wilding, S. S. Davis, F. Pozzi, P. Furlani, A. Gazzaniga, Enteric coated timed release for colonic targeting, *Int J Pharmaceutics 111*:99–102 (1994).
40. T. Levien, D. Baker, Reviews of naproxen enteric coated and varicella vaccine; *Hosp Pharm 30*:1011–1024 (1995).
41. K. Lehmann, Formulation of controlled release tablets with acrylic resins *Acta Pharm Fenn 93*:55–74 (1984).
42. J. J. Orozco-Alcala, J. Baum, Regular and enteric coated aspirin: a revolution, *Arthritis Rheum 22(9)*:1034–1037 (1979).
43. E. B. Bogentoft, P.-O. Lagerström, Comparison of two enteric-coated acetylsalicylic acid preparations by monitoring steady state levels of salicylic acid and its metabolites in plasma and urine, *Biopharm Drug Dispos 5*:251–260 (1984).
44. M. Marvola, J. Heinämaki, E. Westermarck, J. Hepponen, The fate of single-unit enteric-coated drug products in the stomach of the dog, *Acta Pharm Pennica 95*: 59–70 (1986).
45. V. Vilivalam, L. L. Illum, K. K. Iqbal, Starch capsules: an alternative system for oral drug delivery, *Pharm Sci Technol Today 3(2)*:64–69 (2000).
46. B. Brögmann, K. Lehmann, Stability of enteric gelatin and starch capsules with aqueous dispersions of methacrylic acid copolymers; AAPS Annual Congress, San Diego, 1994.
47. I. Cherretté, J. Plaizier-Vercammen, Evaluation of a water dispersion of aquateric for its enteric coating properties of hard gelatine capsules manufactured with a fluidized bed technique, *Pharm Acta Helv 67*:227–230 (1992).
48. L. A. Felton, M. M. Haase, N. H. Shah, G. Zhang, M. H. Infeld, A. W. Malick, J. W. McGinity, Physical and enteric properties of soft gelatin capsules coated with Eudragit L30 D-55, *Int J Pharm 113*:17–24 (1995).
49. G. Digenis, The in vivo behaviour of multiparticulate versus single unit dose formulations, Capsugel Symposium, Seoul and Tokyo, 1990.
50. H. Bechgaard, K. Ladefoged, Gastrointestinal transit time of single-unit tablets, *J Pharm Pharmacol 33*:791–792 (1981).
51. H. Bechgaard, Critical factors influencing gastrointestinal absorption—What is the role of pellets? *Acta Pharm Tech 28*:149–157 (1982).
52. T. Beckert, K. Lehmann, P. C. Schmidt, Compression of enteric-coated pellets to disintegrating tablets, *Int J Pharm 143*:13–23 (1996).
53. T. Beckert, K. Lehmann, P. C. Schmidt, Compression of enteric-coated pellets to disintegrating tablets: Uniformity of dosage units, *Powder Tech 96*:248–254 (1998).

54. K. Walter, G. Weiß, A. Laicher, F. Stanislaus, Pharmacokinetics of ibuprofen following single administration of a suspension containing enteric-coated microcapsules, *Drug Res 45*:886–890 (1995).

55. J. G. Wagner, Disintegration of dosage forms in vitro and in vivo Part IV, *Drug Intell Clin Pharm 3*:324–330 (1969).

56. K. Lehmann, B. Brögmann, C. J. Kenyon, J. R. Wilding, The in vivo behaviour of enteric naproxen tablets coated with an aqueous dispersion of methacrylic acid copolymer, *S.T.P. Pharma Sci 7(6)*:403–437 (1997).

57. W. A. Habib, A. Sakr, Effect of methacrylic acid copolymer USP/NF type C and the release of polypeptide drug nisin, AAPS, Annual Congress Boston, November 1997.

58. J. B. Dressman, G. L. Amidon, Radiotelemetric method for evaluating enteric coatings in vivo, *J Pharm Sci 73(7)*:935–938 (1984).

59. P. Mojaverian, M. L. Rocci, D. P. Conner, W. B. Abreuns, P. H. Vlasses, Effect of food on the absorption of enteric-coated aspirin: Correlation with gastric residence time, *Clin Pharmacol Ther 41(1)*:11–17 (1987).

60. D. Gardner, R. Casper, F. Leith, J. Wilding, Non-invasive methodology for assessing regional drug absorption from the gastrointestinal tract, *Pharm Techn Europe 46–53*, June (1997).

61. M. Marvola, M. Rajamniemi, E. Marttila, K. Vahervuo, A. Sothmann, Effect of dosage form and formulation on the adherence of drugs to the esophagus, *J Pharm Sci 72*:1034–1036 (1983).

62. J.-C. Chaumeil, Y. Piton, Enrobages gastro-résistants à l'acetophthalate de cellulose II.—Relation entre l'épensseur du film et la résitance de l'enrobage in vivo, *Annales Pharmaceutique Françaises 31(11)*:691–704 (1973).

63. M. E. M. van Golderen, M. Oeling, D. M. Barends, J. Meulenbelt, P. Salomons, A. G. Rauws, The bioavailability of diclofenac from enteric coated products in healthy volunteers with normal and artificially decreased gastric acidity, *Biopharm Drug Disp 15*:775–788 (1994).

64. M. S. Gordon, A. Fratis, R. Goldblum, D. Jung, E. Kenneth, K. E. Schwartz, Z. T. Chowhan, In vivo and in vitro evaluation of four different aqueous polymeric dispersions for producing an enteric coated tablet, *Int J Pharmaceut 115*:29–34 (1995).

65. S. A. Qureshi, G. Caillé, Y. Lacasse, I. J. McGilveray, Pharmacokinetics of two enteric coated ketoprofen products in humans with or without coadministration of omeprazole and comparison with dissolution findings, *Pharm Res 11(11)*:1669–1672 (1994).

66. M. Mauro, G. C. Palmien, E. Palazzine, M. Barbanti, F. C. Rindina, M. R. Milani, Pharmacodynamic effects of single and repeated doses of oral sulodexide in healthy volunteers. A placebo-controlled study with an enteric-coated formulation, *Curr Med Res Opin 13*:87–95 (1993).

67. W. Stüber, E. Mutschler, D. Steinbach, The pharmaceutical and biological availability of commercial preparations of furosemide, *Arzneim Forsch/Drug Res 32(I)*, 6, 693–697 (1982).

68. A. Litwin, M. Flanagan, G. Entis, G. Gottschlich, R. Esch, P. Gartside, J. G. Michael, Oral immunetherapy with short ragweed extract employing a novel encapsulated preparation—a double blind study, *J Allergy Clin Immunol 1–34* (1996).

69. J. Leroux, R. Cozens, J. L. Roesel, B. Galli, F. Kubel, E. Doelker, R. Gurny, Pharmacokinetics of a novel HIV-1 protease inhibitor incorporated into biodegradable or enteric nanoparticles following intravenous and oral administration to mice, *J Pharm Sci* *84*(12):1387–1391 (1995).

70. H. Staib, U. Fuhr, Drug absorption differences along the gastrointestinal tract in man: detection and relevance for the development of new drug formulations, *Food-Drug Interactions* (J. Kuhlmann, T. R. Weihrauch, eds.). W. Zuckschwerdt Verlag, München, Germany, 1995, pp. 34–56.

71. A. A. Deshpande, N. H. Shah, C. T. Rhodes, W. Malick, Development of a novel controlled-release system for gastric retention, *Pharm Res* *14*:815–819 (1997).

72. B. C. Lippold, B. H. Lippold, Physibalisch-chemisches Verbalten von polymethacrylsauve derivation, *Pharm Ind* *35*:854ff (1973).

73. P. H. List, G. Kaosio, Water vapor and oxygen permeability of various tablet coatings, *Acta Pharm Techn* *28*(1) 1982.

# 2
# New Experimental Coating Material for Colon-Specific Drug Delivery

**Kurt H. Bauer**
*Freiburg Materials Research Center, Freiburg, Germany*

## I. INTRODUCTION

Colonic targeting has gained increasing interest over the past years. A considerable number of publications dealing with colon targeting, colon-specific drug delivery, and absorption from the cecum and other colon sections indicates a growing focus of research activities in this area.

The scientific rationale of drug absorption from the colon was developed some years ago as exemplified by a series of publications [4–8]. The extent of colonic absorption of drugs has been controversial in the literature, although this only parallels the controversies regarding the bioavailabilities of such drugs in general [10].

In general terms, colonic targeting must be based on:

Physiological realities and requirements (i.e., the function of the small and large intestine)
Colonic absorbability or, alternatively, topical efficacy of the drug of interest
Suitable excipients to release a drug in a targeted fashion in the colon and
The therapeutic usefulness of such systems

## II. PHYSIOLOGICAL REALITIES AND COLONIC ABSORPTION

The most important site of absorption of nutrients and drugs is the small intestine. Its extremely large surface area with many folds, villi, and microvilli, together with the enzymatic intestinal fluids are a prerequisite for digestion and absorption of both nutrients and xenobiotic agents.

In contrast, the function of the large intestine is quite different. Until recently, the large intestine's function was thought to serve solely the consolidation of intestinal contents by withdrawing water and electrolytes. The mucosa of the large intestine exhibits a much smaller surface because of a much smaller number of folds and villi relative to the small intestine. As a result, the absorption conditions are less favorable. In addition, the consistency of the intestinal contents becomes increasingly solid as the material flows in the direction of the ascending, traversing, and descending colon, until the normal consistency of feces is obtained. Clearly, the absorption of drugs will be greatly influenced by this consolidation and decreasing diffusion rate [1–3].

For some drugs that, for physiological and anatomical reasons, mainly follow a passive absorption mechanism, a satisfactory colonic absorption was demonstrated. Similar absorption rates from the small and large intestine were found for oxprenolol, metoprolol, isosorbide-5-mononitrate, and glibenclamide [9]. It has also been known for many years, that some lipophilic vitamins, as well as bile salts and some steroids, that undergo enterohepatic circulation show satisfactory colonic absorption [2].

Obviously, satisfactory colonic absorption of the drug to be delivered is a *sine qua non* for the successful development of a colonic drug delivery system. However, situations exist when even reduced absorption from the large intestine (compared with the small intestine) would still justify colonic delivery, for instance in the case of a peptide drug that would otherwise be efficiently digested in the small intestine.

Satisfactory colonic absorption is also an important prerequisite for the reliable function of prolonged action dosage forms.

Because the large intestine membranes have a much smaller surface area, it is not surprising that many investigators have postulated porous or permeable areas in the colonic membrane, like Peyers' patches or so-called absorption windows to explain the surprisingly good absorption for some drugs (e.g., for verapamil) [11]. The absorption window as a hypothetical model is more or less fictive, whereas Peyers' patches are defined and anatomically discernable lymphatic folliculi aggregates [2, 12, 13]. In addition to absorption through such permeable areas, it cannot be entirely excluded that absorption of water-soluble drugs is also facilitated by the considerable colonic dehydration flux.

## III.  GOALS OF COLON TARGETING

A particular challenge for targeted colon drug delivery is the development of peroral application forms containing peptides and proteins, particularly low–

molecular weight peptide drugs [14–19]. Without protection, such peptide drugs are usually digested within the gastric and small intestinal sections. Therefore, the design of delivery systems must provide reliable protection against gastric and small intestinal digestive secretions and release the peptide unchanged and fully efficacious in predetermined sections of the cecum or colon. Studies with insulin [20, 21] have shown that the large intestine is relatively free of peptidases. Hence, if the delivery system succeeds in transporting the peptide past the small intestine, the probability will be high that the drug will be sufficiently absorbed after peroral application.

Site-specific delivery and release of drugs will not only be useful to achieve systemic therapeutic effects (after absorption), but also topical applications, for instance for the treatment of inflammatory bowel disease, ulcerative colitis, and colon cancer, to name the most prevalent.

Although several clinical trials with macromolecules have been reported, colonic absorption of large molecules still seems problematic. Problems arising with peptide drugs in general are reported in an exemplary fashion for metkephamid, which is normally metabolized before reaching the large intestine [22]. Site-specific degradation along the longitudinal axis of the intestine and biodegradation by luminal gastrointestinal enzymes, mucus binding, and intestinal wall permeability of metkephamid were studied. The results clearly demonstrate how important relevant preformulation studies are [22].

## A. Current Prolonged-Release Materials

The current prolonged-release design in most cases is the use of conventional polymeric coating materials that are available in various types, providing either pH- or diffusion-controlled prolonged drug release [23, 24]. For colonic delivery, these polymers are often applied in thicker coats, assuming that this will provide the delayed dissolution necessary to reach the colon. The simplest case is to use slowly dissolving methacrylic acid copolymers (e.g., EUDRAGIT® S, Röhm, Germany) [25]. Another method uses combinations of methacrylic acid or methacrylate copolymers (EUDRAGIT L, S, or RL) and swelling agents that are biodegradable in the colonic microflora (e.g., guar gum or similar galactomannans) [26]. The strategy behind using such combinations is for the swelling agent to first facilitate slow dissolution of the methacrylate copolymer in the small intestine, and, second, in the colon to enhance the dissolution by enzymatic biodegradation.

This drug design is already pointing to an enzyme-triggered drug delivery system.

A quite similar drug design entails an insoluble or a pH-controlled film-forming polymer with β-cyclodextrin [27]. β-Cyclodextrin is only moderately swelling, but with certain plasticizers, it forms stable mesostructures that should be biodegraded. This has been studied since and the biodegradability of β-cyclodextrin by several amylases was demonstrated [28].

Another possibility to achieve time-dependent release coatings in a simple way is to use two consecutive film layers consisting of different polymer materials. The main excipient of the inner layer will be a water-swelling, highly viscous polymer (e.g., hydroxypropyl-cellulose or hydroxypropylmethylcellulose), and the outer layer will consist of a suitable enteric coating material. By simply using such water-swelling sublayers, time-dependent, chronopharmacokinetic products with encouraging release properties were obtained [29, 30]. A question remains whether the transport times of the intestinal contents are a reliable driving force to accomplish the desired release profiles. A rather complicated colon-targeting drug design was developed by Tanabe Seiyaku, Osaka, Japan [31]. This drug design combined several principles. A hard gelatin capsule contains a powdered or granulated drug together with an organic acid. Subsequently, the capsule is coated with a cationic methacrylate copolymer (EUDRAGIT E). The film former of the third coat is a water-soluble hydroxypropylmethylcellulose (HPMC). Finally, a layer of HPMC acetate succinate (HPMC-AS) is applied. This latter gastric-resistant enteric coat takes the dosage form unchanged into the small intestine, where it rapidly dissolves. The function of the intermediate HPMC layer is first a retarding one, but after the complete dissolution of the outer enteric coat, it should also dissolve quickly. The EUDRAGIT E layer prevents further dissolution. However, as soon as enough water has penetrated into the capsule, the organic acid dissolves and the resulting acidic solution starts to dissolve the EUDRAGIT E layer from the inside out, in a time-dependent fashion as a function of layer thickness.

The underlying problem of all these pH-triggered or time-controlled principles is that they are rather unreliable. Duration of transport across the gastrointestinal tract may differ enormously and is difficult to predict because of the greatly varying length and physiological (or pathological) condition of the individual sections of the gastrointestinal tract. Studies have shown that total transit time from mouth to anus can, under normal conditions, range roughly from 20 to 50 hours, with a mean transit time of 35 hours [32]. These deviations are dependent on several factors, like peristaltic movement and, primarily, simultaneous food intake and the nature and consistency of the ingested food (fraction of liquid and solid, respectively; degree of digestibility).

It was for these reasons that we engaged in a colon-targeted systems design project that was based on a more effective and predictable gradient like microflora differences between small and large intestine to eventually replace simple time- or pH-triggered systems.

## IV. PHYSIOLOGICAL REQUIREMENTS
## FOR COLON TARGETING

Practically no problems exist in developing saliva- or gastric-resistant dosage forms. They can be designed on the basis of the considerable pH gradients between saliva and gastric juice and between gastric and intestinal juice, respectively [23]. Because no such reliable and sufficient pH gradient exists between the small and large intestine, alternative suitable gradients must be found.

The most promising gradient in this regard is the vast difference in the microflora (i.e., in the bacterial counts between the small and large intestine). This is due to a retardation of movement of the contents or substratum within the gastrointestinal tract as a consequence of the widening of the intestinal lumen at the transition from the ileum in the cecum and the subsequent ascending first colon segment. Also, peristalsis is continuously decreasing from the small intestine to the end of the large intestine. These facts and the bagshaped nature of the cecum make this site a favorite region for microbial settlement.

The intestinal microflora in the cecum is highly active. The difference between ileum and cecum in bacterial counts per milliliter is $10^3$ to $10^{12}$ [3, 33–35]. The small intestine is not sterile, but the bacterial gradient, particularly between the last small intestinal section, the ileum, and the cecum is surprisingly high and can reach nine orders of magnitude.

To date, the composition of the intestinal microflora still remains to be definitively described. However, it seems quite clear that it consists of a complex, symbiotic conglomerate of more than 500 different, predominantly anaerobic, strains. The enzymatic equipment of this symbiotically living colon microflora was adapted predominantly because of the changing types of food that were available during the course of evolution [36].

It must be taken into account that the intestinal microflora may change transiently because of illness, medication (e.g., antibiotics), or substantial changes in food intake. However, the microflora has an intrinsic mechanism of regulation that is able to restore normal standards within a relatively short time [37–39]. As mentioned before, although the colonic absorption conditions are relatively good in the colon ascendens, they are, even under normal conditions, not as favorable as in the small intestine and become increasingly less favorable with increasing viscosity and consistency.

When intestinal microflora are obtained from feces, it must be considered that, because of environmental change from the anaerobic *in vivo* to aerobic *in vitro* condition, changes in the activity of the microflora may be encountered. However, in general, the microbiological differences between the small and large intestine seem to be a suitable gradient for the design of drug delivery systems for colon targeting and for the determination of colon-targeting functions.

The most widely used intestinal microbial strains used to test colon targeting and enzymatic biodegradation of the carrier, respectively, are microorganisms with azoreductases and different types of glycosidases.

## V.  TESTING COLONIC BIODEGRADATION

The simplest way to test the properties of a drug or certain excipients with regard to release or biodegradation in the large intestine is to use pharmacopoeial *in vitro* protocols with intestinal fluids for prolonged periods. The USP "dissolution apparatus 3" was used to screen formulations for colonic delivery with simulated colonic fluids [40]. However, as previously mentioned, these methods are only of limited value with respect to simulating the actual environment of the colon. Therefore, it is mandatory to use isolated sections of animal intestines (e.g., rat, rabbit, guinea pig or dog) to test biodegradation or colonic absorption before beginning clinical investigations [10].

Because the physiological conditions in animals differ considerably from those in humans, we developed a special colon microflora test (CMT) for screening purposes [41]. This is an *in vitro* test, based on cecal pig substratum and carried out in an anaerobically ($N_2$ and $CO_2$), mini-fermenter-like device that operates at body temperature (Figure 1). The ceca with the content are obtained from freshly slaughtered pigs and transported in special containers under anaerobic conditions. With exception of primates, the contents of the pig cecum are most similar to that of humans. Hence, this test approaches the human situation as closely as possible and is quite realistic for the testing of drugs, dosage forms, and excipients. The batch-to-batch standardization of pig cecum substrate is problematic because the enzymatic activity can vary, likely caused by the use of different feeding stuff. Some efforts were put into preparing homogenized, freeze-dried extracts or concentrates, but this did not meet with great success.

The experiments can also be performed by means of isolated enzymes that occur in the human cecum and colon [64]. These are, for instance, azoreductases for azo-prodrugs, dextranase, or particularly endo-glucanases that are able to internally cleave polysaccharide chains. The latter seem to be more effective than the azoreductases in metabolizing drugs, prodrugs, and drug-carrier materials.

Other reliable testing methods include scintigraphic evaluations in humans and animals [42–44].

Caco-2 cells have been developed as a useful *in vitro* model to study absorption mechanisms of a variety of drugs. They are also used to measure the relative rates of drug and prodrug permeation across a model colonic membrane (Caco-2 monolayers) [45].

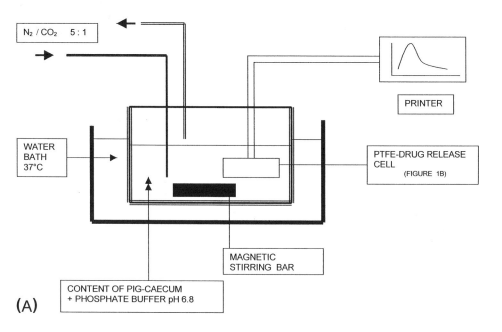

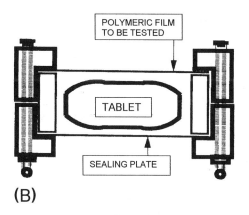

**Figure 1** (A) Colon Microflora Test (CMT) device, an anaerobically working minifermenter consisting of a glass container thermostated by a water bath and hermetically closed by a metal cover plate with several holes for supply pipes. (B) Polytetrafluoroethylene (PTFE) drug release cell. In the case of only small amounts of film-coating materials, this design allows testing under simulated conditions.

## VI.  COLON TARGETING SYSTEMS AND DOSAGE FORMS

Two main strategies accomplish colon targeting or reliable drug release into the cecum or colon [46,47,48]. One approach is to design prodrugs that release the effective drug after enzymatic cleavage at certain predetermined sites of the intestine. Other means include coatings or embedding of the drug with film or matrix-forming polymeric materials that are biodegradable by enzymes of the microflora in the cecum or colon. These materials must, in addition, facilitate retention of the drug's efficacy during gastrointestinal passage. Independent of whether these formulations are designed as coated or embedded dosage forms, mucoadhesive excipients that become available on reaching the colon may play an important role [49]. An instructive example for colon-targeted matrix dosage forms are guar gum tablets with embedded dexamethasone [42]. Guar gum seems to be a suitable colon delivery excipient, because it is satisfactorily biodegradable in the colon.

### A.  Prodrug Systems

1.  Azo-Prodrugs

One of the first prodrugs on the market that used colonic enzymatic biodegradation as its active principle was sulfasalazine (salazosulfapyridine). More modern successor products are olsalazine, balsalazine, and ipsalazine [50], all of them used for the treatment of inflammatory bowel diseases and containing two molecules linked by an azo bond (Figure 2). The prodrugs pass the stomach and the small intestine unchanged and unabsorbed. Reaching the cecum, they are reduced and cleaved by specific azoreductases of the microflora [50, 51]. During this process of enzymatic decomposition, the drug is released in a micro-fine physical state that promotes rapid and extensive dispersion, and thus guarantees maximal topical and systemic activity.

2.  Glycoside-Prodrugs

Alternative systems are glycoside prodrugs, including steroids and narcotics such as dexamethasone-β-D-glucoside, prednisolone-β-D-glucoside, naloxone-β-D-glucoside, and nalmefene-β-D-glucoside [52]. The steroid glycoside prodrugs are evidently better targeted to the colon. They show a more favorable anti-inflammatory effect in the large intestine, a better bioavailability, and potentially reduced side effects, because the effective dose can be reduced. The narcotic prodrug naloxone-β-D-glucoside also passes the small intestine unabsorbed and is enzymatically biodegraded after reaching the cecum.

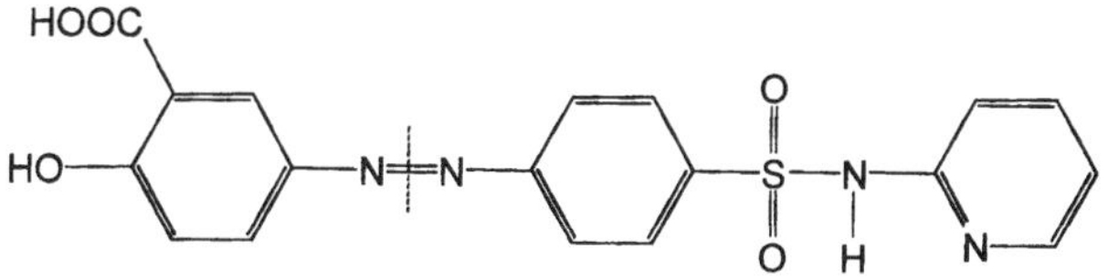

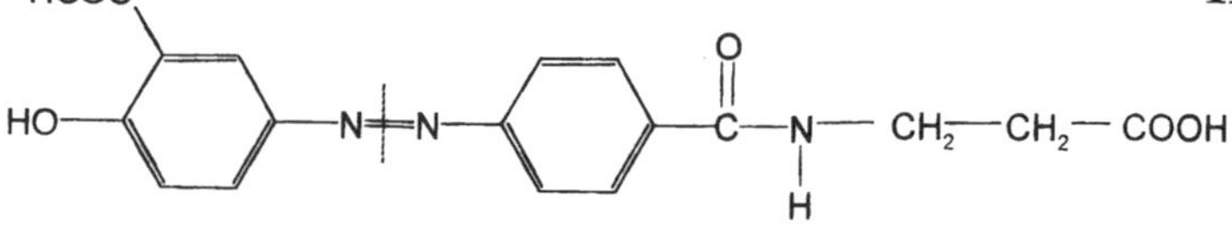

**Figure 2** Three examples of 5-amino salicylic acid prodrugs: sulfasalazine (I), olsalazine (II), and balsalazine (III).

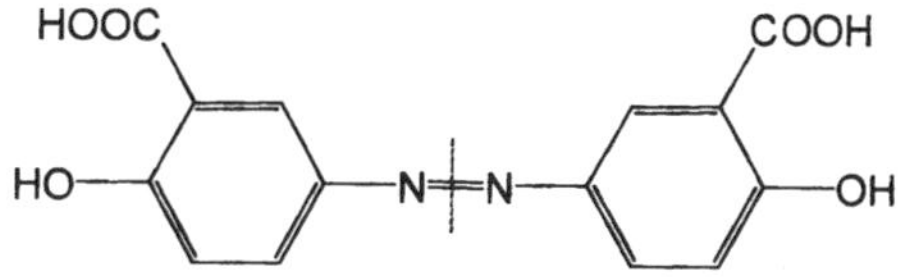

## 3. Glucuronide Prodrugs

Similar to glycoside prodrugs are the glucuronide prodrugs containing corticosteroids [53–55]. They also show improvements in the therapeutic effect and in the reduction of side effects.

## 4. Dextran Prodrugs

Another prodrug group is composed of drug-dextran conjugates [56]. These are synthesized by direct attachment of a drug with a carboxylic group to dextran. If necessary, spacer groups can be used as linkers. Originally, these types of prodrugs were developed for parenteral application. However, they became feasible candidates for colonic delivery, because it has subsequently been demonstrated that they remain unchanged and unabsorbed in the gastrointestinal tract until they reach the cecum. Once there, dextranases of the colonic microflora cleave the ester bond, converting the prodrug to the effective drug. A series of drugs of various pharmacological classes were linked with dextran and modified dextrans and tested in animals. The results indicate that the breakdown of the conjugated prodrugs is mainly mediated by the colonic microflora. An instructive example is the prodrug naproxen-dextran [57].

## B.   Biodegradable Coating and Embedding Excipients

Special coating or embedding excipients for colonic-targeting dosage forms were developed and tested separately or simultaneously with the previously described prodrug systems.

### 1.   Azopolymers

In an effort to develop a peroral application system for insulin, Saffran was the first to use azopolymers as protective colon delivery coating, by use of the afore-mentioned strategy of azo-linked prodrugs [58, 59]. Originally, he used azo-linked copolymers of styrene and hydroxyethylmethacrylate to coat and protect the insulin. However, this approach is not restricted to hydroxyethylmethacrylate and can indeed be accomplished in the same way by an almost unlimited number of similar polymer types.

With his studies, Saffran heralded two interesting areas of research: novel coating systems for colonic delivery coating and novel strategies for the peroral administration of digestible peptide drugs.

Initially, an obvious disadvantage was the relatively slow dissolution rate. Therefore, other azo-linked polymers with a more favorable dissolution profile have been synthesized and tested [60, 61], although their degradation rates evidently were still too slow.

### 2.   Oligosaccharides and Polysaccharides

More recently, polymers or copolymers with oligosaccharides or polysaccharide groups have been used, mainly because of the relatively high activity of glycosidases in the cecum, particularly the endoglycosidases, which cleave and degrade polysaccharides in the interior of the polymer chain, whereas exoenzymes only cleave one polymer unit at a time. Thus, in the presence of endoglycosidases, the coating materials lose their mechanical qualities and their film-forming properties much more rapidly.

One of the first experiments in this direction in our laboratories was the development of copolymers containing polyurethanes (PU) and maltose, oligosaccharides, or polysaccharides. The intention was to use the film-forming qualities of PU and its insolubility in water, whereas maltose, or the polysaccharides or oligosaccharides should function as a ''colon-biodegradable estimated site'' of separation. Maltose was used as the simplest building block, but it proved too short for an enzymatic attack. It was found that a straight, but not a branched, chain of at least five glucose or hexose units is required for a satisfactory biodegradation [62]. Such short, straight chains of oligosaccharides are, for instance, derived from natural galactomannans by complete acetylation or ethylation, purification, and subsequent hydrolysis. In this way oligomeric acetyl- or ethylgalactomannans with terminal hydroxy groups that can be changed into PU by means of isocyanate reaction are obtained. However, these initial products, thought of as raw materials for the synthesis of PU, were found to be excellent film-forming excipients *per se* by use of these production and purification methods [63]. This was a good starting point for the next step, because a major disadvantage of the PU project, despite encouraging results, certainly lies in toxicological problems that must be expected.

Investigation of the acetyl- and ethylgalactomannans initially showed a surprising result. Both products were insoluble in water and sufficiently resistant against gastric and intestinal fluids. But only ethylgalactomannan was biodegraded by the colonic microflora, not acetylgalactomannan. Evidently, the acetylation is complete, whereas ethylation is incomplete, which facilitates biodegradation by means of residual hydroxy groups necessary to allow an enzymatic attack. In contrast, acetylgalactomannan is too hydrophobic and essentially nonswelling in aqueous fluids. Further important findings were that films made of carubin (galactomannan from carob beans) were better biodegraded by the colonic microflora than films with tara or guar galactomannans (guaran). This is because the D-mannose/D-galactose ratio in carubin is $4:1$, whereas it is $3:1$ and $2:1$, respectively, in the latter two. Hence, carubins contain statistically significant longer and unbranched chains and, therefore, have a higher probability for biodegradation. The required water insolubility can also be obtained by cross-linking of galactomannans with 1,4-butandiglycidyl ether or similar cross-linking agents.

With regard to the ecology and economy of the synthesis, film formation quality, stability in gastric, and small intestinal fluids, we found the most suitable colon-biodegradable excipients to be dextran fatty acid esters [64, 65]. The advantage of these colonic-targeting excipients is also due to their manufacture from nontoxic, naturally growing raw materials. Dextrans with molecular weights between 40,000 and 60,000 are widely used as blood plasma expanders. Pure, unsubstituted dextrans with molecular masses more than 1,000,000 exhibit sufficient film-forming properties, yet are still water soluble. Therefore, it is necessary to mask enough hydroxy groups by substitution (e.g., by acetylation). But, as previously mentioned, this reaction usually covers too many hydroxy groups of a polysaccharide. Alternately, hydrophobization by partial alkylation with hydrocarbons having longer chains (e.g., fatty acids) is a simple, straightforward reaction. However, hydrophobization is only one factor and is not able to solve the problem completely. To obtain satisfactory film-forming qualities, the molecular weights of the dextrans substituted with medium chain lengths must be at least

200,000. It was found that the lipophilic properties of the acetyl group were insufficient, and also chain lengths up to eight C-atoms still required unattainably high degrees of substitution. Therefore, fatty acids with more than nine C-atoms were used to determine the optimal ratio of hydrophobic chain length and amount of hydrophilic hydroxyl groups necessary with respect to stability and biodegradation. It was found empirically that the optimum chain length was within the range represented by caprylic acid stearic acid. The best results were obtained with lauroyl substituents. Optimal products with regard to the film-forming and stability qualities resulted from dextrans with molecular weights (MW) about 250,000 and with average degrees of substitution (DS) between 0.1 and 0.24 when tested with the colon microflora test. A survey of small intestinal resistances, colonic biodegradability, and relevant properties of some dextran fatty acid esters is shown in Table 1.

Recently L. Hovgaard and H. Brøndsted published a critical review in which they stated that polysaccharides appear as promising compounds for use in developing colon-specific drug delivery systems [66].

In the field of colonic-targeting excipients, the Kopecek group in Salt Lake City investigated water-soluble N-(2-hydroxypropyl)methacrylamide (HPMA) copolymers and tailor-made derivatives thereof as site-specific drug delivery systems. They developed lysosomotropic carriers for anticancer drugs. They further studied the degradability of oligopeptide side chains in HPMA copolymers by gastrointestinal enzymes isolated from the brush border membranes and lumen and the biorecognition of peptides by certain HPMA copolymer conjugates. Most

**Table 1**   Relevant Properties of Some Dextran Fatty Acid Esters ($-$ = Negative, (+) = Moderate, + = Positive, ++ = Very Positive; Bold Face = Optimal)

| Dextran fatty acid ester | MW | DS | Film formation | Water solubility | Biodegradability (CMT) |
|---|---|---|---|---|---|
| Acetyl- | 500,000 | <1.2 | − | sol. | − |
| | 3.3 Mio | 1.2–3.0 | +/weakly | insol. | − |
| Caproyl- | 250,000 | 0.62–1.7 | −/weakly | insol. | − |
| | 3.3 Mio | 0.08–0.13 | weakly | swelling | − |
| Lauroyl- | 66,000 | 0.3 | − | insol. | (+) |
| | 146,000 | 0.19 | + | insol. | (+) |
| | 146,000 | 0.28 | + | insol. | (+) |
| | **250,000** | **0.11** | **+** | **insol**. | **++** |
| | 250,000 | 0.24 | + | insol. | + |
| Stearoyl- | 250,000 | 0.32 | − | insol. | (+) |
| | 250,000 | 0.5 | neg. | insol. | (+) |
| | 250,000 | 1.16 | weakly | insol. | (+) |

important in our opinion are the investigations concerning colon-specific drug delivery systems combining the concept of colon-specific drug release by the activity of enzymes of the colonic microflora and the colon-specific binding of certain polymeric carriers to colonic mucosal lectins [67, 68, 69].

For many years, Rubinstein and co-workers in Jerusalem, Israel, have been engaged in the investigation and development of microbially controlled drug delivery systems to the colon, extensively publishing studies with coatings and matrix and hydrogel formulations with cross-linked chondroitin, pectins, pectates, and cross-linked guar with several drugs [70, 71, 72]. Calcium pectinate has shown promising results in the colon targeting of insulin, whereas amidated pectins were unsatisfactory. Obviously, the influence of the degree of amidation needs to be assessed [73].

Interesting are amylose coatings for colon-specific drug delivery. It was found that some physical forms of starch are resistant to digestion by pancreatic amylase but are biodegraded by the enzymes of colonic microflora. This led to a particular coating of solid dosage forms, predominantly pellets, with glassy amylose and various amounts of ethylcellulose. Amylose is a natural, nontoxic, and readily available raw material. The products have been tested both *in vitro* and *in vivo* [74, 75]. Attention should also be paid to new matrix systems with cross-linked amylose [76] and cross-linked dextrans that also show interesting results as embedding or coating excipients [77, 78, 79].

## 3. Hydrogels

Novel developments are matrix dosage forms that are "squeezing" hydrogels, which are distinguished by certain swelling and deswelling properties [80]. These properties could become quite interesting in release and diffusion control as a function of moisture binding.

## 4. Chitosan

Newer still is the use of chitosan in colon-specific drug delivery. Chitosan is a partly or largely deacetylated chitin, a basic polymer with a number of free amino groups. Chitosan swells and dissolves in diluted acids and in gastric juice, but it is resistant to intestinal fluids. Chitosan capsules, microcapsules, or chitosan-coated solid dosage forms must be additionally coated with gastric-resistant excipients. Reaching the large intestine, they are biodegraded [81, 82, 83, 84]. Experiments with insulin suggested that coated chitosan capsules may be useful for colon-specific peptide drug delivery after peroral ingestion.

A unique idea is the use of a pressure-controlled capsule in colonic targeting. The inner surface of this capsule is coated with ethylcellulose. Capsules prepared in this way do not disintegrate in the stomach and small intestine but disintegrate in the colon as a result of the colonic peristalsis, where they release their drug content [85].

A controversial problem is still the selection of suitable enhancing agents to overcome the often limited absorption from the colon. In a survey a number of commonly used enhancers, bile salts, surfactants, fatty acids, glycerides, etc. were also reported as effective in the colon [86].

## VII. CONCLUSION

Over the past years, considerable and promising progress has been made in colonic delivery and uptake. Despite significant success, however, a number of problems are still awaiting pharmaceutically acceptable and therapeutically useful solutions. First, more reliable *in vitro* and *in vivo* tests are necessary. Targeted biodegradation must be improved, as well as the enhancement of colonic absorption. As a prerequisite for rational design, this requires intensive investigations of living and activity conditions of the microflora (e.g., optimal conditions at the targeting site), the search for microbial activators, and the search for suitable absorption enhancers. Furthermore, all materials used should be ecologically and economically acceptable. Finally, long-awaited clinical investigations in humans will provide unequivocal proof of the usefulness of colonic drug delivery systems.

## REFERENCES

1.  G. Thews, E. Mutschler, and P. Vaupel, *Anatomie, Physiologie, Pathophysiologie des Menschen, Gastrointestinal Tract and Digestion*, Scientific Publishers, Stuttgart, 1991.
2.  T. T. Kararli, Gastrointestinal absorption of drugs, *Crit. Rev. Ther. Drug Carrier Systems 6*:39–86 (1989).
3.  G. T. MacFarlane, and J. H. Cummings, The colonic flora, fermentation and large bowel digestion function, Chapter 4, *The Large Intestine: Physiology, Pathophysiology and Disease* (S. F. Phillips, J. H. Pemberton, and R. G. Shorter, eds.), Raven Press, New York, 1991.
4.  D. Brockmeier, H. G. Grigoleit, and H. Leonhardt, Absorption of glibenclamide from different sites of the gastrointestinal tract, *Eur. J. Clin. Pharmacol. 29*:193– 197 (1985).
5.  A. H. Staib, D. Loew, S. Harder, E. H. Grayl, and R. Pfab, Measurement of theophylline absorption from different regions of the gastrointestinal tract using a remote controlled drug delivery device, *Eur. J. Clin. Pharmacol. 30*:95–97 (1986).
6.  A. H. Staib, B. G. Woodcock, D. Loew, and O. Schuster, Chapter 8: Remote control of gastrointestinal drug delivery in man (L. F. Prescott, and W. S. Nimmo, eds.), *Novel Drug Delivery and Ist Therapeutic Application*, J. Wiley & Sons, Chichester, UK, 1989, pp. 79–88.

7. H. Laufen, and A. Wildfeuer, Absorption of isosorbide-5-nitrate from the gastrointestinal tract (N. Riethbrock, B. G. Woodcock, A. H. Staib, and D. Loew, eds.), *Methods in Clinical Pharmacology*, Vol. 7, Vieweg Verlag, Braunschweig, Wiesbaden, 1987.

8. K. H. Antonin, P. R. Bieck, M. Scheurlen, M. Jedrychowski, and H. Malchow, Oxprenolol absorption in man after single bolus dosing into two segments of the colon compared with that after oral dosing, *Br. J. Clin. Pharmacol. 19*:137S–142S (1985).

9. P. R. Bieck, Arzneimittelabsorption aus dem menschlichen Colon, *Acta Pharm. Technol. 33*:109–114 (1987).

10. H. Yuasa, Y. Kimura, K. Hamamoto, and J. Watanabe, Evaluation of large intestine as an absorption site for colonic drug delivery, *Proceed. Intern. Symp. Control. Rel. Bioact. Mater. 23*:547–548 (1996).

11. M. Marvola, H. Aito, P. Pohto, A. Kannikoski, S. Nykänen, and P. Kokkonen, Gastrointestinal transit and concomitant absorption of verapamil from a single-unit sustained-release tablet, *Drug Dev. Ind. Pharm. 13*:1593–1609 (1987).

12. D. E. Bockman, and M. D. Cooper, Pinocytosis by epithelium associated with lymphoid follicles in the bursa of Fabricius, appendix, and Peyer's patches, *Am. J. Anat. 136*:455 (1973).

13. D. J. Keljo, and J. R. Hamilton, Quantitative determination of macromolecular transport rate across intestinal Peyer's patches, *Am. J. Physiol. 244*:G637 (1983).

14. B. C. Lippold, Neue Darreichungsformen für Peptidarzneistoffe, *Pharm. Ztg. 136*: 9–18 (1991).

15. R. T. Borchardt, N. A. Mazer, J. H. Rytting, E. Shek, E. Ziv, E. Touitou, and W. I. Higuchi, The delivery of peptides, *J. Pharm. Sci. 78*:883–892 (1989).

16. M. E. K. Kraeling, and W. A. Ritschel, Development of a colonic release capsule dosage form and the absorption of insulin, *Meth. Find. Exp. Clin. Pharmacol. 14*: 199–209 (1992).

17. S. S. Rao, and W. A. Ritschel, Development and *in vitro/in vivo* evaluation of a colonic release capsule of vasopressin, *Int. J. Pharmaceut. 86*:35–41 (1992).

18. S. S. Rao, and W. A. Ritschel, Colonic drug delivery of small peptides, *S.T.P. Pharma Sciences 5*:19–29 (1995).

19. A. Rubinstein, B. Tirosh, M. Baluom, T. Nassar, A. David, R. Radai, I. Gliko-Kabir, and M. Friedman, The rationale for peptide drug delivery to the colon and the potential of polymeric carriers as effective tools, *J. Controlled Release 46*:59–73 (1996).

20. N. Gardner, W. Haresign, R. Spiller, N. Faraj, J. Wiseman, H. Norbury, and L. Illum, Development and validation of a pig model for colon specific drug delivery, *J. Pharm. Pharmacol. 48*:689–693 (1996).

21. A. Manosroi, and K. H. Bauer, The entrapment of a human insulin-DEAE-dextran complex in different compound liposomes, *Drug Dev. Ind. Pharm. 15*:2531–2546 (1989).

22. P. Langguth, H. P. Merkle, and G. L. Amidon, Site-dependent intestinal metabolism and absorption of the pentapeptide metkephamid, *Proceed. Intern. Symp. Control. Rel. Bioact. Mater. 20*:180–181 (1993).

23. K. H. Bauer, K. Lehmann, H. P. Osterwald, and G. Rothgang, *Coated Pharmaceutical Dosage Forms*, MEDPHARM Scientific Publishers, Stuttgart, Germany, 1997.

24. J. W. McGinity, *Aqueous Polymeric Coatings for Pharmaceutical Dosage Forms*, Marcel Dekker Inc., New York, 1996.
25. M. Ashford, J. T. Fell, D. Attwood, H. Sharma, and P. J. Woodhead, An *in vivo* investigation into the suitability of pH-dependent polymers for colonic targeting, *Int. J. Pharmaceut. 95*:193–199 (1993).
26. K. Lehmann, and K. D. Dreher, Methacrylate-galactomannan coating for colon-specific drug delivery, *Proceed. Intern. Symp. Control. Rel. Bioact. Mater. 18*:331–332 (1991).
27. V. Siefke, H. P. Weckenmann, and K. H. Bauer, Colon-targeting with CD-matrix films, *Proceed. Intern. Symp. Control. Rel. Bioact. Mater. 20*:182–183 (1993).
28. A. Fetzner, L. Pfeuffer, and R. Schubert, Enzymatic degradation studies of β-cyclodextrin with regard to the development of colonic specific release formulations, *Proceed. Intern. Symp. Control. Rel. Bioact. Mater. 24*:801–802 (1997).
29. A. Gazzaniga, M. E. Sangalli, G. Maffione, and P. Iamartino, Time-dependent oral delivery system for colon-specific release, *Proceed. Intern. Symp. Control. Rel. Bioact. Mater. 20*:318–319 (1993).
30. A. Gazzaniga, C. Busetti, M. E. Sangalli, R. Orlando, S. Silingardi, A. Bettero, and F. Giordano, *Proceed. Intern. Symp. Control. Rel. Bioact. Mater. 23*:571–572 (1996).
31. T. Ishibashi, H. Hatano, Y. Matukawa, M. Mizobe, M. Kobayashi, and H. Yoshino, Design and evaluation of a new capsule-type dosage form for colon targeted delivery of drugs, *Proceed. Intern. Symp. Control. Rel. Bioact. Mater. 23*:549–550 (1996).
32. A. M. Metcalf, S. F. Phillips, A. R. Zinsmeister, L. MacCart, R. W. Beart, and B. G. Wolff, Simplified assessment of segmental colonic transit, *Gastroenterology 29*:40–47 (1987).
33. G. L. Simon, and S. H. Gorbach, Intestinal flora in health and disease, *Gastroenterology 86*:174 (1984).
34. A. A. Salyers, Bacteroides of the human lower intestinal tract, *Ann. Rev. Microbiol. 38*:293 (1984).
35. W. E. C. Moore, E. P. Cato, and L. V. Holdeman, Anaerobic bacteria of the gastrointestinal flora and their occurrence in clinical infections, *J. Infect. Disease 119*:641–649 (1969).
36. W. D. Keidel, *Kurzgefaßtes Lehrbuch der Physiologie*, Thieme Verlag, Stuttgart, 1979.
37. O. M. Wrong, A. J. Vince, and J. C. Waterlow, The contribution of endogenous urea to faecal ammonia in man, determined by [15]N labelling of plasma urea, *Clin. Sci. 68*:193–199 (1985).
38. G. L. Simon, and S. L. Gorbach, The human intestinal microflora, *Digest. Dis. Sci. 31(suppl.)*:147S–162S (1986).
39. R. Prizont, N. Konigsberg, Identification of bacterial glycosidases in rat cecal contents, *Digest. Dis. Sci. 26*:733–777 (1981).
40. D. Wong, S. Larrabee, K. Clifford, J. Tremblay, and D. Friend, USP apparatus 3 for in vitro screening of colonic delivery formulations, *Proceed. Intern. Symp. Control. Rel. Bioact. Mater. 23*:555–556 (1996).
41. A. W. Sarlikiotis, J. Betzing, Chr. Wohlschlegel, and K. H. Bauer, A new in-vitro method for testing colon targeting drug delivery systems or excipients, *Pharmacol. Lett. 2*:62–65 (1992).

42. C. Kenyon, R. Nardi, D. Wong, G. Hooper, I. Wilding, and D. Friend, Colonic delivery of dexamethasone: A pharmacoscintigraphic clinical evaluation, *Proceed. Intern. Symp. Control. Rel. Bioact. Mater. 23*:553–554 (1996).

43. I. R. Wilding, Scintigraphic evaluation of colonic delivery, *S.T.P. Pharma Sciences 5*:13–18 (1995).

44. A. N. Fisher, L. Illum, S. S. Davis, W. Haresign, J. Jabbal-Gill, and M. Hinchcliffe, *Use of a Pig Model for Colon Specific Delivery of Peptides, Proteins and Other Drugs, European Symposium on Formulation of Poorly-Available Drugs for Oral Administration,* Editions de Santé, Paris, 1996, pp. 183–186.

45. W. Rubas, J. Villagran, M. Cromwell, A. McLeod, J. Wassenberg, and R. Mrsny, Correlation of solute flux across Caco-2 monolayers and colonic tissue in vitro, *S.T.P. Pharma Sciences 5*:93–97 (1995).

46. D. R. Friend, *Oral Colon-specific Drug Delivery,* CRC Press, Boca Raton, Ann Arbor, London, Tokyo, 1992.

47. A. Rubinstein, and D. R. Friend, Specific delivery to the gastrointestinal tract, (A. Domb, ed.), *Polymer Site-Specific Pharmacotherapy,* John Wiley & Sons, New York, 1994.

48. P. J. Watts, and L. Illum, Colonic drug delivery, *Drug Dev. Ind. Pharm. 23*:893–913 (1997).

49. B. Tirosh, and A. Rubinstein, The varied mucus secretory response of the rat GI tract to cholinergic stimulus, *Proceed. Intern. Symp. Control. Rel. Bioact. Mater. 23*:186–187 (1993).

50. M. A. Peppercorn, and P. Goldman, *J. Pharmacol. Exp. Ther. 181*(3), 555–562 (1972); and M. A. Peppercorn, Sulfasalazine: pharmacology, clinical use, toxicity and related new drug development, *Ann. Intern. Med. 3*:377–386 (1984).

51. R. P. Chan, D. J. Pope, A. P. Gilbert, P. J. Sacra, J. H. Baron, and J. E. Lennard-Jones, Studies of two novel sulfasalazine analogs, ipsalazide and balsalazide, *Digest. Dis. Sci. 28*:609–716 (1983).

52. D. R. Friend, Glycosides in colonic drug delivery, Chapter 6 (D. R. Friend, ed.), *Oral Colon-specific Drug Delivery,* CRC Press, Boca Raton, 1992, pp. 153–187.

53. B. Haeberlin, L. Empey, R. Fedorak, H. Nolen III, and D. Friend, In vivo studies in the evaluation of glucuronide prodrugs for novel therapy of ulcerative colitis, *Proceed. Intern. Symp. Control. Rel. Bioact. Mater. 20*:174–175 (1993).

54. B. Haeberlin, H. Nolen III, W. Rubas, and D. Friend, In vitro studies in the evaluation of glucuronide prodrugs for novel therapy of ulcerative colitis, *Proceed. Intern. Symp. Control. Rel. Bioact. Mater. 20*:172–173 (1993).

55. H. Nolen III, R. Fedorak, and D. Friend, Glucuronide prodrugs for colonic delivery: Steady-state kinetics in conventional and colitis rats, *Proceed. Intern. Symp. Control. Rel. Bioact. Mater. 23*:61–62 (1996).

56. A. D. McLeod, Dextran prodrugs for colon-specific drug delivery, Chapter 8 (D. R. Friend, ed.), *Oral Colon-specific Drug Delivery,* CRC Press, Boca Raton, 1992, pp. 214–231.

57. C. Larsen, E. Harboe, M. Johansen, and H. P. Olesen, Macromolecular prodrugs: Naproxen-dextran esters, *Pharm. Res. 6*:919–923, 995–999 (1989).

58. M. Saffran, D. C. Neckers, G. S. Kumar, C. Savariar, and J. C. Burnham, Oral administration of peptide drugs, *Polymer Reprints 27*:23–29 (1986).

59. M. Saffran, G. S. Kumar, C. Savariar, J. C. Burnham, F. Williams, and D. C. Neckers, A new approach to the oral administration of insulin and other peptide drugs, *Science 233*:1081–1084 (1986).

60. S. I. Kim, M. Yamamoto, H. Terashima, H. Tozaki, A. Yamamoto, S. Muranishi, and Y. Kimura, Improvement of oral bioavailability of insulin by colon targeting system using azopolymer, *Proceed. Intern. Symp. Control. Rel. Bioact. Mater. 24*: 377–378 (1997).

61. G. Van den Mooter, M. Offringa, W. Kalala, C. Samyn, and R. Kinget, Synthesis and evaluation of new linear azo-polymers for colonic targeting, *S.T.P. Pharma Sci. 5*:36–40 (1995).

62. A. W. Sarlikiotis, and K. H. Bauer, Synthese und Untersuchung von Polyurethanen mit Galactomannan-Segmenten als Hilfsstoffe zur Freisetzung von Peptidarzneistoffen im Dickdarm, *Pharm. Ind. 54*:873–880 (1992).

63. J. Betzing, and K. H. Bauer, Synthese von substituierten Galactomannanen als Hilfsstoffe zur Herstellung von dickdarmabbaubaren Arzneiformen, *Pharm. Ztg. Wiss. 5/137*:131–134 (1992).

64. J. F. Kesselhut, and K. H. Bauer, Herstellung und Untersuchung von wasserunlöslichen Dextranfettsäureestern als Umhüllungsmaterialien für das Dickdarmtargeting, *Pharmazie 50*:263–269 (1995).

65. S. Hirsch, V. Binder, V. Schehlmann, K. Kolter, and K. H. Bauer, Lauroyldextran and crosslinked galactomannan as coating materials for site-specific drug delivery to the colon, *Eur. J. Pharm. Biopharm. 47*:61–71 (1999).

66. L. Hovgaard, and H. Brøndsted, Current applications of polysaccharides in colon targeting, *Crit. Rev. Ther. Drug Carrier Systems 13*:185–223 (1996).

67. J. Kopecek, P. Kopeceková, *N*-(2-hydroxypropyl)methacrylamide copolymers for colon-specific drug delivery, Chapter 7 (D.R. Friend, ed.), *Oral Colon-specific Drug Delivery*, CRC Press, Boca Raton, 1992, pp. 189–211.

68. H. Ghandehari, P. Kopeceková, P. Y. Yeh, H. Ellens, P. L. Smith, and J. Kopecek, Oral colon-specific protein and peptide delivery: Polymer system and permeability characteristics, *Proceed. Intern. Symp. Control. Rel. Bioact. Mater. 23*:59–60 (1996).

69. V. Omelyanenko, P. Kopeceková, R. K. Prakash, C. M. Clemens, C. D. Ebert, and J. Kopecek, Biorecognition of HPMA copolymers mediated by synthetic receptor binding epitopes, *Proceed. Intern. Symp. Control. Rel. Bioact. Mater. 24*:51–52 (1997).

70. A. Rubinstein, and A. Sintov, Biodegradable polymeric matrices with potential specificity to the large intestine, Chapter 9 (D.R. Friend, ed.), *Oral Colon-specific Drug Delivery*, CRC Press, Boca Raton, 1992, pp. 233–257.

71. R. Radai, and A. Rubinstein, In vitro and in vivo analysis of colon specificity of calcium pectinate formulations, *Proceed. Intern. Symp. Control. Rel. Bioact. Mater. 20*:330–331 (1993).

72. A. Rubinstein, I. Gliko-Kabir, A. Penhasi, and B. Yagen, Enzyme dependent release of budenoside from crosslinked guar hydrogels, *Proceed. Intern. Symp. Control. Rel. Bioact. Mater. 24*:839–840 (1997).

73. Z. Wakerly, J. Fell, D. Attwood, and D. Parkins, Studies on amidated pectins as potential carriers in colonic drug delivery, *J. Pharm. Pharmacol. 49*:622–625 (1997).

74. S. Milojevic, J. M. Newton, J. H. Cummings, G. Gibson, R. L. Botham, S. G. Ring, M. Allwood, and M. Stockham, In vitro and in vivo evaluation of amylose coated pellets for colon specific drug delivery, *Proceed. Intern. Symp. Control. Rel. Bioact. Mater. 20*:288–289 (1993).

75. S. Milojevic, J. M. Newton, J. H. Cummings, G. Gibson, R. L. Botham, S. G. Ring, M. Allwood, and M. Stockham, Amylose, the new perspective in oral drug delivery to the human large intestine, *S.T.P. Pharma Sciences 5*:47–53 (1995).

76. M. A. Mateescu, Y. Dumoulin, G. Delmas, V. Lenaerts, and L. Cartilier, Crosslinked amylose—A new matrix for drug controlled release, *Proceed. Intern. Symp. Control. Rel. Bioact. Mater. 20*:290–291 (1993).

77. H. Brøndsted, and L. Hovgaard, A novel hydrogel system designed for controlled drug delivery to the colon, *Proceed. Intern. Symp. Control. Rel. Bioact. Mater. 20*: 178–179 (1993).

78. L. Hovgaard, and H. Brøndsted, Dextran hydrogels for colon-specific drug delivery, *J. Controlled Release 36*:159–166 (1995).

79. H. Brøndsted, L. Hovgaard, and L. Simonsen, Dextran hydrogels for colon-specific drug delivery, *S.T.P. Pharma Sciences 5*:60–69 (1995).

80. A. Gutowska, J. S. Bark, I. C. Kwon, Y. H. Bae, Y. Cha, and S. W. Kim, Squeezing hydrogels for controlled oral drug delivery, *J. Controlled Release 48*:141–148 (1997).

81. H. Tozaki, T. Fujita, A. Yamamoto, S. Muranishi, T. Sugiyama, A. Terabe, T. Matsumoto, and T. Suzuki, Chitosan capsules for colon-specific drug delivery: Improvement of insulin absorption from the rate colon, *Proceed. Intern. Symp. Control. Rel. Bioact. Mater. 23*:551–552 (1996).

82. T. Odoriba, H. Tozaki, A. Yamamoto, A. Terabe, T. Suzuki, and S. Muranishi, Colon delivery of anti-inflammatory drugs accelerates healing of TNBS induced-colitis in rats, *Proceed. Intern. Symp. Control. Rel. Bioact. Mater. 24*:345–346 (1997).

83. K. Aiedeh, E. Gianas, I. Orienti, V. Zecchi, Chitosan microcapsules as controlled release systems for insulin, *J. Microencapsulation 14*:567–576 (1997).

84. H. Tozaki, J. Komoike, C. Tada, T. Maruyama, A. Terabe, and T. Suzuki, Chitosan capsules for colon-specific drug delivery: Improvement of insulin absorption from the rat colon, *J. Pharm. Sci. 86*:1016–1021 (1997).

85. T. Takaya, K. Niwa, K. Matsuda, N. Danno, and K. Takada, Evaluation of pressure-controlled colon delivery capsule made of ethylcellulose, *Proceed. Intern. Symp. Control. Rel. Bioact. Mater. 23*:603–604 (1996).

86. B. J. Aungst, H. Saitoh, D. L. Burcham, S. M. Huang, S. A. Mousa, and M. A. Hussain, Enhancement of the intestinal absorption of peptides and non-peptides, *J. Controlled Release 41*:19–31 (1996).

# 3
# Pharmacokinetic Considerations in the Design of Pulmonary Drug Delivery Systems for Glucocorticoids

**Cary Mobley**
*Nova Southeastern University, Fort Lauderdale, Florida*

**Günther Hochhaus**
*University of Florida, Gainesville, Florida*

## I. SUMMARY

Inhalation therapy is gaining importance in the therapy of a number of pulmonary diseases. This site-specific targeted delivery aims to achieve high local activity with reduced systemic side effects.

Although numerous review articles have described the clinical efficacy of these inhalation agents [1–8], detailed reviews on the underlying pharmacokinetics and their relevance for targeted drug therapy are rare [3, 9–11]. This chapter evaluates the importance of pharmacokinetic factors for pulmonary targeting and reviews modern pulmonary drug delivery approaches to achieving pulmonary targeting.

## II. INTRODUCTION

Inhalation therapy with its benefit of reduced systemic side effects should be used for the therapy of pulmonary diseases whenever local therapy is able to achieve its therapeutic goals. Drug targeting should reduce the dose required to achieve a desired pharmacological effect and consequently the systemic load of the drug. Although initially localized drug delivery to the lung through inhalation seems trivial, effective pulmonary therapy is rather complex and difficult to achieve. This is because for all pulmonary forms of administration (aerosols, dry powder inhalers, and nebulizers) only a certain portion of the dose is delivered directly to the lung, whereas, most will be deposited in the oropharynx and consequently swallowed (Figure 1). The swallowed portion of the dose, depending on the oral bioavailability of the drug, is potentially available for systemic absorption and will directly contribute to the systemic side effects of the drug. In addition, pulmonary deposited drug is often absorbed very fast from the lung or removed from the upper portion of the lung by way of the mucociliary transporter. Thus, high levels of drug in the lung are difficult to maintain. On the other hand, systemically available drug (having entered through oral or pulmonary absorption) will induce systemic effects and, therefore, should be removed from circulation as efficiently as possible to achieve pulmonary selectivity. The local and systemic properties of a drug or drug delivery system are important factors in achieving pulmonary selectivity (Table 1, see also [12]). These relevant key features are discussed in the following chapter, followed by a review of new delivery approaches to pulmonary delivery for realizing these optimal targeting conditions.

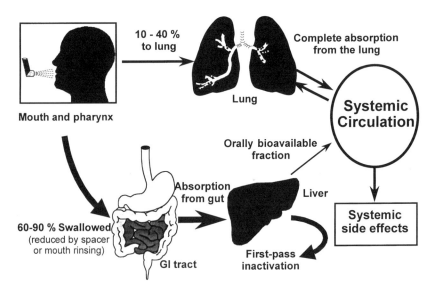

**Figure 1**  The fate of inhaled drugs.

**Table 1**   List of Factors Important for Pulmonary Targeting

| Pharmacodynamic drug characteristics | |
|---|---|
| *Pulmonary PK components* | *Systemic PK components* |
| Pulmonary deposition efficiency | Oral bioavailability |
| Location of pulmonary deposition | Degree of oral deposition |
| Pulmonary residence time (dissolution rate and other factors) | Systemic clearance |
| | Volume of distribution |
| | Protein binding |

## III.  PHARMACOKINETIC/PHARMACODYNAMIC EVALUATION OF PULMONARY SELECTIVITY

### A.  Pharmacodynamic Drug Characteristics

Many articles published over the last decade used the pharmacodynamic activity at the site of action, (e.g., the receptor binding affinities [RBA] of inhaled gluco-corticoids or of β-adrenergic drugs) as parameters for judging the topical activity and the targeting potential of inhalation therapy. A significant body of literature suggested that the receptor-binding affinities of glucocorticoids and $\beta_2$-adrenergic drugs correlates with their activities at the sites of action [13–18]. It has been suggested that high-affinity species achieve pulmonary targeting advantages over lower affinity substances, because the RBAs of inhaled glucocorticoids vary significantly (Table 2). However, theoretical analysis showed that differences in glucocorticoid-receptor affinity can be overcome by selecting the appropriate dose for drugs that induce effects and side effects by means of the same receptors.

**Table 2**   Relative Binding Affinities (RBA)
of Commercially Available Glucocorticoids
to the Glucocorticoid Receptor[a]

| | RBA |
|---|---|
| Fluticasone propionate | 1800 |
| Beclomethasone monopropionate | 1022 |
| Budesonide | 935 |
| Triamcinolone acetonide | 233 |
| Flunisolide | 190 |
| Dexamethasone | 100 |

(RBA of dexamethasone set as 100) As reviewed in Ref. 12.

This assumes that the desired and undesired effects are mediated through the same receptors and that the degree of effects and side effects are directly determined by the number of activated receptors [12]. However, certain glucocorticoids have been reported to show a dissociation between the activity for specific genomic and nongenomic effects (transactivation and AP-1 transrepression [19]). Thus, it might be possible to improve pulmonary selectivity by designing glucocorticoids with favored pharmacodynamic profiles. Similarly, the selectivity ratio between $\beta_1$- and $\beta_2$-receptors, but not the absolute receptor-binding affinity, is important for modulating the pulmonary selectivity of a drug by reducing the systemic $\beta_1$-adrenergic effect [20]. It is, however, clear that both the pharmacokinetic properties of the involved drug and the pharmacokinetic modulation induced by a delivery system are crucial for achieving targeted delivery to the lung (see Table 1).

## B. Clearance

Even when the inhaled stable drug is deposited in the lung, it can be absorbed systemically, consequently inducing systemic side effects. This is the case for all relevant antiasthma drugs. Once absorbed, drug molecules have to be inactivated as quickly as possible to reduce systemic side effects and ensure pulmonary selectivity. The main factor determining the systemic exposure of an absorbed drug molecule is clearance (not half-life). Thus, although drug molecules with the highest possible clearance value should be used for inhalation therapy [12, 21], drugs with long half-lives might still be optimal drug candidates because of large volumes of distribution [12]. Currently available glucocorticoids are inactivated through hepatic metabolism; therefore, maximum systemic clearance is close to the hepatic blood flow. One significant way of improving drugs for pulmonary inhalation is to further increase systemic clearance. Such developments need to concentrate on the extrahepatic inactivation mechanism for the efficient clearance of the drug. Preliminary results with butixocort [22] or fluocortin butylester [23] showed promising increases in clearance. However, to be a successful inhaled glucocorticoid, such derivatives also need to show sufficient pulmonary activity and stability.

## C. Oral Bioavailability

The systemic load of an inhaled agent is the combined result of drug absorbed by way of the pulmonary and the oral routes. Considering the high degree of orally impacted and swallowed drug (commonly >70% of the dose), a significant amount of the swallowed drug is available for oral absorption. Ultimately, this will reduce the pulmonary selectivity of the drug because of the induced systemic effects. Thus, optimal inhaled drugs should consequently show minimal oral deposition and/or low oral bioavailabilities. Because most glucocorticoids have hepatic clearances close to the liver blood flow, oral bioavailabilities of most commercially available glucocorticoids are low. Fluticasone propionate [24, 25] currently shows the lowest oral bioavailability among the commercially available inhaled glucocorticoids, not only because of its high first-pass effect but also because of poor nonlinear absorption from the gastrointestinal tract [24, 25]. Slightly higher oral bioavailabilities have been reported for budesonide (11%, [26]), flunisolide (7–20% [27, 28]), and triamcinolone acetonide (23%, [29]). Within this context, it has been stated that from a practical clinical point of view, oral bioavailabilities of 25% should not induce clinically relevant systemic side effects, as long as the pulmonary deposition is large [30]. For short acting $\beta_2$-adrenergic drugs, oral bioavailabilities range from 1.5% for fenoterol to about 45% for salbutamol, whereas the oral bioavailability for the long-acting agonist salmeterol seems to be more pronounced [31].

## D. Delivery Systems, Deposition Ratio, and Regional Lung Deposition

The delivery system plays a key role in determining the overall pulmonary deposition (percent of drug deposited in the lung), the amount of swallowed drug, and the regional deposition within the lung. All three factors are important for the degree of pulmonary selectivity.

Generally, the pulmonary selectivity of inhaled drugs will increase with increased pulmonary deposition, because the component of drug orally absorbed is reduced. Therefore, a high pulmonary deposition is advantageous. However, the benefits of improved pulmonary deposition are more important for substances with significant oral bioavailabilities, whereas drugs with low oral bioavailabilities will not benefit as much, because the drug does not enter the systemic circulation by the oral route [12, 32]. Thus, in the latter case, reduced pulmonary deposition can be overcome by increasing the dose without losing pulmonary selectivity.

Pulmonary deposition efficiency depends on physicochemical characteristics, such as density of the aerosol or dry powder particles [33–35]. Generally, particle diameters less than than 5 µm are required for efficient pulmonary delivery [36, 37]. Pulmonary deposition also depends on the nature of the delivery device and differs between metered dose inhalers (MDIs). For example, pulmonary deposition expressed as the ratio of pulmonary versus total (pulmonary + oral) absorbed drug, ranged from 15–55% for a number of salbutamol devices and from 66–85% for drugs with lower oral bioavailabilities such as budesonide.

Traditionally, pulmonary deposition of MDIs has been in the range of 10–20% [38–40]. An increase in pulmonary deposition efficiency of MDIs has been achieved with the use of spacer devices [41–46]. Aerosol deposition in the human lung has also been optimized after administration from a microprocessor-controlled pressurized MDI [47, 48]. Improvement of pulmonary deposition of up to 40%

has been reported for newer MDIs that provide smaller aerosol droplet size and lower aerosol velocity [49–53]. Independent of the device, it is crucial that the patient is properly trained in the inhalation technique. This is because pulmonary deposition of conventional (MDIs) depends on patient-related factors, such as the coordination of inhalation and activation by the patient [54]. Breath-activated MDIs have been shown to improve pulmonary deposition characteristics [55].

Inhalation of drug by dry powder inhalers (DPIs) is breath activated, and pulmonary deposition efficiencies are often higher than those measured for MDIs [56, 57]. However, deposition efficiencies also differ among different DPI systems [58]. Because of differences in the physicochemical properties of drugs delivered as MDI or DPI, delivery of the same drug through both devices might result in differences not only of the deposition efficiency but also pulmonary absorption rates [56]. Such differences should be considered in the evaluation of therapeutic benefits of inhalation devices (see later). Similarly, further research is needed to assess the effects of the regional deposition pattern in both the upper and lower part of the lung and the degree of pulmonary targeting, especially because the newer devices are able to deliver a larger portion of the respirable dose into the alveolar region. The pulmonary deposition of inhaled drugs as assessed either by gamma scintigraphy or by pharmacokinetic studies [54, 59–61] will be necessary to assess differences in regional deposition and absorption kinetics, especially when decisions about bioavailability and the equivalency of the regional deposition of formulations have to be made.

## E. Pulmonary Residence Time

It has been suggested by Gerhard Levy that the absorption rate from the target organ is essential for successful drug targeting. It is now recognized that a distinct pulmonary residence time of inhaled drugs is not only beneficial for a prolonged activity but also for increased pulmonary targeting [12, 21]. By use of an integrated PK/PD model of pulmonary targeting, it could be shown that for the upper portion of the lung, an optimal release rate (dissolution rate, release rate from a carrier) exists for which optimal pulmonary targeting is observed [12]. Very fast pulmonary absorption, such as that observed for a glucocorticoid in solution does not result in any pulmonary targeting. The sole reason for this is that fast pulmonary absorption (equivalent to a short pulmonary residence time) results in identical free drug levels in both the systemic circulation and the lung after the termination of the short absorption phase. For delivery systems with reduced pulmonary absorption (e.g., because of a slow pulmonary dissolution rate of the deposited glucocorticoid), the lung is continuously supplied with drug over a prolonged period, resulting in pulmonary drug concentrations being higher than those in the systemic circulation. This results in improved pulmonary targeting. If the release or dissolution rate is too slow, the mucociliary transporter of the upper respiratory tract removes undissolved drug before it is able to interact with the receptor. Consequently, this portion of the pulmonary deposited drug is removed from the lung before it can induce the desired pulmonary effects, and pulmonary selectivity is reduced. Thus, theoretically an optimal pulmonary release or dissolution rate can be defined for the upper respiratory tract [12]. However, a long pulmonary residence time is not easy to achieve because of the physiological characteristics of the lung, including high blood flow, which favors efficient absorption. At present, the following mechanism might be used to prolong the pulmonary residence time: (1) slow dissolution rate of drug particles (e.g., drug lipophilicity or crystal structure), (2) slow release from drug delivery systems (e.g., liposomes or microcapsules), (3) initiation of a biological interaction resulting in prolonged pulmonary residence time (e.g., ester formation or "capturing" in membrane structures).

## 1. Dissolution Rate

The pulmonary absorption rate and, consequently, the degree of pulmonary targeting will depend on the dissolution rate of a given inhalation drug. Once dissolved, most low–molecular weight drugs will be absorbed relatively quickly [62–66], especially from the alveolar region [66]. Indeed, by use of an animal model for pulmonary targeting, it could be shown that the pulmonary targeting achieved with triamcinolone acetonide (TA) (delivered intratracheally to rats) will differ when the pulmonary dissolution rates differ. Figure 2 compares the pulmonary and systemic receptor occupancy for a TA solution, a micronized TA dry powder, and a TA crystal suspension used for treatment of arthritis. The comparison shows that the degree of pulmonary targeting (difference between pulmonary and systemic receptor occupancy) increases from solutions to micronized particles to crystal suspension, which is in in agreement with its anticipated dissolution behavior. This indicates that the biopharmaceutical properties of MDIs and DPIs are important determinants of pulmonary selectivity. Similarly, pharmacokinetic profiles obtained for beclomethasone dipropionate (BDP) DPI and MDI devices differed significantly, with a significantly longer terminal half-life obtained for the DPI device. This argued for a much slower pulmonary dissolution from the dry powder–delivered drug, whereas BDP delivered through a MDI resulted in a faster drug dissolution in the lung [56]. The identification of drug preparations with optimized physicochemical properties (e.g., slow dissolution rate caused by the selection of certain crystal modifications) and the selection of the right device might therefore be one way of further improving inhalation performance. By the same token, it will be crucial to assess the pharmacokinetic absorption profiles of drugs in the developmental stage, because studies for commercially available drugs have shown significant differences in the absorption profiles [67]. In addition, with newer devices delivering more to the alveolar

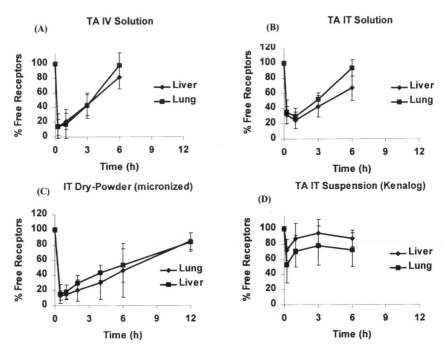

**Figure 2**   Lung and liver glucocorticoid receptor occupancy after administration of 100 µg/kg triamcinolone acetonide (TA): intravenous administration (A) and intratracheal (B) administration of a triamcinolone acetonide solution, micronized TA powder (C) and Kenalog TA crystal suspension (D). Data taken from Ref. (133) and (134).

region of the lung, the goal of ensuring a slow absorption from these regions of the lung will be even more challenging.

## F.  Modulation of Pulmonary Selectivity by Sustained-release Drug Delivery Systems

As discussed later, a number of drug delivery systems, including liposomes and microspheres, have been evaluated for increased pulmonary residence time and, as a consequence, increased pulmonary selectivity. These delivery systems need to ensure a slow release of the drug. For example, a liposomal formulation of TA showed high encapsulation efficiency and shelf stability but failed to provide a sustained drug release, because the lipophilicity of the steroid molecule favored fast release of the drug under sink conditions of the lung. As a consequence, such formulations did not show any pulmonary selectivity (Figure 3). On the contrary,

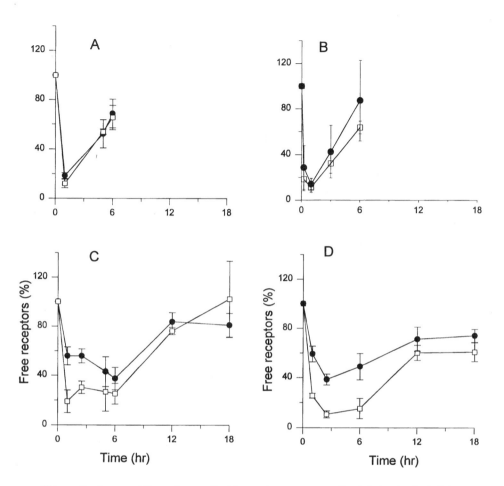

**Figure 3** Lung and liver glucocorticoid receptor occupancy after intratracheal administration of 100 μg/kg triamcinolone acetonide phosphate solution (TAP, A), liposomal preparation of triamcinolone acetonide (B) and intratracheal administration of 200 nm TAP liposomes (C, intermediate release liposomes) and 800 nm TAP liposomes (D, slow release liposomes). Data taken from Ref. 69.

liposomal formulations of TA phosphate (TAP) (Figure 3), showed pulmonary targeting directly related to the stability of the formulations, because the water-soluble drug was captured within the liposome and thereby increased the pulmonary residence time of the formulation [68, 69]. The slow release of TAP in the lung slowly supplies the receptors with drug over a prolonged time period, resulting in an increased pulmonary targeting. If the release rate is further sustained, an even more pronounced pulmonary targeting was observed (Figure 3). Other approaches used highly lipophilic prodrugs of budesonide for achieving stable liposome preparations. In addition, liposomal-encapsulated budesonide esters, but not budesonide, showed significant targeting of the alveolar region of the lung [70]. Alternative approaches for increasing pulmonary targeting, based on low-density microspheres [34, 35] or nanofunctional coatings [71], have been described with significant increases in pulmonary residence times and lung selectivity. These will be discussed later in this chapter.

## G.  Biological Interaction Leading to Prolonged Pulmonary Residence Time

Long-acting $\beta$-2-adrenergic drugs show a prolonged pulmonary residence time because of a specific interaction of these drug molecules with their cellular targets. It has been shown that salmeterol binds reversibly to an active site on the $\beta_2$-receptor and irreversibly to an exosite, which may be a domain adjacent to the active site within the $\beta_2$-receptor in the lipid bilayer of the cell membrane [72]. Similar interactions of formoterol with membrane components have also been attributed to its long action [73]. Not only is the membrane retention of formoterol and salmeterol the reason for the long action of this class of drugs, but it also might be responsible for improved pulmonary selectivity.

Another potential mechanism for prolonged pulmonary residence time (Figure 4) has been the formation of fatty acid esters of the glucocorticoid molecule in the lung [74–76]. These conjugates are unable to cross pulmonary membranes and are consequently trapped in the lung as inactive ''prodrugs,'' thereby increasing the pulmonary residence time of such steroids. Slow activation of these esters by esterases provides a slow release of active species. Although these findings describe a new mechanism for attaining a prolonged pulmonary residence

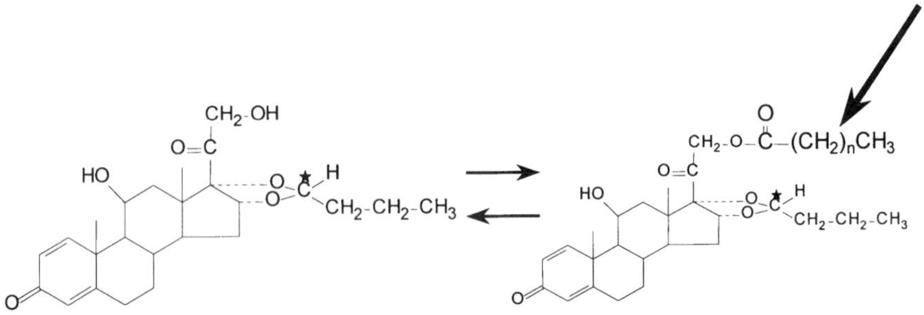

**Figure 4**   Reversible pulmonary esterification of 21-OH glucocorticoids.

time and might be relevant for explaining increased pulmonary selectivity, future studies are needed to show that, indeed, a clinically relevant portion of the overall deposited drug is captured in the lung tissue.

## H. Specific Uptake into Cells

Pulmonary targeting of a number of mainly hydrophilic or high–molecular weight substances is hampered by the lack of cellular uptake. Drug delivery systems such as liposomes, with or without cellular recognition elements such as antibodies, are suitable as cellular delivery devices if they are taken up by cells by phagocytic or receptor-mediated mechanisms. Examples of successful approaches include amikacin liposomes. The use of liposome-encapsulated amikacin allowed the realization of a pronounced cellular drug concentration (Figure 5). Because the uptake is energy dependent and capacity limited, this strategy is only beneficial if cellular drug levels are increased by the delivery systems, or a cellular depot with slow release characteristics is realized (Figure 6). However, this approach may not be successful for all drugs. For example, liposomally encapsulated TA phosphate resulted in a right shift of the dose-response curve,

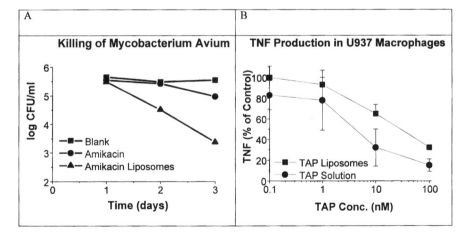

**Figure 5** Pharmacological effects of liposome and liposome-free incubations for amikacin (**A** [132]) and for triamcinolone acetonide phosphate (**B**, unpublished observation) in cell culture. A represents situation **A** in figure 3. **B**, represents situation **B** in Figure 3. Note the right shift for the liposomal preparation, indicating a less favorable uptake of liposomal encapsulated drug. Such formulations are only working as slow-release depot formulations of the drug.

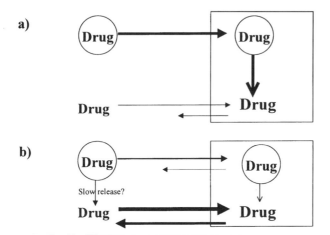

**Figure 6**  Possible scenarios for drug delivery systems. **A**, The drug carrier is able to achieve much higher intracellular levels, because the free drug is hardly able to enter the cell. In this case pulmonary targeting is increased as drug levels in pulmonary cells are increased over free drug. **B**, Intracellular drug levels will not be as high with the drug carrier than with free drug, because free drug is able to enter the cell without any problems, whereas the uptake of the carrier is capacity limited. In this case, the carrier might be suitable for realizing a slow release of drug, thereby improving pulmonary targeting.

because cellular levels with liposomal TA were lower than with free drug (Figure 5).

## I.  Conclusions

The pharmacokinetic/dynamic parameters involved in pulmonary drug delivery of glucocorticoids have been reviewed. Among these, low oral bioavailability, high pulmonary deposition, pronounced clearance, and sustained pulmonary release are the most important parameters to be considered.

## IV.  NEW DELIVERY FORMS

This section will review potential delivery strategies to achieve pulmonary targeting. Several drug delivery systems are discussed here (see Table 3).

## A.  Liposomes

Among those systems, liposomes are the most commonly studied systems for pulmonary targeting, and their usefulness for pulmonary delivery is the subject of several reviews [77–82]. Studies carried out over the past three decades have demonstrated a capacity for liposomes to favorably alter the pharmacokinetic/ pharmacodynamic profile of free drug by prolonging drug release, limiting systemic exposure, or enhancing drug uptake into cellular targets.

In an early study, McCullough and Juliano [83] demonstrated both a pharmacokinetic and pharmacodynamic improvement conferred by liposomes to cytosine arabinoside, a water-soluble antineoplastic compound. They administered the drug, in solution and formulated in egg phosphatidylcholine/cholesterol (PC/ CH) stearylamine (25:7:1 w/w) multilamellar vesicles (MLVs) to rats by means of intratracheal instillation. They showed that the drug persisted in the lungs longer when it was administered in the liposomal formulation. Also, less drug entered the systemic circulation when administered in the liposomal formulation compared with the drug solution. Furthermore, although the drug administered as a solution suppressed the incorporation of thymidine and DNA synthesis at sites distant from the lungs, the drug administered in the liposomal formulation had localized activity within the lungs. Thus, McCullough and Juliano demonstrated that liposomes can be effective for pulmonary drug targeting.

Ten years later, Taylor et al. [84] published a study on the effects of liposome encapsulation on the pharmacokinetics in humans of sodium cromoglycate, a water-soluble antiasthmatic/antiallergenic compound. The drug was formulated in dipalmitoylphosphatidylcholine/cholesterol (DPPC/CH) (1:1 molar ratio) liposomes and administered as a nebulized aerosol. The researchers demonstrated that, compared with drug administered in solution, the liposomally encapsulated drug achieved a lower $C_{max}$ and prolonged plasma half-life, indicating that the liposomes were controlling drug delivery.

Schreier et al. [85] examined the effects of liposome encapsulation on the pharmacokinetics in sheep of amikacin, a water-soluble aminoglycoside. The drug was formulated in 200 nm liposomes and administered by means of intratracheal instillation. The liposome formulations were soy PC/phosphatidyl glycerol (PG) (7:3 molar ratio) and soy PC/PG/CH (4:3:3). They found that both liposome formulations reduced plasma $C_{max}$ and prolonged the plasma half-life of the amikacin compared with the drug administered as a solution, once again indicating that liposomes were controlling drug delivery in the lungs. The inclusion of cholesterol in liposomes more than tripled the plasma half-life for the drug compared with the liposomes without cholesterol. Cholesterol reduces the fluidity and permeability of liposomes in their liquid crystalline phase.

Fielding and Abra systematically examined factors affecting the release rate of $^3$H-terbutaline sulfate from liposomes after intratracheal instillation in guinea pigs [86]. Terbutaline sulfate is a water-soluble bronchodilator. The researchers monitored the clearance of the radioactive label from the lungs and found half-lives ranging from 1.4 to 18 h. Liposome composition had a significant effect on the half-life. On the low end were DPPC dipalmitoyl phosphatidyl glycerol (DPPG), 95:5 and egg phosphatidylcholine/egg phosphatidylglycerol/

**Table 3**  Selected Studies of Sustained Drug Delivery Systems for Pulmonary Delivery

| System | Drug/marker | Type of study | Parameter(s) studied | References |
|---|---|---|---|---|
| *Liposomes* | Cytosine arabinoside | *In vivo*—rats | Pharmacokinetics, pharmacodynamics | [83] |
| | Sodium cromoglycate | *In vivo*—humans | Pharmacokinetics | [84] |
| | Amikacin | *In vivo*—sheep | pharmacokinetics, effect of lipid composition | [85] |
| | ³H-terbutaline sulfate | *In vivo*—guinea pigs | Pharmacokinetics, effect of lipid composition | [86] |
| | Tobramycin | *In vivo*—rats | Pharmacokinetics | [88] |
| | | *In vivo*—rats | Pharmacokinetics, eradication of *P. aeruginosa* | [87] |
| | | *In vitro* | Drug release, effect of lipid composition | [90] |
| | | *In vivo*—mice | Pharmacokinetics, effect of lipid composition | |
| | | *In vivo*—rats | Pharmacokinetics, eradication of *P. aeruginosa* | [89] |
| | Amphotericin B | *In vivo*—mice | Treatment of pulmonary and systemic *Cryptococcus neoformans* infections | [91] |
| | | *In vivo*—mice | Prophylaxis of pulmonary aspergillosis | [92] |
| | | *In vivo*—mice | Pharmacokinetics, effect of lipid composition | [93] |

| | | | | |
|---|---|---|---|---|
| | Beclomethasone dipropionate | In vivo—humans | Tolerance | [95] |
| | | In vivo—humans | Pulmonary distribution and clearance of radiolabeled phospholipid in healthy volunteers | [96, 131] |
| | | In vivo—humans | Pulmonary distribution and clearance of radiolabeled phospholipid in asthmatic patients | [97] |
| | Triamcinolone acetonide phosphate | In vitro | Drug release, | [68] |
| | | In vivo—rats | Pharmacokinetics, pharmacodynamics | |
| | | In vitro | Drug release, | [69] |
| | | In vivo—rats | Pharmacodynamics | |
| | 4-hydroxy-L-proline (cHyp) | In vivo—rats | Pharmacokinetics, pharmacodynamics | [104] |
| Microspheres/ microparticles | Isoproteranol | In vivo—rats | Pharmacodynamics | [99] |
| | Tobramycin | In vivo—rats | Pharmacokinetics | [88] |
| | Albuterol | In vivo—guinea pigs | Pharmacodynamics | [33] |
| | Testosterone | In vitro | Disposition in a cascade impactor | [35] |
| | Insulin | In vivo—rats | Pharmacokinetics, pharmacodynamics | [35] |
| Coated particles | Budesonide | In vivo—rats | Pharmacodynamics | [101] |
| | Pentamidine | In vitro | Drug release | [102] |
| | Fluorescein | In vitro | Drug release, pharmacokinetics | [102] |
| | | In vivo—dogs | | |
| Polymer carriers | Peroxidase | In vitro | Macrophage uptake | [103] |
| | 4-hydroxy-L-proline (cHyp) | In vivo—rats | Pharmacokinetics, pharmacodynamics | [105] |

cholesterol (EPC/EPG/CH), 55:5:40, which gave half-lives of 1.4 h and 4.8 h, respectively. On the high end were DPPC/DPPG/CH, 55:5:40 and DSPC/ DSPG/CH, 55:5:40, which gave half-lives of 17.5 h and 17.9 h, respectively. Thus, the addition of cholesterol to the DPPC/DPPG liposomes significantly improved their ability to control terbutaline release, suggesting that cholesterol not only improves the ability of low-transition temperature liposomes to control drug release (as found by Schreier et al. [85]), but it also improves control for high-transition temperature liposomes such as DPPC/DPPG. Fielding and Abra also found in this study that liposomal $^3$H-terbutaline pulmonary half-life increased according to the degree of hydrogenation of EPC in EPC/EPG/CH liposomes. This improvement conferred by hydrogenation was likely the result of the decrease in membrane fluidity that accompanied the increase in hydrocarbon saturation.

Tobramycin, a water-soluble aminoglycoside has been found to exhibit pulmonary targeting when encapsulated in liposomes [87–90]. Poyner et al. instilled 720 nm EPC/PA/CH (29.7:3.3:33) tobramycin liposomes into the lungs of rats [88]. They found that 55% of the liposomal tobramycin remained in the lungs after 24 h, compared with only 5% of the drug instilled as a solution. Furthermore, the authors showed that less drug was present in both the blood and kidney at 6 and 24 h when administered in the liposomal form. In the same study, similar results to liposomes were found with microspheres. Thus, liposomes help to selectively retain tobramycin in the lungs for a prolonged period.

Omri et al. instilled 400 nm DSPC/DMPG (10:1) tobramycin liposomes into the tracheas of rats infected with *Pseudomonas aeruginosa* [87]. Compared with the drug solution, liposomal tobramycin stayed in the lung longer, exhibiting a relatively constant amount in the lung over the study period of 16 h. However, the animals remained infected after the administration of both free and encapsulated drug. For the free drug, this lack of efficacy may be explained by the rapid elimination of the drug from the lung. For liposomal tobramycin, this may be explained as an insufficient quantity of drug released from the liposomes. Evidence for this latter explanation was found in a study in which they examined the release of tobramycin from liposomes *in vitro* and following *in vivo* intratracheal instillation in mice [90]. They examined several combinations of DSPC, DPPC, DMPC, and DMPG, with each combination possessing different membrane phase transition temperatures ($T_c$). They found that combinations based on DSPC ($T_c = 40–46°C$) retained tobramycin *in vitro* better than combinations based on DPPC ($T_c = 29.5–35°C$). Similarly, in the *in vivo* portion of study, the DSPC-based formulation retained tobramycin better than the DPPC-based formulations.

In another study, the same researchers found that tobramycin encapsulated in the more fluid DPPC-based liposomes essentially eradicated mucoid *P. aeruginosa* from the lungs of rats [89]. Thus, the researchers ultimately discovered that the liposome formulation can have a dramatic effect on the efficacy of encapsulated drug. In the same study, they found that a lower fraction of tobramycin reached the kidneys when the drug was administered in liposomes compared with its administration as a solution. Thus, the researchers were able to achieve a therapeutically effective dose of tobramycin with a measurable degree of pulmonary targeting.

In these studies of tobramycin liposomes, the authors attributed differences in tobramycin retention to differences in liposome membrane fluidity or charge [90], and they also attributed the effectiveness of DPPC-based liposomal tobramycin to a greater membrane fluidity [89]. However, it must also be considered that the DPPC-based liposome formulations that they used ($T_c$ = 29.5–35°C) underwent phase transitions during *in vitro* and *in vivo* incubations. It is well known that liposomes lose a portion of encapsulated solutes during phase transitions. Thus, the release of free tobramycin during phase transitions could be, in part, responsible for both the lesser retention [90] and greater anti-*P. aeruginosa* activity [89] of the DPPC-based formulations compared with the DSPC-based formulations. Finally, it should be remembered that the DPPC-based liposomes used in these studies possessed lower transition temperatures than a pure DPPC bilayer, which has a $T_c$ of 42°C. Thus, the results for DPPC-based liposomes might differ if the mixtures include phospholipids with longer chains than DMPC and DMPG.

Amphotericin B (AmB) is a water-insoluble fungicidal polyene antibiotic that is effective against several of the common pulmonary fungal pathogens, including *Cryptococcus*, *Aspergillus*, and *Histoplasmosis*. Because of its insolubility and systemic side effects, amphotericin B can benefit from local delivery to the lungs. Studies have found aerosolized liposomal AmB to be effective in the treatment [91] or prophylaxis [92] of fungal pathogens in the lungs of mice.

Lambros et al. studied the disposition of aerosolized liposomal amphotericin B in mice as a function of liposome charge [93]. They produced sonicated AmB-containing liposomes composed of DMPC/AmB (neutral), DMPC/CH/stearylamine, 1:1:0.2 (positively charged), and DMPC/dicetylphosphate, 14:1 (negatively charged). They found that the negatively charged formulation was eliminated from the lungs with a half-life of 4.5 days, whereas the positively charged and neutral formulations exhibited biexponential elimination with terminal half-lives of 15 and 22 days, respectively. Thus, charge of liposomes can influence their capacity for pulmonary targeting.

In the same study, the researchers did not detect AmB in serum or other organs, indicating that the liposomes achieved pulmonary targeting for the drug. However, it has also been observed in other studies that aerosolized liposomal AmB can be effective against systemic *Cryptococcus* and *Candida* infections, indicating that the aerosolized drug can be absorbed sufficiently to reach therapeutic systemic concentrations [91, 94].

In several studies, liposomal formulations of beclomethasone dipropionate (Bec) were administered as nebulized aerosols to humans. The drug is a poorly water-soluble glucocorticoid used in asthma. Waldrep et al. [95] found formulations of the drug in MLVs composed of DLPC to be well tolerated by humans on inhalation. Vidgren et al. [96] measured the pulmonary distribution and clearance of radiolabeled ($^{99m}$Tc) Bec-DLPC MLVs on administration to healthy volunteers. They found that more than 80% of the deposited radiolabel was retained in the lung 3 h after inhalation. Saari et al. studied the same formulation in mildly and severely asthmatic patients [97]. They found that although mildly asthmatic patients exhibited a uniform distribution of the radiolabel, severely asthmatic patients had a greater fraction of counts associated with the central airways. The severely asthmatic patients also cleared the radiolabel more rapidly, retaining 54% of the initial dose at 24 h compared with a retention of 72% in the mildly asthmatic patients. The faster clearance in the severely asthmatic patients was possibly the consequence of its greater deposition in the upper airways, where the mucociliary escalator operates to clear foreign material. Also, the greater inflammation experienced by the severely asthmatic patients may have played a role in radiolabel clearance.

It should be noted that in theses studies of beclomethasone diproprionate liposomes, only the fate of radioactively labeled lipid was measured. Although the deposited lipid showed prolonged residence in the lungs, the fate of the drug was not ascertained and is uncertain, because it is possible that drugs may leak out of liposomes after administration. This leakage has been demonstrated for the lipophilic glucocorticoid TA by Schreier et al., who found that the drug was rapidly released from liposomes on dilution or administration [98]. One of the reasons for the efficacy of glucocorticoids is that they are readily able to pass through cell membranes to reach their intracellular target. Drugs with moderate octanol/water partition coefficients, like some glucocorticoids, are usually able to pass through lipid bilayer systems like liposomes and are thus less likely to be retained within the bilayer.

The inability of liposomes to retain TA caused Gonzalez-Rothi et al. to shift their research focus to the water-soluble prodrug, TA phosphate (TAP) [68]. They encapsulated the drug in 200-nm extruded DSPC/DSPG (9:1) liposomes and found that the liposomes retained about 85% of the drug when incubated in lung lavage fluid for 24 h. On intratracheal administration of the liposomal TAP to rats, they found that the liposomes enabled sustained glucocorticoid receptor occupancy in the lungs, as well as receptor occupancy that was more pronounced for the lung than for the liver. Thus, they were able to demonstrate that compared with TAP administered as a solution, TAP liposomes showed a preferential sustained drug activity in the lungs, (i.e., liposomes conferred a measurable degree of lung targeting for the drug). In a later study, the same group of researchers showed data (from 800-nm TAP liposomes administered intratracheally to rats) that suggested that there was a drug dose optimum for TAP liposomes to achieve maximum pulmonary selectivity [69]. This finding prompted them to advise that drug release rate and dose should be considered in drug development.

## B. Microspheres and Microparticles

Polymeric microspheres have also been examined as particulate drug carriers for pulmonary delivery. Lai et al. measured pharmacological responses to isoproterenol formulated in poly(glycolide-co-lactide) (PLGA) 50:50 microspheres, after endotracheal instillation into rats [99]. They found that a formulation consisting of 70% free drug and 30% in microspheres (mean diameter, 4.5 μm) ameliorated serotonin-induced bronchoconstriction for at least 12 h. This was considerably longer than the amelioration conferred by the drug administered as a solution, which was less than 30 min. Along with their examination of liposomally encapsulated tobramycin (previously discussed), Poyner et al. examined the capacity of poly lactic acid (PLA) microspheres to control tobramycin delivery in rats [88]. They administered the microspheres (c.a. 740 nm mean diameter) by way of endoteracheal instillation to rats. Twenty-four hours after instillation, a much larger fraction of the drug was retained in the lungs when the drug was administered in microspheres (46% of the total dose retained) compared with its administration as a solution (5% of the total dose retained). Although PLA microspheres appear to control drug delivery in the lung, there is evidence that they may be toxic to the lungs. Armstrong et al. demonstrated that PLA microspheres evoked a significant inflammatory response in the lungs of rabbits after their administration by nebulization [100]. Thus, other polymer formulations may be more suitable for pulmonary delivery.

Recent studies with microspheres have also demonstrated the importance of the distinction between aerodynamic and true particle diameter. Edwards et al. examined large porous particles (diameter = 8.5 μm, density = 0.1 g/cm$^3$) of testosterone-containing 50:50 poly(lactic acid-co-glycolic acid) (PLGA) and compared them to nonporous particles (diameter = 3.5 μm, density = 0.8 g/cm$^3$) [35]. Even though the true mean diameter of the porous particles was greater, the mean aerodynamic diameters of the porous and nonporous particles were nearly identical because of differences in particle density. On aerosolization through a cascade impactor, the authors found that the porous particles exhibited a greater respirable fraction (50%) than the nonporous particles (20%). Because the aerodynamic diameters were nearly identical, they attributed the greater respirable fraction of the porous particles to their lesser tendency to aggregate, thus a lesser deposition by gravity and inertia. The same authors further applied the concept of large respirable porous particles for drug delivery by examining serum levels of insulin and glucose after the inhalation of insulin-containing PLGA particles by rats. Once again they compared porous particles (diameter >5 μm) with nonporous particles (diameter <5 µm), both possessing an aerodynamic diameter of about 2 µm. The authors found that after an initial peak in serum insulin, the inhaled large porous particles conferred relatively constant elevated serum insulin levels for up to 96 h. Concurrently, the authors showed that after inhalation of porous insulin particles, serum glucose levels fell during the first 10 h and remained low for at least 96 h, whereas they stated that glucose levels rose 24 h often inhalation of the nonporous particles. Besides the importance of particle mass in pulmonary deposition, the authors also demonstrated the importance of particle size in achieving sustained residence for the particles in the respiratory region of the lungs. They showed that because of the large diameter of the porous particles (greater than 3 µm), they were better able than the smaller nonporous particles to escape phagocytosis and removal by macrophages. In summary, the authors showed that the large porous particles possess a relatively high respirable fraction because of their low mass density and low tendency to aggregate, and they can achieve sustained drug delivery, in part because of their ability to avoid phagocytosis.

Ben-Jebria et al. further demonstrated the potential efficacy of large porous particles for pulmonary targeting, but this time with particles not prepared from polymers [33]. They produced large porous particles (diameter = 10 µm) by spray-drying a formulation of human serum albumin (18%), lactose (18%), DPPC (60%), and albuterol (4%). They measured the ability of these particles to protect against carbachol-induced bronchoconstriction after their intratracheal instillation into guinea pigs. For comparison, they used "fast-release" spray-dried nonporous particles (diameter = 3 µm) composed of albuterol (4%) and lactose (96%). It was discovered that the inhaled large porous particles protected against bronchoconstriction for at least 16 h, whereas the smaller nonporous formulation protected for less than 8 h. This prolonged drug effect conferred by the large porous particles was once again attributed to their low density ($0.06$ g/cm$^3$), their lesser tendency to aggregate, and their avoidance of macrophage uptake. The authors stated that they examined this particular composition of large porous particles (DPPC/albumin/lactose), in part, because the particle ingredients are either endogenous to the lungs or are Food and Drug Administration (FDA)-approved for inhalation and because the particle production method (spray-drying) is relatively simple to scale up for commercialization.

## C. Coated Particles

A recent novel approach uses the PLGA polymer as a sustained-release coating on budesonide dry powders. Talton et al. showed that these coated particles conferred a statistically significant increase in pulmonary targeting in rats compared with uncoated budesonide powders [101]. Another approach uses a wax coating to control particle dissolution. Pillai et al. produced paraffin wax–coated fluorescein and pentamidine particles by drying, then coating aerosolized droplets of the compounds [102]. The apparent mass median aerodynamic diameters of the coated particles ranged from 2.5–4.1 μm. *In vitro* dissolution profiles demonstrated that wax coating caused an increase in the duration of total release for the fluorescein and pentamidine. The wax-coated fluorescein particles were administered by aerosol to the lungs of dogs, and it was demonstrated that the wax coating conferred more than a threefold increase in absorption half-life time for the marker, reflecting the longer duration of marker release afforded by the coating. In another approach to prolong drug release, Evora et al. coated PLGA, 50/50 microspheres with DPPC [103]. They found that the DPPC coating reduced the microsphere uptake by alveolar macrophages in cell culture. By avoiding macrophage uptake and subsequent removal from the lungs, the microspheres can remain longer to prolong the time period of controlled drug release.

## D. Polymer Carriers

Another agent that has benefited from liposome encapsulation is 4-hydroxy-L-proline (cHyp), which has been examined for its ability to prevent bleomycin-induced pulmonary fibrosis. Poiani et al. encapsulated cHyp in liposomes composed of DPPC/cholesterol/stearylamine, 14:7:4, and administered them to rats by intratracheal instillation [104]. They found that liposomal cHyp suppressed collagen accumulation in bleomycin-induced pulmonary fibrosis and exhibited prolonged residence in the lungs with minimal uptake by nonlung tissue. In another approach for sustained-release cHyp [105], the researchers produced a polymeric carrier for cHyp. The polymer had a mean molecular mass of 21,700 and had a backbone of alternating poly(ethylene glycol) and lysine molecules with cHyp linked as a pendant of the carboxylate group of each lysine residue. The researchers instilled polymeric cHyp, alone and encapsulated in liposomes, into rat lungs. They found that free polymeric-cHyp exhibited prolonged residence in the lungs (with greater amounts remaining 7 days after instillation) than when administered within liposomes. Furthermore, polymeric cHyp produced sustained inhibition of collagen accumulation in bleomycin-treated rats. The results of this study indicate that polymeric carriers are viable alternatives to liposomes and microspheres for pulmonary targeting.

## E. Pulmonary Epithelial Cell Targeting

The lung possesses a variety of epithelial cells, with different types found in different regions. Alveolar type II cells represent a potentially important target for the delivery of intracellular enzymes, such as superoxide dismutase (SOD) and catalase, which do not reach their full complement until the final 10–15% of gestation. Until the enzymes do reach their full complement, the premature baby is left without the capacity to withstand oxygen-mediated lung damage. If delivered into the lungs, the enzymes are too large and water-soluble to pass into cells sufficiently to boost the intracellular quantities. Liposomes, which can be taken up intact into type II cells [106], have been shown in a number of studies to deliver protective amounts of antioxidant enzymes after intratracheal administration [107–111]. Although studies show that the enzymes are delivered to type II cells, macrophages and other cells in the alveoli can also take up the enzymes [112]. To improve the selective uptake into type II cells, researchers have included the surfactant protein A (SP-A) in liposome formulations in cell culture studies [113, 114]. The oxidant-sensitive type II cell has a high-affinity specific receptor for SP-A [115]. Walther et al. showed that the addition of SP-A to antioxidant-phospholipid liposomes essentially doubled the type II cellular uptake of encapsulated SOD and catalase [113]. Briscoe et al. also found a near doubling of type II cellular uptake for SOD with the inclusion of SP-A in the liposome formulation [114]. In the same study, they also used a pH-sensitive liposome formulation that was designed to destabilize at a pH of 5. The intent here was to allow the liposome to release the enzymes at low endosomal pH, thus further enhancing intracellular delivery.

## F.  Pulmonary Macrophage Targeting

Pulmonary macrophages are promising targets for drug delivery. In the nonciliated region of the lung, they are the major line of defense, responsible for the phagocytosis of inhaled particles less than 5 μm in diameter that penetrate that far into the lungs. The macrophages function to eliminate ingested particles (including pathogens) by producing substances such as proteases and oxygen radicals and by the transport of the particles to mucociliary escalator, which enables the macrophages to be eventually swallowed. In some cases, the macrophages may penetrate the alveolar wall and translocate their contents to the bronchial lymph nodes. Although macrophages function to eliminate inhaled pathogens, they can also serve as a host to certain facultative and obligate intracellular pathogens, such as *Mycobacterium avium intracellulare* and *Mycobacterium tuberculosis*. It is worthwhile to attempt to target infected macrophages to either improve the uptake of antibiotics or to minimize antibiotic side effects. The potential for using liposomes to target infected tissue macrophages has been reviewed [116]. A survey of approaches for targeting alveolar macrophages is given in the following.

Passive targeting of alveolar macrophages can be achieved with liposomes by taking advantage of the proclivity of macrophages to engulf particulates. It has been shown in both *in vitro* [117] and *in vivo* [118] studies that liposomes are taken up by alveolar macrophages. This uptake has been put to a practical use by using liposomes to deliver encapsulated dichloromethylene diphosphonate to deplete alveolar macrophages [119, 120]. The potential of liposomes to passively target antibiotics to alveolar macrophages has been demonstrated *in vitro* against *M. avium-intracellulare* by Wichert et al., who found that treating the infected cells with liposome-encapsulated amikacin enabled a 100-fold improvement in apparent killing efficacy for the drug [121]. In an *in vivo* study, Conley et al. aerosolized liposome-encapsulated ciprofloxacin to mice infected with *Francisella tularensis*, which is a facultative intracellular bacterium that uses macrophages as a host [122]. The researchers found that all mice treated with the liposomal ciprofloxacin survived the infection, whereas all mice treated with free ciprofloxacin died. The researchers suggested that the efficacy of liposomal ciprofloxacin was possibly due to the enhanced delivery of the drug to macrophages and the prolonged drug retention in the lungs afforded by liposomes.

To improve the selectivity of macrophage targeting, receptors on the macrophage may be exploited. These include receptors for terminal mannose moieties, and receptors for nonimmune opsonins (e.g., vitronectin, fibronectin, and surfactant protein A) and immune opsonins (e.g., IgG). Several *in vitro* studies have tested the principle of actively targeting pulmonary macrophages. Shao and Ma investigated the ability of mannose-bearing liposomes to deliver fluorescein to alveolar macrophages [123]. They found that the mannose-bearing liposomes delivered a significantly greater amount of fluorescein to the macrophages than did liposomes without mannose. Thus, surface mannose residues can improve alveolar macrophage targeting of liposomes.

Mannose can also improve targeting for polymeric drug carriers. Liang et al. produced a complex of mannosylated polylysine, electrostatically coupled to a fluorescent oligonucleotide [124]. On incubation with alveolar macrophages, they found that the mannosylated complex caused a significantly greater macrophage uptake of the oligonucleotide compared with the complex without mannose.

Polylysine can also be coupled with other ligands to target alveolar macrophages. Rojanasakul et al. linked the immune opsonin IgG to polylysine in an effort to selectively deliver a plasmid to rat alveolar macrophages [125]. The researchers in this study found a significantly greater gene expression in the macrophages treated with the polylysine-IgG-plasmid complex compared with those treated with plasmid alone, plasmid-IgG, or plasmid-polylysine.

Targeting ligands can also be used in the absence of drug carriers to target agents to alveolar macrophages. Harrison et al. conjugated IgG to catalase, which can be used to mitigate injury caused by oxidants produced by alveolar macrophages [126]. They found that the IgG-catalase conjugate (and not free catalase) inhibited an induced intracellular oxidation and injury in rat alveolar macrophages. Thus, a particulate or polymeric carrier was not needed to enable uptake of the enzyme into the macrophages.

The foregoing *in vitro* studies demonstrate that active macrophage targeting is possible using a variety of ligand-based approaches. However, when performing *in vivo* studies of targeting alveolar macrophages, there are certain issues to consider. Lung macrophages are phenotypically and morphologically diverse and are found throughout the lungs [127, 128]. They are present not only in the alveoli, but also in the airways, the connective tissue, the pleura, and the pulmonary vasculature. Even within the alveoli, there exist subpopulations with different properties such as receptor expression and phagocytic capacity. Furthermore, the subpopulations may shift during an inflammatory response, which may occur during the studies [128]. For example, there may be an increase in the relative number of monocytes, which are the precursors of alveolar macrophages.

The route of pulmonary administration is an additional concern, because there may be differences in the delivery of intratracheally instilled material versus the more clinically relevant inhaled material [129]. For example, inhalation can produce a more homogeneous distribution among lung macrophages, whereas an instilled dose may result in significant fraction of empty macrophages. The large fraction of empty macrophages may be due to a failure of the instilled dose to reach the peripheral macrophages that are part of the measured population [129], or it may reflect the heterogeneity of phagocytic activity of macrophages that are recruited to the conducting airways [130]. In summary, a variety of phenomena can influence pulmonary macrophage targeting. It may also be worth considering that these phenomena may affect pulmonary delivery studies in which macrophages are not the target, but are still influential in the fate of the drug delivery system.

## ABBREVIATIONS

CH, cholesterol; DLPC, dilaurylphosphatidylcholine; DMPC, dimyristoylphosphatidylcholine; DMPG, dimyristoylphosphatidylglycerol; DPPC, dipalmitoylphosphatidylcholine; DPPG, dipalmitoylphosphatidylglycerol, DSPC, distearoylphosphatidylcholine; DSPG, distearoylphosphatidylglycerol; MLV, multilameller vesicle; PA, phosphatidic acid; PC, phosphatidylcholine; TA, triamcinolone acetonide.

## REFERENCES

1. R. N. Brogden, R. C. Heel, T. M. Speight, and G. S. Avery, Beclomethasone dipropionate. A reappraisal of its pharmacodynamic properties and therapeutic efficacy after a decade of use in asthma and rhinitis, *Drugs 28*:99 (1984).
2. R. N. Brogden, and D. McTavish, Budesonide. An updated review of its pharmaco-

logical properties, and therapeutic efficacy in asthma and rhinitis [published errata appear in Drugs 1992 Dec;44(6):1012 and 1993 Jan;45(1):130], *Drugs 44*:375 (1992).

3. S. M. Holliday, D. Faulds, and E. M. Sorkin, Inhaled fluticasone propionate. A review of its pharmacodynamic and pharmacokinetic properties, and therapeutic use in asthma, *Drugs 47*:318 (1994).

4. G. E. Pakes, R. N. Brogden, R. D. Heel, T. M. Speight, and G. S. Avery, Flunisolide: a review of its pharmacological properties and therapeutic efficacy in rhinitis, *Drugs 19*:397 (1980).

5. C. M. Spencer, and D. McTavish, Budesonide. A review of its pharmacological properties and therapeutic efficacy in inflammatory bowel disease, *Drugs 50*:854 (1995).

6. B. J. Lipworth, Clinical pharmacology of corticosteroids in bronchial asthma, *Pharmacol. Ther. 58*:173 (1993).

7. H. W. Kelly, Establishing a therapeutic index for the inhaled corticosteroids: part I. Pharmacokinetic/pharmacodynamic comparison of the inhaled corticosteroids, *J. Allergy Clin. Immunol. 102*:S36 (1998).

8. A. Kamada, S. Szefler, R. Martin, H. Boushey, V. Chinchilli, J. Drazen, J. Fish, E. Israel, S. Lazarus, R. Lemanske, and A. C. R. Group, Issues in the use of inhaled glucocorticoids, *Am. J. Respir. Crit. Care Med. 153*:1739 (1996).

9. S. Edsbaecker, and S. J. Szefler, Glucocorticoid pharmacokinetics—Principles and clinical applications, *Inhaled Glucocorticoids in Asthma*, (R. P. Schleimer, W. W. Busse, and P. M. O'Byrne, eds.), Marcel Dekker, New York, 1997, p. 381.

10. M. Johnson, Pharmacodynamics and pharmacokinetics of inhaled glucocorticoids, *J. Allergy Clin. Immunol. 97*:169 (1996).

11. I. Pavord, and A. Knox, Pharmacokinetic optimisation of inhaled steroid therapy in asthma, *Clin. Pharmacokinet. 25*:126 (1993).

12. G. Hochhaus, H. Moellmann, H. Derendorf, and R. Gonzalez-Rothi, Pharmacokinetic/pharmacodynamic aspects of aerosol therapy using glucorticoids as a model, *J. Clin. Pharmacol. 37*:881 (1997).

13. G. Hochhaus, and H. Moellmann, Beta-agonists: terbutaline, albuterol, and fenoterol. *Handbook of Pharmacokinetic/Pharmacodynamic Correlations* (H. Derendorf and G. Hochhaus). CRC, New York, 1995, p. 299.

14. E. Dahlberg, A. Thalen, R. Brattsand, J.-A. Gustafsson, U. Johansson, K. Roemke, and T. Saartrok, Correlation between chemical structure, receptor binding, and biological activity of some novel, highly active, 16a, 17a- acetal- substituted- glucocorticoids, *Mol. Pharmacol. 25*:70 (1984).

15. G. Hochhaus, P. Rohdewald, H. Moellmann, and D. Grechuchna, Identification of glucocorticoid receptors in normal and neoplastic human lungs, *Respir. Exp. Med. (Berlin) 182*:71 (1983).

16. P. Rohdewald, H. Möllmann, and G. Hochhaus, Affinities of glucocorticoids for glucocorticoid receptors in the human lung, *Agents Action 17*:290 (1985).

17. P. Druzgala, G. Hochhaus, and N. Bodor, Soft drugs 10: Blanching activity and receptor binding affinity of a new type of glucocorticoid: Ioteprednol etabonate, *J. Steroid Biochem. 38*:149 (1991).

18. M. Beato, G. G. Rousseau, and P. Feigelson, Correlation between glucocorticoid

binding to specific liver cytosol receptors and enzyme induction, *Biochem. Biophys. Res. Commun. 47*:1464 (1972).

19.  B. M. Vayssiere, S. Dupont, A. Choquart, F. Petit, T. Garcia,C. Marchandeau, H. Gronemeyer, and M. Resche-Rigon, Synthetic glucocorticoids that dissociate transactivation and AP-1 transrepression exhibit antiinflammatory activity in vivo, *Mol. Endocrinol. 11*:1245 (1997).

20.  G. Hochhaus, and H. Derendorf, Dose optimization based on pharmacokinetic/ pharmacodynamic modeling, *Handbook of Pharmacokinetic/Pharmacodynamic Correlations* (H. Derendorf and G. Hochhaus, eds.). CRC, New York, 1995, p. 79.

21.  R. Brattsand, and B. I. Axelsson, Basis of airway selectivity of inhaled glucocorticoids, *Inhaled Glucocorticoids in Asthma*, (R. P. Schleimer, W. W. Busse, and P. M. O'Byrne, eds.), Marcel Dekker, New York, 1997, p. 351.

22.  F. Chanoine, C. Grenot, P. Heidmann, and J. L. Junien, Pharmacokinetics of butixocort 21-propionate, budesonide, and beclomethasone dipropionate in the rat after intratracheal, intravenous, and oral treatments, *Drug Metab. Dispos. 19*:546 (1991).

23.  P. S. Burge, J. Efthimiou, M. Turner-Warwick, and P. T. Nelmes, Double-blind trials of inhaled beclomethasone diproprionate and fluocortin butyl ester in allergen-induced immediate and late asthmatic reactions, *Clin. Allergy 12*:523 (1982).

24.  G. Ventresca, A. Mackie, J. Moss, J. McDowall, and A. Bye, Absorption of oral fluticasone propionate in healthy subjects, *Am. J. Respir. Crit. Care Med. 149*:A214 (1994).

25.  C. Falcoz, A. Mackie, J. McDowall, J. McRae, L. Yogendran, G. Ventresca, and A. Bye, Oral bioavailability of fluticasone propionate in healthy subjects, *Br. J. Clin. Pharmacol. 41*:459P (1996).

26.  A. Ryrfeldt, P. Andersson, S. Edsbaecker, M. Tonnesson, D. Davies, and R. Pauwels, Pharmacokinetics and metabolism of budesonide, a selective glucocorticoid, *Eur. J. Respir. Dis. 63 (Suppl. 122)*:86 (1982).

27.  M. D. Chaplin, W. D. Rooks, E. W. Swenson, W. C. Cooper, C. Nerenberg, and N. I. Chu, Flunisolide metabolism and dynamics of a metabolite, *Clin. Pharmacol. Ther. 27*:402 (1980).

28.  G. Dickens, D. Wermeling, C. Matheney, W. John, W. Abramowitz, S. Sista, T. Foster, and C. S., Flunisolide administered via metered dose inhaler with and without spacer and following oral administration, *J. Allergy Clin. Immunol. 103, S135 (1999) 103*:S132 (1999).

29.  H. Derendorf, G. Hochhaus, S. Rohatagi, H. Möllmann, J. Barth, and M. Erdmann, Oral and pulmonary bioavailability of triamcinolone acetonide, *J. Clin. Pharmacol. 35*:302 (1995).

30.  S. Rohatagi, G. R. Rhodes, and P. Chaikin, Absolute oral versus inhaled bioavailability: significance for inhaled drugs with special reference to inhaled glucocorticoids, *J. Clin. Pharmacol. 39*:661 (1999).

31.  G. R. Manchee, A. Barrow, S. Kulkarni, E. Palmer, J. Oxford, P. V. Colthup, J. G. Maconochie, and M. H. Tarbit, Disposition of salmeterol xinafoate in laboratory animals and humans, *Drug Metab. Dispos. 21*:1022 (1993).

32.  G. Hochhaus, S. Suarez, R. J. Gonzales-Rothi, and H. Schreier, Pulmonary targeting of inhaled glucocorticoids: How is it influenced by formulation?, *Respiratory Drug Delivery VI* (R. Dalby, P. Byron, and S. J. Farr, eds.), Interpharm Press, 1998, p. 45.

33. A. Ben-Jebria, D. Chen, M. L. Eskew, R. Vanbever, R. Langer, and D. A. Edwards, Large porous particles for sustained protection from carbachol-induced bronchoconstriction in guinea pigs, *Pharm. Res. 16*:555 (1999).

34. D. A. Edwards, A. Ben-Jebria, and R. Langer, Recent advances in pulmonary drug delivery using large, porous inhaled particles, *J. Appl. Physiol. 85*:379 (1998).

35. D. A. Edwards, J. Hanes, G. Caponetti, J. Hrkach, A. Ben-Jebria, M. L. Eskew, J. Mintzes, D. Deaver, N. Lotan, and R. Langer, Large porous particles for pulmonary drug delivery, *Science 276*:1868 (1997).

36. A. Adjei, and J. Garren, Pulmonary delivery of peptide drugs: effect of particle size on bioavailability of leuprolide acetate in healthy male volunteers, *Pharm. Res. 7*:565 (1990).

37. I. Gonda, A semi-empirical model of aerosol deposition in the human respiratory tract for mouth inhalation, *J. Pharm. Pharmacol. 33*:692 (1981).

38. S. P. Newman, D. Pavia, F. Moren, N. F. Sheahan, and S. W. Clarke, Deposition of pressurised aerosols in the human respiratory tract, *Thorax 36*:52 (1981).

39. S. P. Newman, Aerosol deposition considerations in inhalation therapy, *Chest 88*: 152S (1985).

40. S. P. Newman, Aerosol physiology, deposition, and metered dose inhalers, *Allergy Proc. 12*:41 (1991).

41. L. Agertoft, and S. Pedersen, Importance of the inhalation device on the effect of budesonide, *Arch. Dis. Child. 69*:130 (1993).

42. L. Agertoft, and S. Pedersen, Influence of spacer device on drug delivery to young children with asthma, *Arch. Dis. Child. 71*:217 (1994).

43. R. Ahrens, C. Lux, T. Bahl, and S. H. Han, Choosing the metered-dose inhaler spacer or holding chamber that matches the patient's need: evidence that the specific drug being delivered is an important consideration, *J. Allergy Clin. Immunol. 96*: 288 (1995).

44. P. W. Barry, and O. C. C, The effect of delay, multiple actuations and spacer static charge on the in vitro delivery of budesonide from the Nebuhaler, *Br. J. Clin. Pharmacol. 40*:76 (1995).

45. S. P. Newman, F. Moren, D. Pavia, F. Little, and S. W. Clarke, Deposition of pressurized suspension aerosols inhaled through extension devices, *Am. Rev. Respir. Dis. 124*:317 (1981).

46. O. Selroos, and M. Halme, Effect of a volumatic spacer and mouth rinsing on systemic absorption of inhaled corticosteroids from a metered dose inhaler and dry powder inhaler, *Thorax 46*:891 (1991).

47. S. J. Farr, A. M. Rowe, R. Rubsamen, and G. Taylor, Aerosol deposition in the human lung following administration from a microprocessor controlled pressurised metered dose inhaler., *Thorax 50*:639 (1995).

48. M. E. Ward, A. Woodhouse, L. E. Mather, S. J. Farr, J. K. Okikawa, P. Lloyd, J. A. Schuster, and R. M. Rubsamen, Morphine pharmacokinetics after pulmonary administration from a novel aerosol delivery system, *Clin. Pharmacol Ther 62*:596 (1997).

49. S. Newman, K. Steed, S. Reader, G. Hooper, and B. Zierenberg, Efficient delivery to the lungs of flunisolide aerosol from a new portable hand-held multidose nebulizer, *J. Pharm. Sci. 85*:960 (1997).

50.  S. P. Newman, J. Brown, K. P. Steed, S. J. Reader, and H. Kladders, Lung deposition of fenoterol and flunisolide delivered using a novel device for inhaled medicines: Comparison of RESPIMAT with conventional metered-dose inhalers with and without spacer devices, *Chest 113*:957 (1998).
51.  C. Leach, Targeting inhaled steroids, *Int. J. Clin. Pract. Suppl. 96*:23 (1998).
52.  C. L. Leach, P. J. Davidson, and R. J. Boudreau, Improved airway targeting with the CFC-free HFA-beclomethasone metered-dose inhaler compared with CFC-beclomethasone, *Eur. Respir. J. 12*:1346 (1998).
53.  S. P. Newman, K. P. Steed, G. Hooper, J. I. Jones, and F. C. Upchurch, Improved targeting of beclomethasone diprorionate (250 micrograms metered dose inhaler) to the lungs of asthmatics with the Spacehaler [in process citation], *Respir. Med. 93*:424 (1999).
54.  R. Pauwels, S. Newman, and L. Borgstrom, Airway deposition and airway effects of antiasthma drugs delivered from metered-dose inhalers, *Eur. Respir. J. 10*:2127 (1997).
55.  B. J. Lipworth, and D. J. Clark, Lung delivery of salbutamol given by breath activated pressurized aerosol and dry powder inhaler devices, *Pulm. Pharmacol. Ther. 10*:211 (1997).
56.  M. Hill, Effect of delivery mode on pharmacokinetics of inhaled drugs: Experience with beclomethasone, *Respiratory Drug Delivery VI.* (R. Dalby, P. Byron, and S. J. Farr, eds.), Interpharm Press, Englewood, CO, 1998, p. 53.
57.  L. Thorsson, S. Edsbacker, and T. B. Conradson, Lung deposition of budesonide from Turbuhaler is twice that from a pressurized metered-dose inhaler P-MDI, *Eur. Respir. J. 7*:1839 (1994).
58.  B. J. Lipworth, and D. J. Clark, Comparative lung delivery of salbutamol given via Turbuhaler and Diskus dry powder inhaler devices, *Eur. J. Clin. Pharmacol. 53*:47 (1997).
59.  S. Newman, K. Steed, G. Hooper, A. Kallen, and L. Borgstrom, Comparison of gamma scintigraphy and a pharmacokinetic technique for assessing pulmonary deposition of terbutaline sulphate delivered by pressurized metered dose inhaler, *Pharm. Res. 12*:231 (1995).
60.  L. Borgstroem, and M. Nilsson, A method for determiniation of the absolute pulmonary bioavailability of inhaled drugs: Terbutaline, *Pharm. Res. 7*:1068 (1990).
61.  L. Borgstrom, Local versus total systemic bioavailability as a means to compare different inhaled formulations of the same substance, *J. Aerosol. Med. 11*:55 (1998).
62.  R. A. Brown, Jr., and L. S. Schanker, Absorption of aerosolized drugs from the rat lung, *Drug. Metab. Dispos. 11*:355 (1983).
63.  J. A. Burton, and L. S. Schanker, Absorption of sulphonamides and antitubercular drugs from the rat lung, *Xenobiotica 4*:291 (1974).
64.  J. A. Burton, T. H. Gardiner, and L. S. Schanker, Absorption of herbicides from the rat lung, *Arch. Environ. Health 29*:31 (1974).
65.  J. A. Burton, and L. S. Schanker, Absorption of antibiotics from the rat lung, *Proc. Soc. Exp. Biol. Med. 145*:752 (1974).
66.  J. A. Burton, and L. S. Schanker, Absorption of corticosteroids from the rat lung, *Steroids 23*:617 (1974).
67.  B. Meibohm, H. Moellmann, M. Wagner, G. Hochhaus, A. Moellmann, and H.

Derendorf, The clinical pharmacology of fluticasone propionate, *Rev. Contemp. Pharmacother. 9*:535 (1998).

68. R. Gonzales-Rothi, S. Suarez, G. Hochhaus, H. Schreier, A. Lukyanov, H. Derendorf, and T. Dalla Costa, Pulmonary targeting of liposomal triamcinolone acetonide phosphate, *Pharm. Res. 13*:1699 (1996).

69. S. Suarez, R. Gonzalez-Rothi, H. Schreier, and G. Hochhaus, The effect of dose and release rate on pulmonary targeting of of liposomal triamcinolone acetonide phosphate, *Pharm. Res. 15*:461 (1998).

70. R. Brattsand, and B. I. Axelsson, Basis of airway selectivity of inhaled glucocorticoids. *Inhaled Glucocorticoids in Asthma.* (W. W. B. R. P. Schleimer, and P. M. O'Byrne, eds.), Marcel Dekker, New York, 1997, p. 351.

71. J. Talton, G. Hochhaus, J. Fitz-Gerald, and R. Singh. Pulsed laser deposited polymer films onto pulmonary dry powders for improved drug delivery. MRS 1998 Fall Meeting, Boston, 1998, p. 597.

72. S. A. Green, A. P. Spasoff, R. A. Coleman, M. Johnson, and S. B. Liggett, Sustained activation of a G protein-coupled receptor via "anchored" agonist binding. Molecular localization of the salmeterol exosite within the 2-adrenergic receptor, *J. Biol. Chem. 271*:24029 (1996).

73. G. P. Anderson, A. Linden, and K. F. Rabe, Why are long-acting beta-adrenoceptor agonists long-acting? *Eur. Respir. J. 7*:569 (1994).

74. A. Tunek, K. Sjodin, and G. Hallstrom, Reversible formation of fatty acid esters of budesonide, an antiasthma glucocorticoid, in human lung and liver microsomes, *Drug Metab. Dispos. 25*:1311 (1997).

75. A. Miller-Larsson, H. Mattsson, E. Hjertberg, M. Dahlback, A. Tunek, and R. Brattsand, Reversible fatty acid conjugation of budesonide. Novel mechanism for prolonged retention of topically applied steroid in airway tissue, *Drug Metab. Dispos. 26*:623 (1998).

76. E. Wieslander, E. L. Delander, L. Jarkelid, E. Hjertberg, A. Tunek, and R. Brattsand, Pharmacologic importance of the reversible fatty acid conjugation of budesonide studied in a rat cell line in vitro, *Am. J. Respir. Cell. Mol. Biol. 19*:477 (1998).

77. B. E. Gilbert, and V. Knight, Pulmonary delivery of antiviral drugs in liposome aerosols, *Semin. Pediatr. Infer. Dis. 7*:148 (1996).

78. R. J. Gonzalez-Rothi, and H. Schreier, Pulmonary delivery of liposome encapsulated drugs in asthma therapy, *Clin. Immunother. 4*:331 (1995).

79. K. M. G. Taylor, and S. J. Farr, Liposomes for drug delivery to the respiratory tract, *Drug. Dev. Ind. Pharm. 19*:123 (1993).

80. H. Schreier, R. J. Gonzalez-Rothi, and A. A. Stecenko, Pulmonary delivery of liposomes, *J. Control. Rel. 24*:209 (1993).

81. D. Meisner, Liposomes as a pulmonary drug delivery system, *Pharmaceutical Particulate Carriers*, (A. Rolland, ed.), Marcel Dekker, New York, 1993, p. 31.

82. I. W. Kellaway, and S. J. Farr, Liposomes as drug delivery systems to the lungs. *Adv. Drug Del. Rev. 5*:149 (1990).

83. H. N. McCullough, and R. L. Juliano, Organ-selective action of an anti-tumor drug: Pharmacologic studies of liposome-encapsulated β-cytosine arabinoside administered via the respiratory system of rats, *J. Natl. Cancer Inst. 3*:727 (1979).

84. K. M. G. Taylor, G. Taylor, I. W. Kellaway, and J. Stevens, The influence of liposome encapsulation on sodium cromoglycate pharmacokinetics in man, *Pharm. Res. 6*:633 (1989).
85. H. Schreier, K. J. McNicol, M. Ausborn, D. M. Soucy, H. Derendorf, A. A. Stecenko, and R. J. Gonzalez-Rothi, Pulmonary delivery of amikacin liposomes and acute liposome toxicity in sheep, *Int. J. Pharm. 87*:183 (1992).
86. R. M. Fielding, and R. M. Abra, Factors affecting the release rate of terbutaline from liposome formulations after intratracheal instillation in the guinea pig, *Pharm. Res. 9*:220 (1992).
87. A. Omri, C. Beaulac, M. Bouhajib, S. Montplaisir, S. Sharkawi, and J. Lagace, Pulmonary retention of free and liposome-encapsulated tobramycin after intratracheal administration in uninfected rats and rats infected with Pseudomonas aeruginosa, *Antimicrob. Agents Chemother. 38*:1090 (1994).
88. E. A. Poyner, H. O. Alpar, A. J. Almeida, M. D. Gamble, and M. R. Brown, Comparative study on the pulmonary delivery of tobramycin encapsulated into liposomes and PLA microspheres following intravenous and endotracheal delivery, *J. Contr. Rel.* 41 (1995).
89. C. Beaulac, S. Clement-Major, J. Hawari, and J. Lagace, Eradication of mucoid Pseudomonas aeruginosa with fluid liposome-encapsulated tobramycin in an animal model of chronic pulmonary infection, *Antimicrob. Agents Chemother 40*:665 (1996).
90. C. Bealulac, S. Clement-Major, J. Hawar, and J. Lagace, In vitro kinetics of drug release and pulmonary retention of microencapsulated antibiotic and liposomal formulations in relation to lipid composition, *J. Microencapsulation 14*:335 (1997).
91. B. E. Gilbert, P. R. Wyde, and S. Z. Wilson, Aerosolized liposomal amphotericin B for treatment of pulmonary and systemic Cryptococcus neoformans infections in mice, *Antimicrob. Agents Chemother. 36*:1466 (1992).
92. S. D. Allen, K. N. Sorensen, M. J. Nejdi, C. Durrant, and R. T. Proffit, Prophylactic efficacy of aerosolized liposomal (AmBisome) and non-liposomal (Fungizone) amphotericin B in murine pulmonary aspergillosis, *J. Antimicrob. Chemother. 34*:1001 (1994).
93. M. P. Lambros, D. W. A. Bourne, S. A. Abbas, and D. L. Johnson, Disposition of aerosolized liposomal amphotericin B, *J. Pharm. Sci. 86*:1066 (1997).
94. B. E. Gilbert, P. R. Wyde, G. Lopez-Berestein, and S. Z. Wilson, Aerosolized amphotericin B-liposomes for treatment of systemic Candida infections in mice, *Antimicrob. Agents Chemother. 38*:356 (1994).
95. J. C. Waldrep, B. E. Gilbert, C. M. Knight, M. B. Black, P. W. Scherer, V. Knight, and W. Eschenbacher, Pulmonary delivery of beclomethasone liposome aerosol in volunteers: tolerance and safety, *Chest 111*:316 (1997).
96. M. Vidgren, J. C. Waldrep, J. Arppe, M. Black, J. A. Rodarte, W. Cole, and V. Knight, A study of [99m]technetium-labelled beclomethasone dipropionate dilauroylphosphatidlycholine liposome aerosol in normal volunteers, *Int. J. Pharm. 115*:209 (1995).
97. S. M. Saari, M. T. Vidgren, M. O. Koskinen, V. M. Turjanmaa, J. C. Waldrep, and M. M. Nieminen, Regional lung deposition and clearance of 99mTc-labeled

beclomethasone-DLPC liposomes in mild and severe asthma, *Chest 113*:1573 (1998).

98.　H. Schreier, A. N. Lukyanov, G. Hochhaus, and G.-R. R. J., Thermodynamic and kinetic aspects of the interaction of triamcinolone acetonide with liposomes, *Proc. Intern. Symp. Control. Rel. Bioact. Mater. 21*:228 (1994).

99.　Y. Lai, R. C. Mehta, A. A. Thacker, S. Yoo, P. J. McNamara, and P. P. DeLuca, Sustained bronchodilation with isoproterenol poly (glycolide-co-lactide) microspheres, *Pharm. Res. 10*:119 (1993).

100.　D. J. Armstrong, P. N. Elliott, J. L. Ford, D. Gadsdon, G. P. McCarthy, C. Rostron, and M. D. Worsley, Poly-(D,L-lactic acid) microspheres incorporating histological dyes for intra-pulmonary histopathological studies, *J. Pharm. Pharmacol. 48*:259 (1996).

101.　J. D. Talton, J. M. Fitzgerald, R. Singh, and G. Hochhaus, Pulmonary targeting of poly(lactic-co-glycolic acid)-coated budesonide dry powders vs. uncoated powders using anex vivoreceptor binding assay in rats, *AAPS Pharm. Sci. 1*: (1999).

102.　R. S. Pillai, D. B. Yeates, I. F. Miller, and A. J. Hickey, Controlled dissolution from wax-coated aerosol particles in canine lungs, *J. Appl. Physiol. 81*:1878 (1998).

103.　C. Evora, I. Soriano, R. A. Rogers, K. M. Shakesheff, J. Hanes, and R. Langer, Relating the phagocytosis of microparticles by alveolar macrophages to surface chemistry: the effect of 1,2-dipalmitoylphosphatidylcholine, *J. Contr. Rel. 51*:143 (1998).

104.　G. J. Poiani, M. Greco, J. K. Choe, J. D. Fox, and D. J. Riley, Liposome encapsulation improves the effect of antifibrotic agent in rat lung fibrosis, *Am. J. Respir. Crit. Care Med. 150*:1623 (1994).

105.　M. J. Greco, J. E. Kemnitzer, J. D. Fox, J. K. Choe, J. Kohn, D. J. Riley, and G. J. Poiani, Polymer of proline analogue with sustained antifibrotic activity in lung fibrosis, *Am. J. Respir. Crit. Care Med. 155*:1391 (1997).

106.　W. J. Muller, K. Zen, A. B. Fisher, and H. Shuman, Pathways for uptake of fluorescently labeled liposomes by alveolar type II cells in culture, *Am. J. Physiol. 269*: L11 (1995).

107.　B. J. Buckley, A. K. Tanswell, and B. A. Freeman, Liposome-mediated augmentation of catalase in alveolar type II cells protects against $H_2O_2$ injury, *J. Appl. Physiol. 63*:359 (1987).

108.　A. K. Tanswell, D. M. Olson, and B. A. Freeman, Liposome-mediated augmentation of antioxidant defenses in fetal rat pneumocytes, *Am. J. Physiol. 258*:L165 (1990).

109.　R. V. Padmanabhan, R. Gudapaty, I. E. Liener, B. A. Schwartz, and J. R. Hoidal, Protection against pulmonary oxygen toxicity in rats by the intratracheal administration of liposome-encapsulated superoxide dismutase or catalase, *Am. Rev. Respir. Dis. 132*:164 (1985).

110.　F. J. Walther, R. David-Cu, and S. L. Lopez, Antioxidant-surfactant liposomes mitigate hyperoxic lung injury in premature rabbits, *Am. J. Physiol. 269*:L613 (1995).

111.　M. L. Barnard, R. R. Baker, and S. Matalon, Mitigation of oxidant injury to lung microvasculature by intratracheal instillation of antioxidant enzymes, *Am. J. Physiol. 265*:L340 (1993).

112.　R. R. Baker, L. Czopf, T. Jilling, B. A. Freeman, K. Kirk, and S. Matalon, Quantities of alveolar distribution of liposome-entrapped antioxidant enzymes, *Am. J. Physiol. 263*:L585 (1992).

113.  F. J. Walther, R. David-Cu, M. C. Supnet, M. L. Longo, B. R. Fan, and R. Bruni, Uptake of antioxidants in surfactant liposomes by cultured alveolar type II cells is enhanced by SP-A, *Am. J. Physiol. 265*:L330 (1993).

114.  P. Briscoe, I. Caniggia, A. Graves, B. Benson, L. Huang, A. K. Tanswell, and B. A. Freeman, Delivery of superoxide dismutase to pulmonary epithelium via pH-sensitive liposomes, *Am. J. Physiol. 268*:L374 (1995).

115.  Y. Kuroki, R. J. Mason, and D. R. Voelker, Alveolar type II cells express a high-affinity receptor for pulmonary surfactant protein A, *Proc. Natl. Acad. Sci. USA 85*:5566 (1988).

116.  I. A. J. M. Bakker-Woudenberg, Delivery of antimicrobials to infected tissue macrophages, *Adv. Drug Del. Rev. 17*:5 (1995).

117.  R. J. Gonzalez-Rothi, J. Cacace, L. Straub, and H. Schreier, Liposomes and pulmonary alveolar macrophages: functional and morphological interactions, *Exp. Lung Res. 17*:687 (1991).

118.  M. A. Myers, D. A. Thomas, L. Straub, D. W. Soucy, R. W. Niven, M. Kaltenbach, C. I. Hood, H. Schreier, and R. J. Gonzalez-Rothi, Pulmonary effects of chronic exposure to liposome aerosols in mice, *Exp. Lung Res. 19*:1 (1993).

119.  J. T. Berg, S. T. Lee, T. Thepen, C. Y. Lee, and M. Tsan, Depletion of alveolar macrophages by liposome-encapsulated dichloromethylene diphosphonate, *J. Appl. Physiol. 74*: 2812 (1993).

120.  S. Worgall, P. L. Leopold, G. Wolff, B. Ferris, N. VanRoijen, and R. G. Crystal, Role of alveolar macrophages in rapid elimination of adenovirus vectors administered to the epithelial surface of the respiratory tract, *Hum. Gene Ther. 8*:1675 (1997).

121.  B. V. Wichert, R. J. Gonalez-Rothi, L. E. Straub, B. M. Wichert, and H. Schreier, Amikacin liposomes: preparation, characterization, and in vitro activity against Mycobacterium avium-intracellulare infection in alveolar macrophages, *Int. J. Pharm. 78*:227 (1992).

122.  J. Conley, H. Yang, T. Wilson, K. Blasetti, V. D. Ninno, G. Schnell, and J. P. Wong, Aerosol delivery of liposome-encapsulated ciprofloxacin: aerosol characterization and efficacy against Francisella tularensis infection in mice, *Antimicrob. Agents Chemother. 41*:1288 (1997).

123.  J. Shao, and J. K. H. Ma, Characterization of a mannosylphospholipid liposome system for drug targeting to alveolar macrophages, *Drug Delivery 4*:43 (1997).

124.  W. W. Liang, X. Shi, D. Deshpande, C. J. Malanga, and Y. Rojanasakul, Oligonucelotide targeting to alveolar macrophages by mannose receptor-mediated endocytosis, *Biochim. Biophys. Acta 1279*:227 (1996).

125.  Y. Rojanasakul, L. Y. Wang, C. J. Malanga, J. K. H. Ma, and J. Liaw, Targeted gene delivery to alveolar macrophages via Fc receptor-mediated endocytosis, *Pharm. Res. 11*:1731 (1994).

126.  J. Harrison, X. Shi, L. Y. Wang, J. K. H. Ma, and Y. Rojanasakul, Novel delivery of antioxidant enzyme catalase to alveolar macrophages by Fc receptor-mediated endocytosis, *Pharm. Res. 11*:1110 (1994).

127.  R. G. Canto, G. R. Robinson, and H. Y. Reynolds, Defense mechanisms of the

respiratory tract, *Pulmonary Infections and Immunity* (H. Chmel, M. Bendinelli, and H. Friedman, eds.), Plenum Press, New York, 1994, p. 1.

128. M. F. Lipscomb, D. E. Bice, C. R. Lyons, M. R. Schuyler, and D. Wilkes, The regulation of pulmonary immunity, *Adv. Immunol. 59*:369 (1995).

129. A. M. Dorries, and P. A. Valberg, Heterogeneity of phagocytosis for inhaled versus instilled material, *Am. Rev. Respir. Dis. 164*:831 (1992).

130. M. Geiser, M. Baumann, L. M. Cruz-Orive, V. I. Hof, U. Waber, and P. Gehr, The effect of particle inhalation on macrophage number and phagocytic activity in the intrapulmonary conducting airways of hamsters, *Am. J. Respir. Cell Mol. Biol. 10*: 594 (1994).

131. M. Saari, M. T. Vidgren, M. Koskinen, O., V. M. H. Turjanmaa, and M. M. Nieminen, Pulmonary distribution and clearance of two beclomethasone liposome formulations in healthy volunteers, *Int. J. Pharm. 181*:1 (1999).

132. B. V. Wichert, R. J. Gonzalez-Rothi, L. E. Straub, B. M. Wichert, and H. Schreier, Amikacin liposomes: preparation, characterization, and in vitro activity against *Mycobacterium avium-intracellulare* infection in alveolar macrophages, *Int. J. Pharmaceut. 78*:227 (1992).

133. G. Hochhaus, R. J. Gonzalez-Rothi, A. Lukyanov, H. Derendorf, H. Schrier, and T. D. Costa, Assessment of glucocorticoid lung targeting by ex-vivo receptor binding studies in rats, *Pharm. Res. 12*:134 (1995).

134. J. D. Talton, S. Suarez, R. Gonzales-Rothi, and G. Hochhaus, Pulmonary targeting of intratracheal triamcinolone acetonide dry-powder using an ex-vivo receptor binding assay, *Pharm. Sci. 1*:S (1998).

# 4

# Lipid-Based Formulations for Oral Administration

## Opportunities for Bioavailability Enhancement and Lipoprotein Targeting of Lipophilic Drugs

**Christopher J. H. Porter and William N. Charman**
*Monash University, Parkville, Victoria, Australia*

## I. INTRODUCTION

The identification of increasingly complex intracellular drug targets, escalating requirements in terms of drug potency and the use of combinatorial chemistry libraries and *in vitro* receptor-based activity assays, has led to a trend toward the identification of increasingly lipophilic lead molecules. The clinical usefulness of highly lipophilic, poorly water-soluble drugs, however, is often limited by their low and variable oral bioavailability, and although co-administration with lipids, lipidic excipients, or fatty meals may improve the bioavailability of lipophilic drugs, the mechanistic aspects of this enhancement are incompletely understood.

The inherent physicochemical properties of lipophilic drug molecules dictate that they become associated with endogenous lipidic microdomains, ranging from lipids and lipid digestion products within the gastrointestinal tract (GIT) to lymph and plasma lipoproteins in the systemic circulation. The affinity with which lipophilic drugs bind to, and interchange between, these carrier systems can have a significant impact on the free drug fraction available for absorption, distribution, metabolism, and excretion and can therefore play a major role in defining both drug pharmacokinetics and therapy.

Importantly, the nature and behavior of these lipid microdomains changes dramatically in response to the ingestion of exogenous lipid (as either food or formulation excipients), thereby influencing the disposition of lipophilic drugs. In this commentary, we will describe the known effects of co-administered lipids on drug absorption from the GIT, drug processing and metabolism within the enterocyte, and drug association with lymph and plasma lipoproteins.

The examples described will illustrate that co-administration of lipophilic drugs with lipids or lipid-based excipients may provide significant opportunities for enhancement of the biopharmaceutical profile of lipophilic drugs. Conversely, it is also apparent that the potential effects of lipids and lipid-based vehicles on drug distribution and metabolism may have an impact on the design and implementation of preclinical and clinical drug progression programs.

## II. LIPID DIGESTION AND ABSORPTION AND THE PREABSORPTIVE PHASE

Lipids, unlike many excipients, whether present in food or as discreet pharmaceutical additives, are processed both chemically and physically within the GIT before absorption and transport into the portal blood (or mesenteric lymph). Indeed, most of the effects mediated by formulation-based lipids or the lipid content of food are mediated by means of the products of lipid digestion—molecules that may exhibit *very* different physicochemical and physiological properties when compared with the initial excipient or food constituent. Therefore, although administered lipids have formulation properties in their own right, many of their effects are mediated by species that are produced after transformation or "activation" in the GIT. An understanding of the luminal and/or enterocyte-based processing pathways of lipids and lipid systems is therefore critical to the effective design of lipid-based delivery systems.

The general process of lipid digestion is well known and well described in a number of recent publications [1–5]. Ingested triglycerides are digested by the action of lingual lipase in the saliva and gastric lipase and the pancreatic lipase/co-lipase complex in the stomach and small intestine, respectively. These sequential processes convert essentially water-insoluble, nonpolar triglyceride into progressively more polar diglycerides, monoglycerides, and fatty acids. The end point (chemically) of digestion of one molecule of triglyceride is the liberation of two molecules of fatty acid and one molecule of 2-monoglyceride.

In addition to the chemical breakdown of ingested lipids, the physical properties of lipid digestion products are markedly altered to facilitate absorption. Initial lipid digestion products become crudely emulsified on emptying from the stomach into the duodenum (because monoglycerides and diglycerides have some amphiphilic, emulsifying properties, and gastric emptying provides sufficient shear to provoke emulsification). The presence of partially digested emulsion in the small intestine leads to the secretion of bile salts and biliary lipids from the gallbladder that stabilize the surface of the lipid emulsion and reduce its particle size, presenting a larger lipid surface area to the pancreatic lipase/co-lipase digestive enzymes. In the presence of sufficient bile salt concentrations, the products of lipid digestion are finally incorporated into bile salt micelles to form a solubilized system consisting of fatty acids, monoglycerides, bile salts, and phospholipid— the so-called intestinal mixed micellar phase. The intestinal mixed micellar phase co-exists with a number of physical species in the small intestine, including multi-lamellar and unilamellar lipid vesicles, simple lipid solutions, and fatty acid soaps [6, 7].

The complexity and dynamism of the postdigestive intestinal contents (in terms of the interconversion and equilibrium-driven transfer of lipids across the various dispersed species) is a likely contributor to the uncertainty in defining the effects of lipids on drug absorption. Conversely, a more complete understanding of this preabsorptive phase and its interaction with lipophilic drugs will enhance appreciation of the effects of lipids on drug absorption and improve the ability to select appropriate lipid excipients.

Solubilization of lipid digestion products in intestinal mixed micelles enhances their dissolution and dramatically increases the GI lumen-enterocyte concentration gradient that drives absorption by means of passive diffusion. Micelles, however, are not absorbed intact [8, 9], and lipids are thought to be absorbed from a monomolecular intermicellar phase in equilibrium with the intestinal micellar phase [10]. The dissociation of monomolecular lipid from the micellar phase appears to be stimulated by the presence of an acidic microclimate associated with the enterocyte surface [11, 12]. In addition to passive diffusion, growing evidence suggests that active uptake processes mediated by transport systems located in the enterocyte membrane are also involved in the absorption of (in particular) fatty acids into the enterocyte [4].

After-absorption, the biological fate of lipid digestion products is defined primarily by the chain length of the absorbed lipid. Short- and medium-chain length fatty acids are much less water insoluble than longer chain lipids, and they diffuse relatively unhindered across the enterocyte into the portal blood [13]. Long-chain lipids, however, are trafficked through the endoplasmic reticulum, re-esterified to triglyceride, assembled into lymph lipoproteins, and secreted into the intestinal lymph [2]. Subsequently, long-chain lipids are transported through the intestinal lymph and into the central lymph, before entering the systemic circulation at the junction of the thoracic lymph duct and the left internal jugular vein in the neck. After entering the systemic circulation, the poor water solubility of lipids dictates their association with endogenous carrier systems such as plasma proteins and plasma lipoproteins. These carrier systems facilitate the distribution of lipids to peripheral tissues, where they are either stored as fat deposits, metabolized as an energy source, or used as a structural building block in lipidic structures such as membranes. The interested reader is directed to the following reviews for more details [14, 15].

## III. THE INTERACTION OF LIPOPHILIC DRUGS WITH THE LIPID DIGESTION/ABSORPTION CASCADE

The inherent physicochemical similarities between many lipophilic drugs and dietary and/or formulation-derived lipids in terms of high partition co-efficients and low water solubilities suggests that the processes that control lipid digestion, absorption, and distribution may similarly affect the disposition of lipophilic drugs. Therefore, the co-administration of lipids might be expected to have an impact on the disposition of lipophilic drugs in the following ways:

1.  By stimulating the release of biliary and pancreatic secretions, thereby providing an intestinal micellar phase into which a poorly water-soluble drug may become solubilized—increasing its effective solubility, dissolution rate, lumen-to-enterocyte concentration gradient and, consequently, extent of absorption. Increasing evidence suggests that co-administered lipids also have significant effects on drug absorption and metabolism at a cellular level through attenuation of enterocyte-based metabolic and antitransport processes.
2.  By enhancing the formation and turnover of lymph lipoproteins through the enterocyte and provoking, or improving, the targeting of orally administered lipophilic drugs to the intestinal lymphatics.
3.  By altering the relative proportions and constituents of plasma lipoproteins and changing the degree of binding of lipophilic drugs to discreet lipoprotein (LP) subclasses. The presence of specific receptors for lipoprotein subclasses such as the low-density lipoprotein receptor suggests that alteration of LP-binding profiles may have a significant impact on both pharmacokinetic issues such as drug clearance and volume of distribution and on pharmacodynamic end points such as toxicity and activity.

Lipids may also have effects on gastric transit (in terms of delaying gastric emptying) and intestinal permeability (enhancing the absorption of poorly permeable compounds). These areas have been well covered in other texts [16–19].

## IV.  LIPIDS AND BIOAVAILABILITY ENHANCEMENT

Armstrong and James have reviewed much of the pre-1980 literature [20], and Humberstone and Charman have recently addressed the post-1980 literature [21] detailing the use of lipid-based dose forms to enhance the bioavailability of lipophilic drugs. Similarly, the bioavailability-enhancing effects of food (in cases where the effect has been attributed to the lipid content of food) have been recently reviewed elsewhere [22].

In this chapter we will provide a brief overview of the early approaches to bioavailability enhancement by use of simple lipid-based delivery systems (lipid solutions, emulsions etc), and then describe recent progress in the application of self-emulsifying- and microemulsion-based formulations. The effects of lipids on the oral bioavailability of co-administered poorly water-soluble drugs may also be classified from a mechanistic (and to a degree, historical) perspective as "physicochemically" mediated effects (solubility, dissolution, surface area) and "biochemically" mediated effects (metabolism, transport related events), and these will be approached separately. It is readily apparent, however, that in many cases physicochemically and biochemically mediated mechanisms will operate side by side. In some instances, bioavailability may also be enhanced by the stimulation of intestinal lymphatic transport, and these studies will be addressed in a separate section.

## A. Bioavailability Enhancement by Means of Physicochemical Mechanisms

Simple suspensions and solutions of drugs in lipids have been shown to enhance the oral bioavailability of a number of poorly water-soluble compounds, including phenytoin, progesterone, and cinarrizine [23–28]. In these examples, bioavailability enhancement appears to have been mediated by way of improved drug dissolution from lipid solutions (compared with aqueous suspensions) and enhanced drug solubility in the lipid/bile salt–rich GI contents. Optimal bioavailability enhancement was generally provided by lipids in which the drug was most soluble, although factors including the solubility of the lipid in the GI fluids (short-chain lipids typically dissolve in the intestinal lumen leading to drug precipitation) and the ability of long-chain lipids to stimulate lymphatic transport complicate choice of the optimal lipid.

As a consequence of the intestinal processing that lipids undergo before absorption, there has been significant interest in assessing the "digestibility" of formulation lipids as a potential indicator of *in vivo* bioavailability enhancement. In this regard, digestible lipids such as dietary fats (triglycerides, diglycerides, fatty acids, phospholipids, cholesterol, etc) are generally more effective in terms of bioavailability enhancement than indigestible oils such as mineral oil [29–31]. However, more complex correlations of lipid chain length (medium chain versus long chain lipids) or lipid class (triglycerides versus diglycerides or monoglycerides) with digestibility and bioavailability enhancement have met with little success.

The degree of dispersion of a lipid-based delivery system appears to have the most marked effect on the bioavailability of a co-administered drug, and this has stimulated many of the most recent articles in the literature. Clearly, by decreasing the particle size of a dispersed formulation, the surface area available for lipid digestion and drug release or transfer is enhanced. In this regard, the bioavailability of griseofulvin [32, 33], phenytoin [23], penclomedine [30], danazol [34], REV 5901 [35], and, more recently, ontazolast [36] has been shown to be enhanced after administration in an emulsion formulation compared with a tablet, aqueous solution, or suspension formulation. It is not clear in these cases how much more efficient the emulsion formulation would have been compared with a simple lipid solution.

In many cases the relatively complex nature of lipid-based formulations in terms of lipid class, chain length, degree of dispersion, and choice of surfactant makes explanation of the mechanistic information difficult. For example, the bioavailability of vitamin E after administration of vitamin E acetate is greater after administration in a medium-chain triglyceride (MCT)–based emulsion compared with a long-chain triglyceride (LCT)–based lipid solution; however, the differential roles of lipid dispersion or lipid class (MCT vs. LCT) cannot be separated [37].

Although emulsion formulations show great promise for the enhancement of lipophilic drug bioavailability, the limited acceptability of oral emulsions has led to the more recent development of self-emulsifying drug delivery systems or SEDDS. These systems are composed of an isotropic mixture of drug, lipid, and surfactant and are generally filled into a soft or sealed hard gelatin capsule. After administration, the capsule ruptures, and an emulsion is spontaneously formed on contact with the intestinal fluids. The optimized interfacial properties (i.e., low interfacial tension) of these systems facilitate spontaneous emulsification and also result in the formation of emulsions with particle sizes that are generally lower than that formed with conventional emulsions ($<1$ μm), providing additional benefits in terms of enhanced surface areas of interaction.

The technology to formulate SEDDS is not new and was developed many years ago in the agrochemical industry. However, the application of SEDDS technology to oral pharmaceuticals has only recently been exploited with the identification of generally recognized as safe (GRAS) status lipids and surfactants with the appropriate physicochemical properties. A large number of publications have subsequently described the factors affecting the formulation of SEDDS and have discussed various methods of *in vitro* assessment of emulsification [38–45]. Descriptions of the *in vivo* bioavailability enhancing effects of SEDDS formulations are limited and include progesterone [46] and various experimental compounds including WIN 54954 (an antiviral) [47], L-365260 (a $CCK_B$ antagonist) [48], Ro 15-0778 (a naphthalene derivative) [49], 8-methoxypsoralen [50], tebufalone [51], L-683,453 [52], and tirlakiren [53].

The most recent development (in terms of physicochemical/particle size approaches) in the design of lipid-based delivery systems has been the use of microemulsions, microemulsion preconcentrates, or self-microemulsifying drug delivery systems (SMEDDS), typified by the Sandimmun Neoral formulation. Microemulsions are defined as isotropic, transparent, and thermodynamically stable (in contrast to conventional emulsions) mixtures of a hydrophobic phase (lipid), a hydrophilic phase (often water), a surfactant, and in many cases a co-surfactant. From a lipid formulation perspective, microemulsions are generally regarded as the ultimate extension of the "decreased particle size/increased surface area" mantra, because emulsion particle sizes are usually less than 50 nm. Microemulsions also have additional pharmaceutical advantages in terms of their solubilizing capacity [54, 55], thermodynamic stability, and capacity for stable, infinite dilution.

The preferred method of delivery of oral microemulsion formulations is either as a combination of drug, lipid, and surfactant/cosurfactant (generally filled into soft or sealed hard gelatin capsules) that spontaneously microemulsify in the GIT, or as a microemulsion preconcentrate in which the dose form contains a small quantity of hydrophilic phase and is in itself a concentrated O/W or W/O microemulsion, which becomes diluted or phase inverted in the GI fluids.

In addition to their usefulness in the enhancement of oral bioavailability of lipophilic drugs, microemulsion formulations have found considerable application as potential delivery systems for peptides whose delivery is often limited by poor GI permeability. W/O microemulsions provide a convenient means of delivery of both permeability-enhancing lipids and water-soluble peptides. The GI permeability-enhancing effects of lipids and their use in the delivery of highly water-soluble compounds are reviewed elsewhere [18, 56, 59].

Perhaps the most well-known example of the bioavailability-enhancing effects of microemulsion formulations is that of cyclosporin. Cyclosporin (CY) is a cyclic undecapeptide with potent immunosuppressive activity, which is lipophilic (log P 2.92) and poorly water soluble. Early cyclosporin studies showed that immunosuppressive effects were evident after administration as a lipid solution in olive oil (a triglyceride consisting primarily of glycerides of long-chain fatty acids) but not after administration in Migliol 812 (a synthetic triglyceride of $C_{10-12}$ fatty acids) [60]. Subsequent studies confirmed these data and showed that the absorption of cyclosporin was significantly improved after administration in a lipid vehicle made up of glycerides of long-chain fatty acids compared with MCT [60]. Attempts at correlation with the relative digestabilities of the long-chain triglycerides (LCT) and the MCT were unsuccessful, suggesting that the rate of digestion of the lipid was not a limiting factor [61]. Behrens et al. subsequently proposed that the improved intestinal absorption of CY from LCT vehicles was a function of the improved capacity of the products of LCT digestion to intercalate into intestinal mixed micelles [62].

The first commercial formulation of cyclosporin (Sandimmun) contained drug, LCT, ethanol, and polyglycolized LCT administered as either an oral solution/dispersion in water or milk or filled into a soft gelatin capsule. The soft-gel formulation produced a crude oil in water emulsion on capsule rupture *in situ*. The biopharmaceutical performance of the Sandimmun formulation, however, was relatively poor, because absorption was variable, incomplete ($\approx 30\%$), and affected by food [63–67].

Subsequent studies showed that absorption could be enhanced by administration of emulsions with smaller particle sizes [68], and this stimulated a number of preclinical [69–71] and clinical [72] studies examining the benefit of microemulsion formulations for the enhancement of cyclosporin bioavailability. With some exceptions (e.g., solid formulations with slow dissolution profiles [72]), and notwithstanding the problems of cross-study and cross-species comparisons, the data showed that the use of microemulsions with smaller particle sizes could increase the oral bioavailability of cyclosporin compared with the Sandimmun formulation. As a caveat, however, a recent report on the bioavailability of cyclosporin after oral administration in either simple lipid solution combinations of long-chain monoglycerides and triglycerides or as a predispersed emulsion-like formulation containing the same lipids has suggested that the particle size of an emulsion formulation or an emulsion formed on dilution/emulsification in the intestine may not be as important in dictating bioavailability as previously thought [73].

The current proprietary cyclosporin formulation, Sandimmun Neoral, is a microemulsion formulation. Although the formulation details of the Neoral formulation are not generally available, the relative bioavailabilities of the Neoral formulation and the initial Sandimmun formulation have been reported. In a dose linearity study, the relative bioavailability of the Neoral formulation compared with the Sandimmun formulation varied from 1.74 at a 200-mg dose to 2.39 at an 800-mg dose, illustrating the usefulness of the microemulsion formulation and suggesting an approximate twofold increase in bioavailability from the microemulsion formulation [74]. Further studies showed that the absorption of cyclosporin from the Neoral formulation was significantly less variable [75] and less dependent on bile flow [76] than oral Sandimmun and that its absorption was unaffected by food [77]. In terms of its apparent lack of reliance on bile for absorption, it is not known whether cyclosporin is absorbed from the formulation directly or just requires much lower bile salt concentrations to facilitate absorption.

Halofantrine (Hf) is a new phenanthrenemethanol antimalarial that shares many physicochemical similarities with cyclosporine and has been the subject of a number of investigations in our laboratory. Hf is orally active, well tolerated, and is finding increasing use in the treatment of malaria associated with multidrug resitant strains of *Plasmodium falciparum*. However, Hf is extremely lipophilic (log P $\approx$ 8) and poorly water soluble (<10 $\mu$g/mL), and the bioavailability of Hf after oral administration of Hf.HCl tablets is low and variable.

Co-administration of Hf with food increases bioavailability up to threefold in humans [78] and 10-fold in beagle dogs [79]; however, the clinical application of co-administration with food is limited because of possible uncontrolled increases in plasma levels, leading to cardiac side effects including a lengthening of the $QT_c$ interval.

In an attempt to increase the oral bioavailbility of Hf, we developed three novel lipid-based formulations, the details of which are given in Table 1. The commercial Hf formulation is formulated as a hydrochloride salt; however, to improve the lipid solubility of Hf in lipid-based formulations, the free base of Hf was used in this study. All three formulations consisted of an isotropic mix of drug, lipids, surfactant, and ethanol filled into soft gelatin capsules. On capsule rupture and interaction with an aqueous environment (*in vitro* screens used distilled water or 0.1 N HCl), the systems designated as self-microemulsifying (SMEDDS) produced a clear/translucent microemulsion (particle size <50 nm) and the self-emulsifying system (SEDDS) produced a bright white emulsion (particle size approximately 200 nm). The three formulations were designed to probe the effects of both particle size (emulsion vs. microemulsion) and lipid class (long chain vs medium chain) on Hf bioavailability. Formulations were optimized by the construction of partial phase diagrams, in which the lipid phase consisted of either a 2/1 w/w mass ratio of Captex 355 to Capmul MCM (for the medium chain system) or a 1/1 mass ratio of soybean oil to Maisine 35/1 in the case of the long-chain systems. The quantity of Hf and ethanol was kept constant.

*In vivo* evaluation was performed as a four-treatment crossover study in four male beagles (three experimental formulations and an IV leg to facilitate bioavailability estimation). Animals were dosed orally with a single capsule and

**Table 1** Formulation Details for Three Novel Self-emulsifying Formulations of Halofantrine

| Components | Medium-chain triglyceride SEDDS (% w/w) | Medium-chain triglyceride SMEDDS (% w/w) | Long-chain triglyceride SMEDDS (% w/w) |
|---|---|---|---|
| Halofantrine | 5 | 5 | 5 |
| Captex 355 | 46.7 | 33.3 | — |
| Capmul MCM | 23.3 | 16.7 | — |
| Soybean oil | — | — | 29 |
| Maisine 35-1 | — | — | 29 |
| Cremophor EL | 15 | 35 | 30 |
| Absolute ethanol | 10 | 10 | 7 |

*Source*: Ref. 80.

50 mL of tap water. The mean plasma levels of Hf and Hfm are shown in Figure 1, and the bioavailability data are summarized in Table 2.

The mean absolute bioavailability of the free base of Hf after administration in the lipid-based formulations ranged from 52–67% [80]. This compares favorably with previous data in which the absolute bioavailability of Hf from the com-

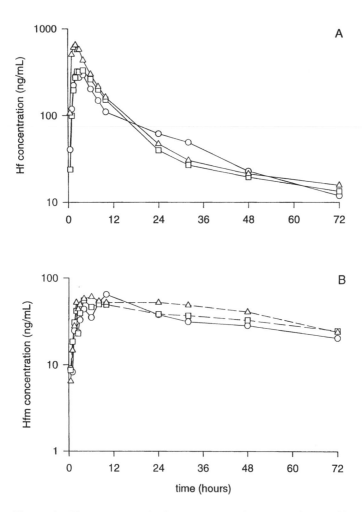

**Figure 1**   The mean ($n = 4$) plasma concentration versus time profiles of Hf (panel A) and Hfm (panel B) after the oral administration of MCT SEDDS (○), MCT SMEDDS (□), and LCT SMEDDS (△) formulations of Hf base (50 mg) to fasted beagles. (From Ref. 80.)

**Table 2** Bioavailability Parameters Obtained in Male Beagle Dogs (mean ± SD, $n = 3$) After Oral Administration of 50 mg of Hf Base Formulated as Either a Long chain Self-microemulsifying Formulation (SMEDDS) or a Medium-chain Self-emulsifying (SEDDS), or Self-microemulsifying Drug Delivery System (SMEDDS)

| Parameter | MCT SEDDS | MCT SMEDDS | LCT SMEDDS |
|---|---|---|---|
| $C_{max}$ (ng/mL) | 363 ± 156 | 374 ± 198 | 704 ± 308 |
| $T_{max}$ (h) | 2.8 ± 0.8 | 4.2 ± 1.5 | 2.3 ± 0.5 |
| $AUC^{(0-inf)}$(ng.h/mL) | 5313 ± 1956 | 5426 ± 2481 | 6973 ± 2388 |
| Absolute BA (%) | 51.6 ± 19.2 | 52.7 ± 24.0 | 67.3 ± 21.0 |

[a] In a previous study in male beagle dogs the absolute BA of Hf HCl was 8.6 ± 5.3% [79].
*Source*: Ref. 80.

mercially available tablet formulation (250 mg Hf HCl) in fasted beagles was 8.6 ± 5.3% [79]. The capacity of dispersed lipid-based formulations to enhance the bioavailability of Hf is further supported by a recent study in fasted beagles in which the absolute oral bioavailability of crystalline Hf base was found to be even lower (2.9 ± 3.3%, unpublished results).

No particle size effect was seen with the medium-chain formulations, suggesting that the differences in particle size of the emulsion and microemulsion formulations were either unimportant or not sufficiently large to result in a difference in intestinal processing. Although not statistically significant (possibly because of the size of the study), there was a trend to higher bioavailability with the long-chain formulation, indicating that for formulations that produce very fine dispersions, lipid class effects may be more important than small variations in particle size. Significant intestinal lymphatic transport of Hf base has been described in the rat (using higher lipid doses), and this could explain the improved performance of the long-chain systems however, it is unclear what role lymphatic transport has in these studies in which the total lipid load is much lower. The trend to higher bioavailability with long-chain systems is in keeping with the results seen with cyclosporin, where the extent of lymphatic transport is low (<1%).

## B. Bioavailability Enhancement by Means of Biochemical/Metabolic Mechanisms

Formulation components may enhance the extent of oral drug bioavailability by altering the extent of absorption (by means of improvements in drug dissolution, GI solubility, GI stability, and intestinal permeability) or by reducing drug metabolism. Historically, the role of the liver in drug metabolism was considered paramount, and alterations in the extent of hepatic presystemic metabolism—the first pass effect—the most likely mechanism to improve drug bioavailability.

Recently, however, interest has increased in the potential role of prehepatic metabolic processes as a major limitation to the oral bioavailability of some drugs [81–83]. It is now accepted that enterocyte-based processes play a central role in the metabolism of ketaconazole [82], midazolam[84], cyclosporin [81, 85–87], tacrolimus [85, 88], erythromycin [89], saquinavir [90], and rifabutin [91], and in most cases this has been attributed to metabolism by cytochrome P450 enzymes and more specifically the CYP3A subfamily of enzymes. Over a similar time frame interest has also increased in the role of intestinal efflux pumps that limit drug transfer through the enterocyte by pumping xenobiotics from inside the cell back into the lumen of the intestine (i.e., a countertransport function). Recent data have implicated P-glycoprotein (P-gp) (the MDR 1 gene product) located in the enterocyte brush border membrane in the limitation of the oral bioavailability of cyclosporin [92], vinblastine [93], verapamil [94], celiprolol [95], etoposine [96], and others [89, 97].

P-gp and CYP3A are co-localized within the apex of absorptive enterocytes, they have similar substrate specificities, and they are co-inducible in response to some xenobiotics [89, 97]. These observations have led to the proposition that p-gp and CYP3A may be functionally linked and act in concert to limit the exposure of xenobiotics/drugs to the small intestine and portal circulation [89, 97].

Because of the relatively recent and growing realization of the importance of enterocyte-based CYP3A and p-glycoprotein in the limitation of drug bioavailability, few published studies have addressed the potential impact of formulation components such as lipids on these processes. However, a recent patent has described in some detail the potential for many lipids to inhibit both CYP3A-based metabolic processes and p-glycoprotein-mediated antitransport processes [98]. The patent covers the use of essential oils to improve bioavailability and presents data detailing the inhibitory capacity of essential oils using *in vitro* drug metabolism screens. Surfactants found in many dispersed lipid formulations have also been shown to inhibit the extent of p-glycoprotein–mediated efflux of a model peptide [99]. However, research in this area is in its infancy and the factors controlling the intracellular fate of lipophilic drugs *vis á vis* metabolism by CYP3A or binding to the p-gp efflux pump are not well known.

One of the major obstacles to definitively examining the role of formulation components on enterocyte-based processes is the possible effect of excipient inclusion on other complicating factors. For example, the inclusion of lipids or surfactants in *in vitro* metabolic or transport screens runs the risk of affecting the thermodynamic activity of the drug in solution, thereby obscuring the role of metabolic and transport processes. Similarly, some surfactants and lipid-surfactant conjugates may cause transient increases in intestinal permeability as a result of cellular damage as opposed to more subtle effects on intracellular transport processes.

Important indirect indications of metabolic activity may be gained from careful evaluation of clinical or preclinical pharmacokinetic profiles obtained in the presence and absence of known inhibitors of CYP3A. By conducting studies after both intravenous and oral administration, the relative contribution of hepatic and prehepatic metabolism may be deconvoluted.

The oral bioavailability of a drug ($F_{oral}$) may be regarded as the product of the fraction of the drug dose absorbed into and through the gastrointestinal membranes ($F_{abs}$), the fraction of the absorbed dose that passes unchanged through the gut into the hepatic portal blood ($F_G$) and the hepatic first pass availability ($F_H$) (Eq. 1)

$$F_{oral} = F_{abs} \cdot F_G \cdot F_H \tag{1}$$

Gut and hepatic availability may be defined as one minus the extraction ratio (ER) at each site, therefore

$$F_{oral} = F_{abs} \cdot (1 - ER_G) \cdot (1 - ER_H) \tag{2}$$

By use of these basic concepts, Wu et al. [87] conducted an examination of the pharmacokinetics of cyclosporine after administration in the presence of two known CYP3A inhibitors, ketaconazole and erythromycin and the CYP3A inducer rifampin. Alterations to the cyclosporin hepatic extraction ratio were measured in the presence of the inducer/inhibitor, and oral cyclosporin biovailability (relative to IV administration in the presence/absence of the interacting species) was calculated. These data are reproduced in Table 3 and provide several impor-

**Table 3** Measures of Cyclosporin Bioavailability and Hepatic Extraction After Oral Administration in the Presence or Absence of the CYP3A Inhibitors Ketoconazole and Erythromycin and the CYP3A Inducer Rifampin

| Study | $n$ | $F_{oral}$ | $ER_H$ |
|---|---|---|---|
| Healthy volunteers | | | |
|    Without ketoconazole | 5 | $0.22 \pm 0.05$ | $0.25 \pm 0.06$ |
|    With ketaconazole | 5 | $0.56 \pm 0.12$ | $0.14 \pm 0.03$ |
| Kidney transplant patients | | | |
|    Without erythromycin | 4 | $0.36 \pm 0.09$ | $0.27 \pm 0.10$ |
|    With erythromycin | 4 | $0.62 \pm 0.17$ | $0.20 \pm 0.04$ |
| Healthy volunteers | | | |
|    Without rifampin | 5 | $0.27 \pm 0.09$ | $0.24 \pm 0.04$ |
|    With rifampin | 5 | $0.10 \pm 0.03$ | $0.33 \pm 0.07$ |

*Source*: Ref. 87.

tant pieces of information. First, in the presence of the metabolic inhibitors, a better indication of the fraction absorbed was obtained. The oral bioavailability of cyclosporin was 56% after ketaconazole co-administration and 62% after erythromycin co-administration. However, these values still underestimate the fraction absorbed, because although hepatic CYP3A elimination is suppressed in the presence of the inhibitors, it is not eliminated entirely. Consideration of the reduced hepatic extraction ratio in the presence of the CYP3A inhibitors suggested that the actual fraction absorbed was at least 65% and 77% in the ketaconazole and erythromycin studies, respectively. Second, with the assumption that rifampin enzyme induction in the gut was similar to the extent of enzyme induction measured in the liver, the authors estimated that 86% of the cyclosporin dose was absorbed in the rifampin study and that the reduced bioavailability seen in the absence of any interacting species (27%) was due to a combination of both prehepatic and hepatic first-pass metabolism. In these studies cyclosporin was administered as the original Sandimmun formulation. Because other data have shown that the bioavailability of cyclosporin is markedly improved (approximately twofold) by the Sandimmun Neoral formulation [74], these data suggest that the bioavailability improvement was not due to improved absorption (because up to 80% of the cyclosporin dose was thought to be absorbed from the Sandimmun formulation) and therefore that lipids, surfactant, or co-surfactants contained in the formulation must improve cyclosporin bioavailability, at least in part, by inhibitory effects on enterocyte-based CYP3A metabolism.

In contrast, data from a recent conference presentation by Choc and Robinson [100], indicated that cyclosporin metabolite ratios were similar after oral administration of either Sandimmun or the Neoral formulation, suggesting minimal changes in cyclosporin first-pass metabolism after Neoral administration. Furthermore, Neoral/Sandimmun area under the curve (AUC) ratios were similar (approximately 2) after cyclosporin administration in either the absence or presence of a CYP3A/P-gp inhibitor, whereas an AUC ratio of approximately 1 would have been expected if bioavailability was limited by metabolism or anti-transport. However, these data have not yet been published, and the relative contributions of enhanced absorption or reduced metabolism in the CY bioavailability enhancement provided by Sandimmun Neoral are yet to be fully defined.

We have recently conducted a similar study with halofantrine and obtained pharmacokinetic data after oral Hf administration in the presence and absence of ketaconazole (KC). The plasma profiles for Hf and Hfm (metabolite) are shown in Figure 2, and the $C_{max}$, $T_{max}$, and AUC values are presented in Table 4. Co-administration of Hf with KC produced nonsignificant changes in the measured plasma $C_{max}$ and $T_{max}$ values of Hf, and the calculated plasma Hf $AUC^{0-72h}$ values, although at 72 h after oral administration, there was a clear and developing trend toward higher plasma Hf concentrations in the KC treated group (which may have reached statistical significance if blood samples were taken for a longer

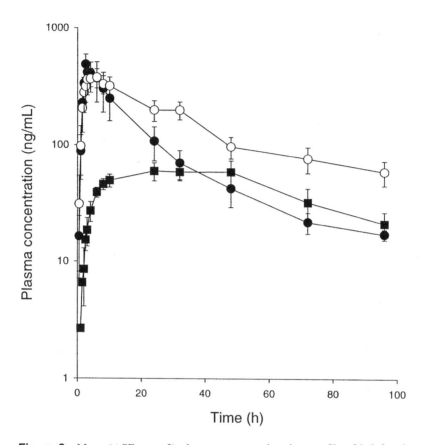

**Figure 2** Mean (±SE, $n = 3$) plasma concentration-time profile of halofantrine (●) and desbutylhalofantrine (■) after fasted oral administration of 150 mg Hf without ketoconazole (closed symbols) or after pretreatment with ketoconazole (open symbols). When Hf was co-administered with ketoconazole, the concentration of Hfm at all time points was below the LOQ of the assay, which was 10 ng/mL.

period). The co-administration of Hf with KC almost completely inhibited formation of Hfm after fasted oral administration and plasma concentrations of Hfm were reduced from approximately 50 to 100 ng/mL (without KC) to below the LOQ of the assay (with KC). The mean (±SD, $n = 3$) ratio of the plasma Hfm/ Hf $AUC^{0-72h}$ values for fasted administration were 0.47 ± 0.08 (without KC) and less than 0.05 (with KC). After intravenous administration (and assuming metabolic contributions primarily from systemic hepatic metabolism), the Hfm/ Hf ratio was higher (0.22). These data suggest that conversion of Hf to Hfm

**Table 4**  Summary Pharmacokinetic Parameters (mean $\pm$ SD, $n = 3$) for Halofantrine (Hf) and Desbutylhalofantrine (Hfm) After Oral Administration of Hf (150 mg) to Fasted Beagles with or without Co-administration of Ketoconazole

| Pharmacokinetic Parameter | Oral Hf (no ketoconazole) | Oral Hf (with ketoconazole) |
|---|---|---|
| Hf | | |
| $C_{max}$(ng/mL) | 516 $\pm$ 148 | 388 $\pm$ 140 |
| $T_{max}$(h) | 3.8 $\pm$ 1.9 | 5.0 $\pm$ 1.7 |
| $AUC^{0-96h}$(ng.h/mL) | 8550 $\pm$ 4286 | 14,308 $\pm$ 5029 |
| Hfm | | |
| $C_{max}$(ng/mL) | 62.8 $\pm$ 22.4 | ND[a] |
| $T_{max}$(h) | 35.1 $\pm$ 12.5 | ND |
| $AUC^{0-96h}$(ng.h/mL) | 4256 $\pm$ 1546 | ND |

[a] ND, not detected.

*in vivo* is likely mediated by means of CYP3A4 and that presystemic CYP3A metabolism limits the bioavailability of Hf after oral administration. Furthermore, on the basis of the localization of functional CYP3A4 in duodenal tissue [97], it is likely that both prehepatic (enterocyte-based) and hepatic CYP3A4 contribute to the presystemic metabolism of Hf.

An interesting feature that arises from comparison of this KC data with previous fed/fasted Hf data [79] is that administration of Hf with a fatty meal produces a similar plasma Hfm/Hf plasma AUC profile to that obtained after co-administration of Hf with KC in the fasted state (and very different to the Hfm/Hf ratio after normal fasted administration). In both the fed/fasted and the KC studies the Hf/Hfm ratio was similar (0.47 and 0.54, respectively) after fasted administration (no food/ketaconazole, respectively). In contrast, in both the fed state and after co-administration with KC, the Hfm/Hf metabolite ratio was significantly reduced (0.08 and <0.05, respectively). Similar alterations in the magnitude of the Hfm/Hf ratio after feeding have also been demonstrated in humans [78]. It is apparent, therefore, that the presystemic metabolism of Hf is reduced in both the fed state and after co-administration with KC. The effects of KC co-administration on hepatic elimination were not studied in the investigation described here, precluding quantification of the relative roles of prehepatic and hepatic presystemic metabolism to the altered metabolic profile seen after KC co-administration. In contrast, recent examination of Hf pharmacokinetics after fed and fasted intravenous administration revealed a decrease in Hf systemic CL of only approximately 15% after fed administration, suggesting that the almost complete inhibition of Hf metabolism in the fed state must be a result of extra hepatic mechanisms. Explanations behind the alterations in fed-state metabolism include alterations to hepatic blood flow, stimulation of lymphatic transport, or the action of one or more components of the diet on gut-based metabolic processes. We have previously shown that Hf is a substrate for intestinal lymphatic transport [100, 101], and this is therefore a likely contributor, considering the plentiful supply of the long-chain triglycerides in the fed state. However, the cyclosporin data also suggest a possible role for the inhibition of prehepatic metabolism by lipids in the diet. Studies to explain these mechanisms are ongoing.

## V. LIPIDS AND TARGETING TO THE INTESTINAL LYMPH

After absorption, most drugs and xenobiotics traverse the enterocyte and are absorbed into the portal blood. A small number of highly lipophilic drugs, however, are transported to the systemic circulation by means of the intestinal lymphatics.

Drug delivery to the intestinal lymphatics confers two primary advantages over conventional absorption by means of the portal blood. First, transport through the intestinal lymph avoids presystemic hepatic metabolism and therefore enhances the concentration of orally administered drugs reaching the systemic circulation. Second, from a site-specific delivery or targeting perspective, the lymphatics (1) contain relatively high concentrations of lymphocytes and therefore provide attractive targets for cytokines such as interferon and immunomodulators in general, (2) serve as the primary conduit for the dissemination of many tumor metastases and therefore show promise as a target for cytotoxics, and (3) may provide an efficient route of delivery to HIV-infected T-cells, because recent findings have suggested that a significant proportion of the HIV viral burden resides in the lymphoid tissue [103, 104].

As described previously, digested dietary lipids (in the form of fatty acids and monoglycerides) are absorbed into the enterocyte, re-esterified to form triglyceride, and subsequently assembled into colloidal lipid aggregates or prelipoproteins. Prelipoproteins then fuse with the basolateral membrane of the enterocyte, facilitating entry into the lamina propria. The colloidal structure and size of intestinal lipoproteins subsequently precludes their absorption into the blood capillaries (because the capillaries of the small intestine have a continuous "tight" endothelial structure). The structure of the intestinal lymphatic vessels, however, is notably different, and lymphatic endothelial cells have relatively open intercellular junctions. Estimates of intercellular junctional distances range from several microns [105, 106] to 15–20 nm [108–110], and, consequently, intestinal lipoproteins are almost exclusively absorbed into the intestinal lymphatics. The collecting lymphatics from the small intestine and the ascending and transverse colon join to form the superior mesenteric lymph duct, which runs by means of the thoracic lymph into the systemic circulation directly, illustrating that drugs that are transported to the systemic circulation by means of the intestinal lymph avoid the first-pass metabolic effects inherent in absorption by means of the portal blood.

The potential for orally administered drugs to enter the intestinal lymphatics is therefore defined by their selectivity for uptake into the intestinal lymphatics as opposed to the blood capillaries in the subepithelial space. Because selectivity for the lymphatics is primarily defined by size, it is apparent that only macromole- cules or colloids will be preferentially absorbed into the intestinal lymphatics. However, the intestine provides a significant barrier to the absorption of both macromolecules and intact colloids, and the most prevalent mechanism for drug delivery to the intestinal lymph is by means of secondary drug association with intestinal lipoproteins [110]. The size of the drug-lipoprotein complex subse- quently dictates absorption into the lymphatic vessels.

Because drug access to the intestinal lymphatics primarily depends on drug association with lymph lipoproteins, compounds that are inherently lymph direct- ing or "lymphotropic" must be extremely lipophilic. This requirement is en- hanced further by appreciation of the relative mass transfer rates between the portal blood and the intestinal lymph. The absorption "sink" associated with the portal blood may be regarded as the mass of blood flowing through the portal blood per unit time. The absorption sink associated with the intestinal lymph, however, is significantly less, because the fluid flow in the portal blood is some 500-fold higher than intestinal lymph flow. Furthermore, drugs must be associ- ated with lymph lipoproteins to access the intestinal lymph, and consequently the lymphatic absorption sink is more correctly calculated as the flow of lipopro- tein mass per unit time. Because the mass of lipoproteins in lymphatic fluid is approximately 1–2% w/w, there is an approximate 50,000-fold higher mass ratio between the portal blood and lymph lipid. By use of these estimates, Charman and Stella predicted that for appreciable lymphatic transport, drug molecules must have a log P of approximately 4.7 or greater [111].

Intestinal lymphatic transport has been shown to contribute to the absorp- tion of a number of lipophilic vitamins and vitamin derivatives [112] and xenobi- otics such as probucol [113]; naftifine [114]; coenzyme-Q [115]; cyclosporin [116]; DDT and related analogues [117–120]; halofantrine [101, 102]; benzo- (a)pyrene [121]; polychlorinated biphenyls [122]; CI 976 [123]; temarotine, etret- inate, and isotretinoin [124]; ontazolast [36], MK 386 [125]. The more recent examples of these data are briefly summarized in Table 5 for examples in which the total extent of lymphatic transport is known. The combination of a high log P and high lipophilicity (as evidenced by >50 mg/mL solubility in long-chain triglyceride oils) is probably the best current physicochemical indicator of the potential for lymphatic transport (reflecting the need for drug molecules to parti- tion into the TG core of lymph lipoproteins), although this is obviously an oversimplification, because compounds such as ontazolast can fulfil both criteria and still exhibit limited lymphatic transport.

With exceptions, including halofantrine, DDT, and the lipophilic vitamins, the extent of lymphatic transport (as a proportion of the dose) is generally low. However, the compounds described in Table 5 are hydrophobic (as evidenced by the high log Ps), and their bioavailability is often low. Therefore, although the absolute extent of lymphatic transport may be low, the lymphatic contribution to the small fraction that is absorbed may be high, and alterations in the extent of lymphatic transport may have a significant effect on the extent of oral bioavailability.

More successful approaches to the enhancement of intestinal lymphatic transport have used various prodrugs (reviewed in 128,129). Two general strategies have been used for lymph directing prodrugs. First, the synthesis of simple lipid ester/ether conjugates that enhance the lipophilicity of the parent molecule, and, more recently, the use of molecules that mimic key components of the lipid digestion absorption pathway such that the prodrug (or a prodrug derivative) becomes intercalated into the natural lipid digestion/absorption pathway. The latter group of compounds can be further split into glyceride mimics, in which the parent drug molecule is linked to a glycerol backbone to form a monoglyceride or triglyceride mimic or a similar strategy that uses phospholipid-based backbones to produce phospholipid mimics.

It is apparent that after absorption, the extent of drug partition into lymph lipoproteins and consequently the extent of intestinal lymphatic drug transport may be enhanced by either increasing the lipophilicity of the drug molecule itself (e.g., using prodrugs) or by increasing the effective lymphatic absorption sink by increasing the throughput of lipid into the lymph (formulation approaches). However, appreciation of the possible increases in lymphatic lipid throughput as a result of lipid absorption (perhaps 2–10 fold), compared with the 50,000-fold difference in portal blood/lymphatic lipid affinity, indicates that the opportunities for the stimulation of lymphatic drug transport using lipid-based formulations are limited. Experimental evidence confirms this hypothesis and suggests that although lipid-based formulations may enhance and optimize the lymphatic transport of lipophilic drugs, there is little chance of stimulating the lymphatic transport of drugs that are not inherantly lymphotropic.

Although lipids and lipid-based formulations cannot promote drug association with intestinal lipoproteins in the absence of the requisite physicochemical drug properties, lipid-based delivery systems can have an appreciable effect on the extent of drug absorption into the enterocyte as described in the previous section. The eventual extent of lymphatic drug transport therefore is the product of the sequential processes of drug diffusion and dissolution in the GIT, drug absorption and metabolism within the enterocyte, and partition of the drug mole-

**Table 5** Summary of Recent Intestinal Lymphatic Transport Data (Lymphotropic Prodrugs Have not Been Included)

| Compound | Log P | Lipid Solubility | Dosing vehicle[a] | Cumulative lymphatic transport (% dose) | Collection period | Model | Ref. |
|---|---|---|---|---|---|---|---|
| DDT | 6.19 | 97.5 mg/mL (peanut oil) | 200 µL oleic acid | 33.5% | 10 h | Anesthetized rat/Mesenteric lymph duct/Intraduodenal dose | 111 |
| MK 386 | N/A | >80 mg/mL (soybean oil) | 5 mL/kg methocel suspension 5 mL/kg peanut oil | 4.8% 0.10% | 6 h | Conscious rat/Mesenteric lymph duct/Oral dosing | 125 |
| HCB | 6.53 | 7.5 mg/mL (peanut oil) | 200 µL oleic acid | 2.3% | 24 h | Anesthetized rat/Mesenteric lymph duct/Intraduodenal dose | 111 |
| Cyclosporin | 2.92 | >30 mg/mL (sesame oil) | 2 mL/kg 8% HCO-60 micellar solution 2 mL/kg sesame oil | 2.14% 0.19% | 6 h | Anesthetized rat/Thoracic lymph duct/Oral dosing | 126 |
| Penclomedine | 5.48 | 177 mg/mL (soybean oil) | 0.5 mL 10% soybean oil emulsion | 1.5% | 12 h | Anesthetized rat/Mesenteric lymph duct/Intraduodenal dosing | 127 |
| Halofantrine | 8.5 | >50 mg/mL (peanut oil) | 50 µL peanut oil | 16.7% | 12 h | Conscious rat/Mesenteric lymph duct/Oral dosing | 102 |
| CI-976 | 5.83 | >100 mg/mL (corn oil) | 1 mL 20% soybean/safflower oil emulsion 1 mL 1% Avicel/0.5% CMC/0.2% PS80/4% glycerin suspension | 0.4% 0.06% | 14 h | Conscious rat/Mesenteric lymph duct/Intraduodenal dosing | 123 |
| Ontazolast | 4.0 | 55 mg/mL (soybean oil) | 10 mL/kg 20% soybean oil emulsion 10 mL/kg 1% Methocel/0.2% PS80 suspension | 1.2% <0.1% | 8 h | Conscious rat/Mesenteric lymph duct/Oral dosing | 36 |

[a] Where many different vehicles were administered, simple lipid- and nonlipid-based formulations have been chosen for comparison.

cule after absorption into either the portal blood or the intestinal lipoproteins. All of these processes can be markedly affected by co-administered lipids, and the differentiation of lipid effects is problematic. For example, although lipid-based formulations generally enhance the lymphatic transport of lymphotropic drug molecules, simple lipid-free micellar delivery systems have been shown to enhance the lymphatic transport of cyclosporine and retinyl palmitate compared with emulsified or lipid solution formulations [126, 130]. Under these circumstances, it appears that the lipid-free formulations enhance the mass of drug available for partition into lymph lipoproteins (by either increasing absorption or reducing enterocyte-based metabolism) and that this effect overrides the increase in the lymphatic lipid turnover provided by the lipid-based formulations.

The lipoproteins secreted into the intestinal lymphatics as part of the lipid transport process are less precisely defined than plasma lipoproteins and vary significantly in response to a change in the nature of the ingested lipids. The major classes of lipoprotein secreted into the lymph are chylomicrons and very low-density lipoproteins (VLDL). High-density lipoproteins are also synthesized in the intestine but do not appear to play a significant role in lymphatic drug transport. Lymph lipoproteins are endogenous colloidal lipid carrier systems that consist of a hydrophobic core of triglyceride and cholesterol ester surrounded by a surface layer of amphiphilic lipids such as phospholipid, free cholesterol, and protein. Although the structure of lymph lipoproteins is similar to plasma lipoproteins, lymph lipoproteins carry much higher exogenous lipid loads. Thus, lymph chylomicrons consist primarily of triglyceride (86–92%) with small amounts of phospholipid (6–8%), free and esterified sterols (2–4%), and protein (1–2%) [131]. Chylomicrons are responsible for most transport of exogenous lipid into the intestinal lymphatics. VLDL may also be regarded as triglyceride-rich lipoproteins; however, they are smaller (20–50 nm), have a higher surface area/mass ratio, and consequently are made up of a higher proportion of surface-based lipoprotein components. The triglyceride load of VLDL is approximately 50%.

Most of the mass of lymphatically transported drugs appears to be contained within the chylomicron and VLDL fraction of the lymph. A small proportion is also associated with the nonlipoprotein fraction because of re-equilibration of drug within the lymphatic compartment; however, the high log P of lymphatically transported drugs dictates that most of the drug remains associated with the lipoproteins. Table 6 summarizes some of the available drug-lymph lipoprotein distribution data and illustrates that most of the drug mass is associated with the chylomicron fraction, probably reflecting the carrier with the largest triglyceride capacity. Interestingly, the proportion associated with the chylomicron fraction also appears to be correlated with the mass of lipid transported into the lymph.

Because lymphatically transported drugs are almost entirely contained within chylomicrons and VLDL, they are effectively introduced into the systemic circulation as a slow intravenous infusion of a drug-lipoprotein delivery system.

**Table 6** Relative Distribution of Some Lipophilic Drugs, Vitamins, and Xenobiotics between Different Lipoprotein Fractions in the Lymph

| Compound | Vehicle | Lipoprotein distribution (%) | Ref. |
|---|---|---|---|
| [3]H retinol | Soybean oil ($\approx$1 mL/kg) | >95% chylomicron | 132 |
| Vitamin D₃ | Soybean oil ($\approx$1 mL/kg) | >95% chylomicron | 132 |
| DDT | Olive oil | 87–92% chylomicron | 120 |
| Co-enzyme Q | Sesame oil | >80% chylomicron <br> <20% VLDL/other | 115 |
| Hexadecane | Safflower oil | 99% chylomicrons | 133 |
| Octadecane | Safflower oil | 98% chylomicrons | 133 |
| DDT | Safflower oil | 97% chylomicrons | 133 |
| Mepitiostane[a] | Aqueous suspension | 30% chylomicron <br> 60% VLDL | 134 |
| | Sesame oil—0.02 mL/kg | 45% chylomicron <br> 45% VLDL | |
| | 0.2 mL/kg | 60% chylomicron <br> 35% VLDL | |
| | 0.8 mL/kg | 80% chylomicron <br> 13% VLDL | |

[a] Data approximated from a graphical representation.

Although there has been little data generated on the topic, this altered method of delivery to the systemic circulation may have profound effects on the subsequent distribution, pharmacokinetics, and pharmacodynamics of the drug molecule and have an impact on the wider area of drug binding to plasma lipoproteins and the potential effects of lipids on this process.

Haus and co-workers [123] recently examined the mechanism of absorption of CI-976, a poorly water-soluble lipid regulator molecule, and found that intestinal lymphatic transport contributed up to 43% and 57% of the eventual CI-976 plasma AUC after administration in an emulsion and suspension formulation, respectively. However, when the relative transport of CI-976 into the lymph was assessed directly (i.e., by means of cannulation of the intestinal lymph duct), the extent of lymphatic transport of CI-976 was seven times greater for the emulsion compared with the suspension formulation. These observations suggested that a significantly larger amount of CI-976 was transported through the intestinal lymph after administration in the lipid-based emulsion formulation (compared with the lipid-free suspension formulation), but that after entering the systemic circulation (presumably in association with lymph lipoproteins) the drug was rapidly cleared and the sevenfold higher levels of lymphatic transport were not reflected in a higher systemic AUC. In support of this hypothesis, subsequent studies that used [14]C CI-

976 showed that compared with the suspension, the emulsion delivery system resulted in 43% greater accumulation of intact CI-976 in perirenal fat, suggesting rapid redistribution of chylomicron-associated CI-976 into the fat.

Current opinion suggests that after emptying into the bloodstream, drugs associated with lymph lipoproteins equilibrate across the various plasma lipoprotein and protein fractions and take on the same clearance properties as drug introduced into the systemic circulation by way of the portal blood. The data of Haus et al. suggest that this is an oversimplification and that the mechanism of interconversion and interaction of drug molecules between lymph and plasma lipoproteins is not clear [123].

The consequences of intestinal lymphatic transport are therefore more wide ranging than the opportunity to target the intestinal and central lymph and to avoid first-pass metabolism. Indeed, drug transport by means of the intestinal lymph may markedly alter the patterns of subsequent systemic drug clearance and disposition, and these profound alterations could be stimulated simply by a change in formulation. Importantly, the common approach of the use of plasma AUC as an indicator of the available fraction may be misplaced when dealing with lipophilic and lymphotropic drug molecules.

## VI. LIPIDS AND DRUG BINDING TO PLASMA LIPOPROTEINS

Most drug molecules are assumed to be ''in solution'' in plasma and consequently available for distribution, metabolism, and elimination and interaction with the appropriate receptor or biological target. Although there are many examples documenting cases in which this situation may be complicated as a result of drug binding to either plasma proteins or red blood cells (resulting in a decrease in the plasma free fraction), the role of drug binding to plasma lipoproteins has been relatively poorly studied.

However, the identification of increasingly lipophilic lead molecules, the physicochemical properties of which (low aqueous solubility and high log Ps) suggest a natural predisposition for increased plasma lipoprotein binding, has increased interest in the possible pharmacokinetic, therapeutic, and toxicological ramifications of drug binding to plasma lipoproteins.

Unlike most plasma proteins and red blood cells, the metabolism and disposition of plasma lipoproteins is highly regulated and mediated in part by specific receptors located at discreet sites around the body. The concentrations of plasma lipoproteins in the blood can also vary significantly (up to tenfold) as a function of diet, disease, and both within and between healthy individuals. The implications, therefore, for the clearance of drug molecules associated with these plasma lipoproteins are considerable but have not been clearly defined.

Plasma lipoproteins are generally classified by their density and separation achieved with ultracentrifugation. According to this density-based classification system, the major lipoprotein classes are chylomicrons (CH), very low-density lipoproteins (VLDL), intermediate-density lipoproteins (IDL), low-density lipoproteins (LDL), and high-density lipoproteins (HDL).

The general structure of lipoproteins is shown schematically in Figure 3. The core of the lipoprotein contains the more hydrophobic lipids namely cholesterol ester (CE) and triglyceride (TG) and is surrounded by a surface monolayer consisting of the more polar phospholipid (PL) and free cholesterol (FC). Apoproteins are associated with the lipoprotein surface. The proportional composition of human plasma lipoproteins is given in Table 7.

The primary function of lipoproteins is to transport insoluble lipids through the aqueous environment of the vascular and extravascular fluids, and the relative

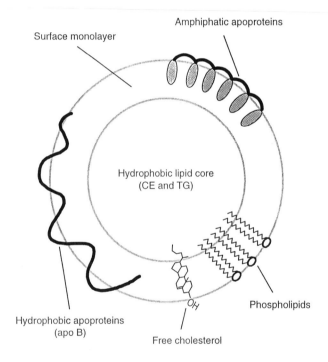

**Figure 3** Schematic representation of a lipoprotein particle. The core in made up of hydrophobic lipids (TG and CE) surrounded by a monolayer of phospholipids and free cholesterol. Large hydrophobic apoproteins (apo-B) bind irreversibly to the surface of CM, VLDL, and LDL, whereas smaller amphipathic apoproteins (apo-A, C, D, and E) are reversibly bound to the surface of the lipoprotein by hydrophobic α-helical domains of the apoprotein.

**Table 7** The Proportional Composition of Human Plasma
Lipoproteins (% of Total Weight)

| Class | Plasma conc (mg/mL$^{-1}$) | TG | CE | FC | PL | Protein |
|-------|------------|----|----|----|----|---------|
| CM   | 0.1–0.3    | 85 | 4  | 2  | 7  | 2  |
| VLDL | 0.7–1.8[a] | 53 | 13 | 7  | 19 | 8  |
| LDL  | 3.4–6.3    | 8  | 40 | 8  | 22 | 22 |
| HDL  | 2.7–4.1    | 3  | 15 | 4  | 28 | 50 |

[a] This includes IDL concentration.
*Source*: Refs. 14, 135, 136.

role of each lipoprotein class in terms of lipid transport is reflected in the structure. Thus, CM and VLDL (which may be collectively referred to as triglyceride-rich lipoproteins, TRL) are primarily responsible for the transport of TG, whereas cholesterol transport is primarily mediated by LDL and HDL.

TG are derived directly from the diet and secreted from the intestines (primarily by way of the lymph) as CM and TRL or synthesized into VLDL in the liver. The net transport of TG is therefore from the intestines and the liver to skeletal and cardiac muscle or to adipose tissue for storage. Cholesterol is used for membrane synthesis and steroid production and is primarily synthesized in extrahepatic tissues. It is continuously transported between the liver, intestines, and extrahepatic tissues, but the net transport of cholesterol is from the extrahepatic tissues to the liver and intestines from where it is eliminated.

CM and VLDL secreted by intestinal cells and VLDL synthesized and secreted in the liver have similar metabolic fates. After secretion into the blood, newly formed CM and VLDL take up apoprotein (apo-C) from HDL and are subsequently removed from the blood (plasma half-life of less than 1 h in humans [137]) primarily by the action of lipoprotein lipase (LPL). Lipoprotein lipase is situated mainly in the vascular bed of the heart, skeletal muscle, and adipose tissue and catalyzes the breakdown of core TG to monoglycerides and free fatty acids, which are taken up into adjacent cells or recirculated in blood bound to albumin. The activity of LPL in the heart and skeletal muscle is inversely correlated with its activity in adipose tissue and is regulated by various hormones. Thus, in the fasted state, TG in CM and VLDL is preferentially delivered to the heart and skeletal muscle under the influence of adrenaline and glucagon, whereas in the fed state, insulin enhances LPL activity in adipose tissue, resulting in preferential uptake of TG into adipose tissue for storage as fat.

The LPL-mediated catabolism of TRL removes about 80–90% of the core TG and results in reduction of the size of the lipoproteins, with concurrent transfer of excess surface materials to HDL. The lipoproteins are now termed remnant particles, and in the case of chylomicron remnants are rapidly cleared by the liver by receptor-mediated uptake. In contrast, although VLDL remnants are also cleared by the liver, most are converted to LDL by means of an IDL intermediate.

The plasma half-life of LDL is about 2 days, and their primary function is the delivery of cholesterol to the tissues. LDL are taken up into cells by two routes, one that is receptor mediated (and is regulated by the cholesterol requirement of the cell) and one that appears to be nonreceptor mediated (and depends entirely on the extravascular concentrations of LDL). The receptor-mediated uptake occurs by binding of apo-B100, which is predominantly present on LDL. Hence these receptors are also known as LDL receptors and have been identified on a variety of cell types. In normal humans, about two thirds of total LDL clearance is mediated by the LDL receptor, and about 80–90% of the receptor-mediated uptake occurs in the liver. However, the relative importance of receptor and nonreceptor-mediated LDL uptake can vary depending on factors including diet and different disease states.

The metabolism of HDL is less well defined than that of the other lipoprotein classes. Reasons for this include the heterogeneous nature of HDL and the interchange of both lipid and protein components between HDL and other lipoproteins. Furthermore, HDL are not removed from the circulation, because whole particles and their different components have different metabolic fates.

The main precursors of plasma HDL are most likely disk-shaped bilayers composed of PL and protein and secreted by the liver and intestine. HDL are also derived from the surplus surface material removed from TG-rich lipoproteins during lipolysis. HDL are involved in the net transfer of cholesterol from peripheral tissues to the liver, where it can be eliminated or recirculated. This process is initiated by the uptake of FC from cell membranes into the HDL. The nature of this uptake is not known but may involve binding of HDL to the membrane.

Once associated with HDL, FC is converted to CE by the action of lecithin-cholesterol acyltransferase (LCAT), which resides on the surface of HDL. This conversion creates a concentration gradient for the further uptake of FC. The newly formed CE enters the core of the HDL because of its highly hydrophobic nature. This process transforms nascent disk-shaped HDL into mature spherical HDL ($HDL_3$).

Mature $HDL_3$ undergo further metabolic changes in plasma. The transfer of surface lipids, proteins, and core TG from TG-rich lipoproteins results in the conversion of $HDL_3$ to $HDL_2$. Hence, the levels of $HDL_2$ are directly related to chylomicron metabolism. Further transfer of TG from LDL in exchange for CE results in formation of TG-rich $HDL_2$ particles, which in turn are reconverted to $HDL_3$ by removal of TG by the action of hepatic or lipoprotein lipase.

The transfer of CE from HDL to TRL and LDL and the transfer of TG back to HDL is facilitated by cholesteryl ester transfer protein (CETP) or lipid transfer protein 1 (LTP 1). CETP also catalyzes the transfer of phospholipids (transfer activity for PL is one order of magnitude lower than for CE) and other nonpolar lipids such as retinyl esters between plasma lipoproteins. The transfer of phospholipids between HDL and other lipoproteins is also facilitated by a specific phospholipid transfer protein.

The general topic of drug binding to plasma lipoproteins has been well reviewed by Urien et al. [138] and most recently by Wasan and Cassidy [139], and the interested reader should consult these texts for a comprehensive review of the documented examples of drug-lipoprotein binding. These authors have also addressed the implications of disease states on lipoprotein levels and the associated changes in drug lipoprotein-binding patterns and have examined the increasing number of studies that suggest a correlation between the extent of lipoprotein binding and therapeutic activity and toxicity. These areas are beyond the scope of this chapter and will not be addressed in detail. The primary thrust of this commentary is to review the potential factors controlling the patterns of lipoprotein binding and to address the implications in terms of the co-administration of drugs with exogenous lipids.

For many compounds, for example, benzpyrene [140], nicardipine, [141] propranolol [142], and a series of anthracycline derivatives [143], the degree of binding to specific lipoprotein (LP) subgroups (VLDL, LDL, and HDL) correlates with the size or total lipid content of the lipoprotein (VLDL > LDL > HDL). Subsequent studies have shown in specific cases that a particular lipid component may be a more important determinant of the extent of drug binding. For example, probucol LP binding is most efficiently correlated with the amount of phospholipid in the LP fraction [144] and nicardipine binding with the LP TG content [145]. In contrast, studies with cyclosporin have shown that drug-LP affinity is highest for HDL and LDL, suggesting that lipoprotein cholesterol content and not total lipid content (because HDL and LDL are the primary cholesterol LP carriers) is the primary determinant of drug LP binding [146, 147].

Data obtained with etretinate and its more polar metabolite acitretin [148] and a series of anthracyline derivatives [143] have also suggested that the primary determinant of LP binding is drug lipophilicity, indicating the presence of a relatively nonspecific partition phenomena driving the LP-binding process. A recent examination of the LP-binding behavior of three basic ligands (binedaline, nicardipine, and darodipine) at different pH values supports this view and demonstrates an increase in LP binding across all LP classes at pHs at which the ligands become increasingly un-ionized [149].

Comparison of drug-LP binding profiles across a series of compounds may uncover significant differences in binding profiles across seemingly similar molecules. Thus, for benzo(a)pyrene, the extent of drug LP uptake (normalized for the lipid volume of each lipoprotein fraction) was found to be essentially identical for each LP fraction (VLDL, LDL, and HDL), with differences in binding reflecting the relative volume of lipid in the fraction [140]. However, for the

3-hydroxy metabolite of benzo[a]pyrene, the uptake was the same in TRL and LDL as for benzo[a]pyrene, whereas in HDL the normalized uptake was more than twofold greater, suggesting an additional interaction of this metabolite with HDL. The normalized uptake of the 7,8-dihydrodiol metabolite was much less in all fractions, possibly as a function of the relatively greater polarity of this compound. These data suggest that for the most hydrophobic compound (benzo [a]pyrene) solubilization in lipoprotein lipids was the major mechanism of uptake into all the lipoproteins, whereas with the increased polarity of the metabolites, other factors were involved [140].

The process of drug-LP binding is therefore primarily one of solubilization (supplemented in certain cases by drug association with specific LP components), and the binding of lipophilic drugs to plasma lipoproteins is generally determined by the quantity of lipid within each lipoprotein fraction. Many of the compounds studied thus far appear to bind in either a nonlipid-specific manner or one favoring lipoprotein core lipids and therefore on a mole-to-mole basis, the highest drug affinities are generally seen for the lipid-rich (and especially triglyceride rich) TRL. Conversely, some molecules (typified by cyclosporin) have a more specific affinity for cholesterol and consequently have less affinity for TRL (which are relatively cholesterol poor) and higher affinity for cholesterol-rich LDL and HDL.

The eventual distribution of drug molecules across plasma, however, is determined not only by the specific affinities and binding constants of drug molecules to isolated lipoprotein fractions but also to the concentration of each lipoprotein fraction in the plasma. Consequently, because the relative concentrations of plasma lipoprotein fractions increase from VLDL ($\approx 0.1$ μM) < LDL ($\approx 1$ μM) < HDL ($\approx 11$ μM) [142, 148, 150], the plasma distribution profiles of lipophilic drugs often reflect the relative lipoprotein concentrations and not the specific binding affinities.

Alterations to both the total concentration of lipoproteins in plasma and the lipid composition of lipoproteins will therefore likely affect the pattern of drug lipoprotein binding and potentially both pharmacokinetic and pharmacodynamic drug end points.

Plasma lipoprotein composition and/or concentration can change both as a function of acute events such as administration of lipids in food (or a dosage form) or more chronically as a function of disease. As described, the focus of this review is to examine the potential effects of co-administration of lipophilic drugs with lipids, and, therefore, the emphasis here is on LP changes secondary to the ingestion of lipid. However, it is instructive to examine the literature describing alterations in drug-LP binding as a function of chronically altered lipid loads in, for example, hypertriglyceridaemia and hypotriglyceridaemia, although these data should be viewed with the caveat that although the end result of both feeding and hypertriglyceridaemia may be similar in terms of lipoprotein profile, it is likely that the mechanisms underlying both events are different and therefore that the impact on drug binding may also alter.

Sgoutas et al. [151] determined the cyclosporine distribution in plasma from fasted and nonfasted patients, and these data are shown in Table 8. An increased proportion of cyclosporine was associated with the TRL in the nonfasted state compared with fasted patients with a corresponding decrease in LDL and HDL CY levels. An extreme example of altered cyclosporine plasma distribution has also been reported in a case study of a patient with severe hypertriglyceridemia [152]. This was characterized by huge increases in plasma chylomicron concentrations (with plasma TG concentrations concentrations up to 264 mg/mL$^{-1}$) and much higher than expected plasma concentrations of cyclosporine (considering the dose), of which up to 83% was associated with the chylomicrons.

Other examples of altered plasma binding of drugs caused by altered plasma lipoprotein profiles include imipramine, where the total plasma binding was higher in hyperlipoproteinemic patients than normal subjects [153], and probucol, where Eder demonstrated that 66% of the probucol plasma concentration in the plasma of fasted rhesus monkeys was associated with LDL, but that after oral administration with a high-fat meal, most of the plasma probucol was associated with the TRL fraction at the peak of hypertriglyceridemia [154].

We have recently examined the pattern of plasma lipoprotein binding of halofantrine in fed and fasted healthy volunteers. Plasma samples were take from fasted subjects and from subjects 4 hr after a standard fatty meal and subsequently incubated *in vitro* with Hf at 37°C for 60 min. After incubation, lipoprotein fractions were separated by single spin density gradient ultracentrifugation, and the mass of Hf and various lipid components in the fractions was determined. (Validation studies indicated that the drug-LP distributions obtained using this *in vitro* method were identical to distributions obtained *in vivo*). The percentage distribution of halofantrine among the plasma lipoprotein fractions is given in Table 9. Significant differences were seen between the drug distributions obtained both fed and fasted and also in distributions obtained in beagles and humans.

**Table 8** Lipoprotein Distribution of Cyclosporine After Spiking Cyclosporine in Serum Samples from Either Fasted or Nonfasted Human Subjects

| Sample | % of total plasma concentration | | | |
|---|---|---|---|---|
| | TRL | LDL | HDL | LPDP |
| Fasted | 8 | 31 | 46 | 15 |
| Nonfasted | 16 | 28 | 39 | 11 |

*Source*: Ref. 151.

**Table 9**  Percentage Distribution of Halofantrine (Hf) between Plasma Lipoprotein Fractions[a] (mean $\pm$ SD, $n = 3$) Obtained after *In Vitro* Incubation of 1000 ng/mL$^{-1}$ Hf.HCl with Plasma Obtained from Preprandial and Postprandial Human Subjects, and Preprandial and Postprandial Beagles

| Lipoprotein fraction | Percentage distribution of Hf in human plasma | | Percentage distribution of Hf in beagle plasma | |
|---|---|---|---|---|
|  | Preprandial | Postprandial | Preprandial | Postprandial |
| TRL | 17.3 $\pm$ 1.2 | 39.7 $\pm$ 1.2[b] | 3.3 $\pm$ 0.9 | 26.6 $\pm$ 2.0[b] |
| LDL | 31.5 $\pm$ 0.2 | 18.6 $\pm$ 1.0[b] | 10.1 $\pm$ 0.8 | 7.5 $\pm$ 1.1[b] |
| HDL | 6.1 $\pm$ 0.6 | 6.5 $\pm$ 2.4 | 51.2 $\pm$ 4.4 | 37.5 $\pm$ 0.9[b] |
| Total LP binding | 54.8 $\pm$ 0.9 | 64.8 $\pm$ 3.2[b] | 64.6 $\pm$ 2.7 | 71.5 $\pm$ 1.4[b] |
| LPDP | 45.2 $\pm$ 0.9 | 35.2 $\pm$ 3.2[b] | 35.4 $\pm$ 2.7 | 28.5 $\pm$ 1.4[b] |
| % recovery[c] | 99.1 $\pm$ 4.4 | 109.1 $\pm$ 1.6 | 93.8 $\pm$ 3.7 | 90.2 $\pm$ 4.0 |

[a] TRL, triglyceride-rich lipoproteins that include VLDL and chylomicrons; LDL, low-density lipoproteins; HDL, high-density lipoproteins; LPDP, lipoprotein-deficient plasma.
[b] Significantly different from the corresponding preprandial value ($p < 0.05$).
[c] Recovery is defined as the percent mass of drug recovered after plasma fractionation divided by the mass of drug present in the original sample.

The proportion of Hf associated with the TRL and LDL fractions was higher in humans compared with beagles, whereas the proportion associated with HDL was higher in beagles. These findings broadly reflect the species difference in lipoprotein profiles between humans and beagles, such that a larger proportion of human plasma lipid is carried by TRL and LDL, whereas HDL are the major lipid carriers in dogs.

In both species, the postprandial state induced significant changes in the distribution of Hf between LP fractions, with increased proportions of Hf in the TRL fractions in both human and beagle plasma at the expense of a decrease in the proportion of Hf in LDL and HDL fractions.

To investigate more closely the basis for the altered interaction of Hf with the different LP fractions, correlations were investigated between the amount of Hf present in the individual fractions and the corresponding mass of total protein, phospholipid, triglyceride, and apolar lipid in human and beagle plasma lipoproteins. The Hf distribution profile was poorly correlated with individual lipoprotein surface components (phospholipid and protein), reasonably correlated with TG profiles in humans (but not beagles), and well correlated with the mass of apolar lipid (defined as the mass of TG plus CE) in both beagles and human lipoprotein fractions. These correlations suggest that the driving force for Hf-LP association was solubilization of Hf within the apolar lipid core of the LP and that this was proportionally altered when the mass of apolar lipid in plasma (and particularly the mass of TG in TRL LP) was increased by feeding. However, close examination of the data revealed some anomalies, for example, the mass of Hf in the postprandial LDL fraction was lower than the mass in preprandial LDL, whereas the mass of apolar lipid in preprandial and postprandial LDL was similar. These data can be explained by appreciation that the effective driving force for the distribution of Hf across the competing plasma LP fractions is not the absolute mass of apolar lipid in each fraction, but the proportional distribution of apolar lipid within the plasma. Therefore, although the mass of apolar lipid was similar between preprandial and postprandial states, the proportion of apolar plasma lipid carried by LDL decreased because of the increased quantity of apolar lipid carried by the TRL fraction. The correlation between the proportional distribution of plasma apolar lipid and Hf across all lipoprotein fractions is given in Figure 4. Interestingly, this correlation was independent of both the extent of feeding and species (human vs. beagle), suggesting that the factors regulating the proportional distribution of Hf were identical in fed and fasted beagles and humans.

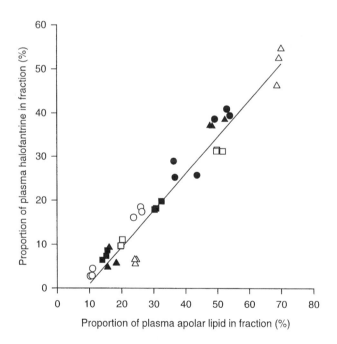

**Figure 4** Combined correlation between the proportion of Hf distributed between lipoprotein fractions (TRL [●], LDL [■], HDL [▲]) in human and beagle plasma and the proportional distribution of apolar lipid (TG + CE) within the individual lipoprotein fractions from preprandial (open symbols) and postprandial plasma (closed symbols). The correlation coefficient was 0.95.

Subsequent experiments in collaboration with Dr. K. Wasan at the University of British Columbia have examined variations in the plasma-LP binding profile of Hf as a function of alterations in plasma lipid load, secondary to underlying dyslipidemia (i.e., in hypo- and hyperlipidaemic patients and normolipidaemic controls). These studies uncovered broadly similar correlations (i.e., positive correlations between the distribution of Hf across plasma LPs and the proportional distribution of apolar lipid within a particular patient group). However, these correlations did not hold true across all the dyslipidemic patient groups, and an inverse correlation between the mass of Hf in the TRL fraction and the mass of Hf in the HDL fraction as a function of the dyslipidemic state was seen. Thus, from hypolipidemic patients through normolipidemic controls to hyperlipidemic patients, there appeared to be a redistribution of Hf from HDL to TRL. These data suggested the presence of additional mechanisms that modulate the distribution of Hf across dyslipidemic plasma in addition to the partition into apolar lipid seen before.

Examination of the activity of lipid transfer protein 1 (LTP 1), a plasma protein that facilitates the transfer of cholesterol ester, triglyceride, and phospholipids between plasma lipoproteins, in each of the dyslipidemic plasmas revealed a significant increase in LTP 1 activity from hypo $<$ normo $<$ hypertriglyceridemic plasma, indicating a possible role for LTP in the altered Hf distribution in dyslipidemic patients. It is unclear at this stage what role LTP 1 may play in acute changes in drug binding because of altered plasma lipoprotein distributions postprandially.

## A. The Impact of Lipid-induced Alterations in Drug–LP Binding Profiles on Drug Pharmacokinetics

Because it is generally assumed that only unbound drug is available for uptake across cell membranes and therefore determines distribution and possibly elimination of a drug, it is likely that changes in drug plasma binding (secondary to alterations in plasma lipoprotein binding) will affect pharmacokinetic parameters such as volume of distribution ($V_D$) and clearance (CL).

Cyclosporine (CY) is an example of a lipophilic drug that is predominantly bound to lipoproteins in plasma, has a low extraction ratio ($E_H < 0.3$), and a volume of distribution of about 3.9 L/kg [155]. Both the CL and $V_D$ of low extraction ratio highly bound drugs such as CY are highly dependent on the unbound fraction present in plasma. It might be expected that an increase in the concentration of plasma lipoproteins would result in a decreased unbound fraction of CY and therefore decreased CL and $V_D$. In support of this hypothesis, an inverse linear relationship has been shown between cyclosporine CL and the level of TRL-TG, TRL-CH, and LDL-TG in uremic patients [155].

Similarly, we have examined the pharmacokinetics of Hf after intravenous administration to fed and fasted beagles and shown that Hf plasma AUCs were significantly higher and CL and $V_{SS}$ significantly lower in the fed state compared with the fasted state [156]. The mean postprandial increase in plasma AUC values was 18%, with corresponding decreases in CL (15%) and $V_{SS}$ (21%). There was also a linear relationship between the increase in postprandial Hf concentrations at specific time points and the corresponding postprandial increase in TG concentrations. Similar increases in the proportion of Hf associated with the postprandial TRLs were seen in this study *in vivo* to those discussed earlier *in vitro*, suggesting that increased TRL LP binding may have resulted in the decrease in CL and $V_{SS}$ presumably mediated by a decrease in the plasma free fraction of Hf.

In contrast to these studies, Gupta and Benet reported an increase in the CL and $V_D$ of cyclosporine when administered to healthy subjects with a high-fat meal compared with administration to the same subjects in the fasted state [157]. They suggested that this unexpected result could be explained if the CL and distribution were not purely a function of the unbound fraction of drug in blood, but that lipoprotein-bound drug could also be taken up into tissues. They suggested that after the fatty meal, an increased proportion of the drug would be associated with the TRL-lipoproteins, and because these lipoproteins are primarily catabolized in the adipose tissues in the postprandial state, these factors would increase distribution of cyclosporine to the adipose tissue, thereby increasing the volume of distribution. Similarly, they suggested that the increased clearance of cyclosporine when administered with a fatty meal could be explained by increased uptake of cyclosporine into hepatocytes because of its association with chylomicron remnants, which are cleared primarily in the liver. This uptake in conjunction with a slow rate of diffusion of free cyclosporine out of cells would result in increased metabolism and, therefore, the observed increase in clearance.

A similar mechanistic basis was proposed for the increased $V_{SS}$ and CL with increased plasma lipid concentrations demonstrated for amphotericin B (which associates with HDL and LDL [158] in a diabetic hyperlipidemic rat model [159]). The increased lipid concentrations in the diabetic compared with the non-diabetic rats were characterized by increased LDL concentrations. It was therefore suggested that increased association of amphotericin B with LDL in the diabetic rats increased LDL-associated uptake in peripheral tissues and in the liver, resulting in the increased $V_{SS}$ and CL.

## B. The Impact of Lipid-induced Alterations in Drug–LP Binding Profiles on Drug Pharmacodynamics

Amphotericin B (AmB) is a polyene macrolide used in the treatment of systemic fungal infections. The clinical use of amphotericin, however, is limited by its dose-dependent nephrotoxicity. A number of studies have shown that AmB is significantly bound to lipoproteins in the plasma and that the toxicity of AmB is altered in patients with altered plasma lipid and lipoprotein levels [160–164]. For example, renal toxicity was decreased after administration to leukemic [160] and immunocompromised patients [161] who exhibited low plasma cholesterol levels, and no measurable toxicity was seen after AmB administration to cancer patients who exhibited hypocholesterolemia [162]. Koldin and co-workers also demonstrated elevated AmB toxicity when LDL-associated AmB was administered to hypercholesterolemic rabbits [164]. Conversely, administration of AmB to mice with elevated serum triglyceride concentrations appeared to protect against renal toxicity, suggesting that triglycerides may provide protection against AmB toxicity [165]. These data are supported by a study by Souza and co-workers, who showed that co-administration of AmB with a TG-rich emulsion (which behaves as a model chylomicron in the systemic circulation) also protects against AmB toxicity [166].

Wasan and co-workers have further shown a correlation between elevated serum LDL cholesterol levels, elevated AmB levels in LDL, and elevated AmB toxicity and have suggested that the greater proportion of the AmB dose associated with LDL in patients with elevated LDL cholesterol levels leads to the increased toxicity of AmB mediated by the LDL receptor [139, 163].

Lemaire and co-workers [167] and Lithell et al. [168] have also suggested that the disposition of cyclosporine is dictated by its lipoprotein-binding profile and that the patterns of LP association determine both its activity and toxicity. In support of these findings, several investigators have shown that the toxicity of CY may be attenuated in patients with hypertriglyceridemia (and presumably increased levels of CY in TRL) [169, 170] and enhanced in patients with hypercholesterolemia [171]. Specifically, Gardier and co-workers showed an increase in LDL-associated CY in heart transplant patients with hypercholesterolemia and a concurrent increase in CY-mediated renal toxicity [172]. Unfortunately, the activity of CY also appears to be enhanced when bound to LDL (and not VLDL or HDL), suggesting that increases in CY LDL may lead to increases in both therapeutic activity and toxicity [173, 174].

The previously observed effects on the pharmacokinetics of Hf, possibly mediated by a decrease in free fraction, also led us to speculate that a decrease in free fraction produced by an increase in LP binding in fed plasma may reduce the access of Hf to the malaria parasite and reduce the antimalarial activity of Hf. This hypothesis led to studies that assessed the effect of using growth media containing 10% serum with varying concentrations of TG (i.e., different degrees of fed serum) on the $IC_{50}$ of Hf by use of a continuous *in vitro* culture of *P. falciparum*. There was a linear and statistically significant 2.5-fold increase in the $IC_{50}$ of Hf across a sixfold range of increasing TG concentrations, with the increased $IC_{50}$ values being ascribed to a decreased free fraction of Hf in the incubation media, caused by sequestration by TG-rich lipoproteins.

## VII. CONCLUSION

It is readily apparent from the data reviewed earlier that the co-administration of lipophilic drugs with lipids, whether in the form of food or lipid-based delivery systems (notwithstanding the likely differences in exogenous lipid loads), may have significant effects on drug absorption, metabolism, and disposition. These observations present a number of attractive propositions for the formulation scientist, ranging from an enhancement of drug dissolution and reduction in prehepatic enterocyte-based metabolism to specific delivery to the intestinal lymph and avoidance of hepatic first-pass metabolism and improvements in therapeutic and toxicological end points as a function of altered plasma lipoprotein–binding profiles.

However, they also indicate that care must be taken when assessing the biopharmaceutical profile of lipophilic drugs, particularly when data are obtained with different delivery systems or animal models. For example, subtle changes in lipid excipient may provoke changes in enterocyte-based metabolism or alter the extent of intestinal lymphatic transport, leading to differences in drug bioavailability and clearance. Furthermore, the markedly different plasma lipoprotein profiles found in common laboratory animals presents concerns for the preclinical progression of toxicity and activity studies from rat or rabbit to dogs and primates in cases in which drug toxicity or pharmacokinetics may be altered by lipoprotein-binding patterns. Advances in our basic understanding of the possible interaction of lipids in food or lipid-based delivery systems with drug absorption, metabolism, and clearance mechanisms is needed before these effects can be predicted with confidence.

## REFERENCES

1. Y.-F. Shiau, Mechanism of intestinal fat absorption, *Am. J. Physiol. 240*:G1–G9 (1981).
2. C. L. Bisgaier and R. M. Glickman, Intestinal synthesis, secretion and transport of lipoproteins, *Annu. Rev. Physiol. 45*:625–636 (1983).
3. A. B. R. Thomson, M. Keelan, M. L. Garg, and M. T. Clandinin, Intestinal aspects of lipid absorption: in review, *Can. J. Physiol. Pharmacol. 67*:179–191 (1989).
4. P. Tso, Intestinal lipid absorption, *Physiology of the Gastrointestinal Tract* (L. R. Johnson, chief ed.), Raven Press, New York, 1994, pp. 1867–1908.

5.  M. C. Carey, D. M. Small, and C. M. Bliss, Lipid digestion and absorption, *Annu. Rev. Physiol. 45*:651–677 (1983).
6.  J. E. Staggers, O. Hernell, R. J. Stafford, and M. C. Carey, Physical-chemical behaviour of dietary and biliary lipids during intestinal digestion and absorption. 1. Phase behaviour and aggregation states of model lipid systems patterned after aqueous duodenal contents of healthy adult human beings, *Biochemistry 29*:2028–2040 (1990).
7.  O. Hernell, J. E. Staggers, and M. C. Carey, Physical-chemical behaviour of dietary and biliary lipids during intestinal digestion and absorption. 2. Phase analysis and aggregation states of luminal lipids during duodenal fat digestion in healthy adult human beings, *Biochemistry 29*:2041–2056 (1990).
8.  N. E. Hoffman, The relationship between uptake in vitro of oleic acid and micellar solubilization, *Biochim. Biophys. Acta 196*:193–203 (1970).
9.  W. J. Simmonds, The role of micellar solubilisation in lipid absorption, *Aust. J. Exp. Biol. Med. Sci. 50*:403–421 (1972).
10. H. Westergaard and J. M. Dietschy, The mechanism whereby bile acid micelles increase the rate of fatty acid and cholesterol uptake into the mucosal cell, *J. Clin. Invest 58*:97–108 (1976).
11. Y-F. Shiau, P. Fernandez, M. J. Jackson and S. McMonagle, Mechanisms maintaining a low pH microclimate in the intestine, *Am. J. Physiol. 248*:G608–619 (1985).
12. Y-F. Shiau, Mechanism of intestinal fatty acid uptake in the rat: the role of an acidic microclimate *J. Physiol. 421*:463–474 (1990).
13. J. Y. Kiyasu, B. Bloom and I. L. Chaikoff, The portal transport of absorbed fatty acids, *J. Biol. Chem. 199*:415–419 (1952).
14. D. K. Spady, Lipoproteins in biological fluids and compartments: Synthesis, interconversions and catabolism, *Lipoproteins as carriers of Pharmacological Agents* (J. M. Shaw, ed.), Marcel Dekker, New York, 1991, pp. 1–44.
15. P. E. Fielding and C. J. Fielding, Dynamics of lipoprotein transport in the circulatory system. *Biochemistry of Lipids, Lipoproteins and Membranes* (D. E. Vance and J. Vance eds.), Elsevier, New York, 1991, pp. 427–459.
16. J. N. Hunt and M. T. Knox, A relation between the chain length of fatty acids and the slowing of gastric emptying, *J. Physiol. 194*:327–336 (1968).
17. E. K. Anderberg, T. Lindmark and P. Artursson, Sodium caprate elicits dilatations in human intestinal tight junctions and enhances drug absorption by the paracellular route, *Pharm. Res. 10*:857–864 (1993).
18. B. J. Aungst, Novel formulation strategies for improving oral bioavailability of drugs with poor membrane permeation or presystemic metabolism, *J. Pharm. Sci. 82*:979–987 (1993).
19. E. S. Swenson, W. B. Milisen, and W. Curatolo, Intestinal permeability enhancement: Efficacy, acute local toxicity, and reversibility, *Pharm. Res. 11*:1132–1142 (1994).
20. N. A. Armstrong and K. C. James, Drug release from lipid based dosage forms. Part 2, *Int. J. Pharmaceut. 6*:195–204 (1980).
21. A. J. Humberstone and W. N. Charman, Lipid-based vehicles for the oral delivery of poorly water soluble drugs *Adv. Drug Deliv. Rev. 25*:103–128 (1997).

22. W. N. Charman, C. J. H. Porter, S. Mithani and J. B. Dressman, Physicochemical and physiological mechanisms for the effects of food on drug absorption: The role of lipids and pH, *J. Pharm. Sci. 86*:269–282 (1997).

23. S. Chakrabarti and F. M. Belpaire, Bioavailability of phenytoin in lipid containing dosage forms in rats, *J. Pharm. Pharmac. 30*:330–331 (1978).

24. J. T. Hargrove, W. S. Maxson and A. C. Wentz, Absorption of oral progesterone is influenced by vehicle and particle size, *Am. J. Obstet. Gynecol. 161*:948–951 (1989).

25. L. S. Abrams, H. S. Weintraub, J. E. Patrick and J. L. McGuire, Comparitive bioavailability of a lipophilic steroid, *J. Pharm. Sci. 67*:1287–1290 (1978).

26. Y. Yamaoka, R. D. Roberts and V. J. Stella, Low-melting phenytoin prodrugs as alternative oral delivery modes for phenytoin: a model for other high-melting sparingly water-soluble drugs. *J. Pharm. Sci. 72*:400–405 (1983).

27. V. Stella, J. Haslam, N. Yata, H. Okada, S. Lindenbaum, and T. Higuchi, Enhancement of bioavailability of a hydrophobic amine antimalarial by formulation with oleic acid in a soft gelatin capsule, *J. Pharm. Sci. 67*:1375–1377 (1978).

28. T. Tokumura, Y. Tsushima, K. Tatsuishi, M. Kayano, Y. Machida, and T. Nagai, Enhancement of the oral bioavailability of cinnarizine in oleic acid in beagle dogs. *J. Pharm. Sci. 76*:286–288 (1987).

29. Y. Yamahira, T. Noguchi, H. Takenaka and T. Maeda, Biopharmaceutical studies of lipid containing oral dosage forms: relationship between drug absorption rate and digestibility of vehicles, *Int. J. Pharmaceut. 3*:23–31 (1979).

30. R. A. Myers and V. J. Stella, Systemic bioavailability of penclomedine (NSC-338720) from oil-in-water emulsions administered intraduodenally to rats, *Int. J. Pharmaceut. 78*:217–226 (1992).

31. K. J. Palin and C. M. Wilson, The effect of different oils on the absorption of probucol in the rat, *J. Pharm. Pharmac. 36*:641–643 (1984).

32. P. J. Carrigan and T. R. Bates, Biopharmaceutics of drugs administered in lipid-containing dosage forms I: GI absorption of griseofulvin from an oil-in-water emulsion in the rat, *J. Pharm. Sci. 62*:1476–1479 (1973).

33. T. R. Bates and P. J. Carrigan, Apparent absorption kinetics of micronized griseofulvin after its oral administration on single- and multiple-dose regimens to rats as a corn oil-in-water emulsion and aqueous suspension, *J. Pharm. Sci. 64*:1475–1481 (1975).

34. W. N. Charman, M. C. Rogge, A. W. Boddy, and B. M. Berger, Effect of food and a monoglyceride emulsion formulation on danazol bioavailability, *J. Clin. Pharmacol. 33*:381–386 (1993).

35. A. T. Serajuddin, P. Sheen, D. Mufson, D. F. Bernstein, and M. A. Augustine, Physicochemical basis of increased bioavailability of a poorly water-soluble drug following oral administration as organic solutions *J. Pharm. Sci. 77*:325–329 (1988).

36. D. J. Hauss, S. E. Fogal, J. V. Ficorilli, C. A. Price, T. Roy, A. A. Jayaraj, and J. J. Keirns, Lipid-based delivery systems for improving the bioavailability and lymphatic transport of a poorly water soluble $LTB_4$ inhibitor, *J. Pharm. Sci. 87*:164–169 (1988).

37. T. Kimura, E. Fukui, A. Kageyu, Y. Kurohara, T. Nakayama, Y. Morita, K. Shibu-

sawa, S. Ohsawa, and Y. Takeda, Enhancement of oral bioavailability of d-a-tocopherol acetate by lecithin-dispersed aqueous preparations containing medium-chain triglyceridies in rats, *Chem. Pharm. Bull. 37*:439–441 (1989).

38. M. J. Groves and de D. A. Galindez, The self-emulsifying action of mixed surfactants in oil, *Acta. Pharm. Suec. 13*:361–372 (1976).

39. M. J. Groves, R. M. A. Mustafa, and J. E. Carless, Phase studies of mixed phosphated surfactants, n-hexane and water, *J. Pharm. Pharmac. 26*:616–623 (1974).

40. M. J. Groves and R. M. A. Mustafa, Measurement of the 'spontaneity' of self-emulsifiable oils, *J. Pharm. Pharmac. 26*:671–681 (1974).

41. C. W. Pouton, Self-emulsifying drug delivery systems: assessment of their efficiency of emulsification, *Int. J. Pharmaceut. 27*:335–348 (1985).

42. M. G. Wakerly, C. W. Pouton, B. J. Meakin, and F. S. Morton, Self-emulsification of vegetable oil-nonsurfactant mixtures: A proposed mechanism of action. *Am. Chem. Soc. Symp. Ser. 311*:242–255 (1986).

43. D. Q. M. Craig, H. S. R. Lievens, K. G. Pitt, and D. E. Storey, An investigation into the physico-chemical properties of self-emulsifying systems using low frequency dielectric spectroscopy, surface tension measurements and particle size analysis, *Int. J. Pharmaceut. 96*:147–155 (1993).

44. D. Q. Craig, S. A. Barker, D. Banning, and S. W. Booth, Investigation into the mechanisms of self-emulsification using particle size analysis and low frequency dielectric spectroscopy, *Int. J. Pharmaceut. 114*:103–110 (1995).

45. C. W. Pouton, Formulation of self-emulsifing drug delivery systems, *Adv. Drug. Deliv. Rev. 25*:47–58 (1997).

46. T. Gershanik, and S. Benita, Positively charged self-emulsifying oil formulation for improving oral bioavailability of progersterone, *Pharm. Dev. Technol. 1*:147–157 (1996).

47. S. A. Charman, W. N. Charman, M. C. Rogge, T. D. Wilson, F. J. Dutko, and C. W. Pouton, Self-emulsifying drug delivery systems: Formulation and biopharmaceutical evaluation of an investigational lipophilic compound, *Pharm. Res. 9*:87–93 (1992).

48. J. H. Lin, I. Chen, and H. Lievens, The effect of dosage forms on the oral absorption of L-365,260, a potent $CCK_B$ receptor antagonist in dogs, *Pharm. Res. 8*:S–272 (1991).

49. N. H. Shah, M. T. Carvajal, C. I. Patel, M. H. Infeld, and A. W. Malick, Self-emulsifying drug delivery systems (SEDDS) with polyglycolyzed glycerides for improving in vitro dissolution and oral absorption of lipophilic drugs, *Int. J. Pharmaceut. 106*:15–23 (1994).

50. R. Kinget, and H. De Greef, Absorption characteristics of novel 8-MOP semi-solid-lipid-matrix formulations: In vitro-in vivo correlation, *Int. J. Pharmaceut. 110*:65–73 (1994).

51. G. R. Kelm, J. V. Penafiel, R. M. Deibel, G. O. Kinnett, and M. P. Meredith, The relative absorption of tebufalone from self-emulsifying solutions in the beagle dog, *Pharm. Res. 10*:S211 (1993).

52. B. Matuszewska, L. Hettrick, J. V. Bondi, and D. E. Storey, Comparative bioavailability of L-683,453, a 5α-reductase inhibitor, from a self emulsifying drug delivery system in beagle dogs, *Int. J. Pharm. 136*:147–154 (1996).

53.  M. J. Gumkowski, L. A. Fournier, N. K. Tierney, and W. J. Curatolo, Improved bioavailability through use of soft gelatin formulations of terlakiren, a tripeptide renin inhibitor, *Pharm. Res. 11*:S286 (1994).

54.  C. Malcolmson, C. Satra, S. Kantaria, A. Sidhu, and M. J. Lawrence, Effect of oil on the level of solubilisation of testosterone propionate into non-ionic oil-in-water microemulsions, *J. Pharm. Sci. 87*:109–116 (1988).

55.  C. Malcolmson, M. J. Lawrence, Comparison of the incorporation of model steroids into non-ionic micellar and microemulsion systems, *J. Pharm. Pharmacol. 45*:141–143 (1993).

56.  J. M. Sarciaux, L. Acar, and P. A. Sado, Using microemulsion formulations for oral drug delivery of therapeutic peptides, *Int. J. Pharm. 120*:127–136 (1995).

57.  P. P. Constantinides, Lipid microemulsions for improving drug dissolution and oral absorption: Physical and biopharmaceutical aspects, *Pharm. Res. 12*:1561–1572 (1995).

58.  R. R. C. New, and C. J. Kirby, Solbilisation of hydrophilic drugs in oily formulations, *Adv. Drug Deliv. Rev. 25*:59–69 (1997).

59.  P. P. Constantinides and J. P. Scalart, Formulation and physical characterisation of water-in-oil microemulsions containing long-versus medium chain glycerides, *Int. J. Pharm. 158*:57–68 (1997).

60.  J. Reymond and H. Sucker, In vitro model for ciclosporin intestinal absorption in lipid vehicles, *Pharm. Res. 5*:673–676 (1988).

61.  J. Reymond, H. Sucker, and J. Vonderscher, In vivo model for ciclosporin intestinal absorption in lipid vehicles, *Pharm. Res. 5*:677–679 (1988).

62.  D. Behrens, R. Fricker, A. Bodoky, J. Drewe, F. Harder, and M. Heberer, Comparison of cyclosporin A absorption from LCT and MCT solutions following intrajejunal administration in conscious dogs, *J. Pharm. Sci. 85*:666–668 (1996).

63.  T. Beveridge, A. Gratwohl, F. Michot, W. Niederberger, E. Nuesch, K. Nussbaumer, P. Schaub, and B. Speck, Cyclosporin A: Pharmacokinetics after a single dose in man and serum levels after multiple dosing in recipients of allogenic bone marrow grafts, *Curr. Ther. Res. 30*:5–18 (1981).

64.  F. J. Frey, F. F. Horber, and B. M. Frey, Trough levels and concentration time curves of cyclosporin in patients undergoing renal transplantation, *Clin. Pharmac. Ther. 43*:55–62 (1988).

65.  R. Ptachcinski, R. Venkataramanan, and G. J. Burckart, Clinical pharmacokinetics of cyclosporin, *Clinical Pharmacokin. 11*:107–132 (1986).

66.  R. J. Ptachcinski, R. Venkataramanan, J. T. Rosenthal, G. J. Burckart, R. J. Taylor, and T. R. Hakala, The effect of food on cyclospoerin absorption, *Transplantation, 40*:174–176 (1985).

67.  S. K. Gupta and L. Z. Benet, High fat meals increase the clearance of cyclosporine, *Pharm. Res. 7*:46–48 (1990).

68.  B. D. Tarr and S. H. Yalkowsky, Enhanced intestinal absorption of cyclosporine in rats through the reduction of emulsion droplet size, *Pharm. Res. 6*:40–43 (1989).

69.  W. A. Ritschel, Microemulsions for improved peptide absorption from the gastrointestinal tract, *Meth. Find. Exp. Clin. Pharmacol 13*:205–220 (1991).

70.  W. A. Ritschel, S. Adolf, G. B. Ritschel, and T. Schroeder, Improvement of peroral

absorption of cyclosporine A by microemulsions, *Meth. Find. Exp. Clin. Pharmacol. 12*:127–134 (1990).

71.  Z. G. Gao, H. G. Choi, H. J. Shin, K. M. Park, S. J. Lim, K. J. Hwang, and C. K. Kim, Physicochemical characterisation and evaluation of a microemulsion system for oral delivery of cyclosporin A, *Int. J. Pharmaceut. 161*:75–86 (1998).

72.  J. Drewe, R. Meier, J. Vonderscher, D. Kiss, U. Posanski, T. Kissel, and K. Gyr, Enhancement of the oral absorption of cyclosporin in man, *Br. J. Clin. Pharmacol. 34*:60–64 (1992).

73.  M. Bojrup, Z. Qi, S. Bjorkman, O. Ostraat, B. Landin, H. Ljusberg-Wahren, and H. Ekberg, Bioavailability of cyclosporine in rats after intragastric administration: a comparitive study of the $L_2$ phase and two other lipid-based vehicles, *Transplant. Immunol. 4*:313–317 (1996).

74.  E. A. Mueller, J. M. Kovarik, J. B. van Bree, W. Tetzloff, J. Grevel and K. Kutz, Improved dose linearity of cyclosporine pharmacokinetics from a microemulsion formulation, *Pharm. Res. 11*:301–304 (1994).

75.  J. M. Kovarik, E. A. Mueller, J. B. van Bree, W. Tetzloff, and K. Kutz, Reduced inter and intraindividual variability in cyclosporine pharmacokinetics form a microemulsion formulation, *J. Pharm. Sci. 83*:444–446 (1994).

76.  A. K. Trull, K. K. C. Tan, L. Tan, G. J. M. Alexander, and N. J. Jamieson, Absorption of cyclosporin from conventional and new microemulsion oral formulations in liver transplant recipients with external biliary diversion, *Br. J. Clin. Pharmacal. 39*:627–631 (1995).

77.  E. A. Mueller, J. M. Kovarik, J. B. van Bree, J. Grevel, P. W. Lucker, and K. Kutz, Influence of a fat rich meal on the pharmacokinetics of a new oral formulatin of cyclosporine in a crossover comparison with the market formulation, *Pharm. Res. 11*:151–155 (1994).

78.  K. A. Milton, G. Edwards, S. A. Ward, M. L. Orme, and A. M. Breckenridge, Pharmacokinetics of halofantrine in man: effects of food and dose size, *Br. J. Clin. Pharmacol. 28*:71–77 (1989).

79.  A. J. Humberstone, C. J. H. Porter, and W. N. Charman, A physicochemical basis for the effect of food on the absolute oral bioavailability of halofantrine, *J. Pharm. Sci. 85*:525–529 (1996).

80.  S-M Khoo, A. J. Humberstone, C. J. H. Porter, G. A. Edwards, and W. N. Charman, Formulation design and bioavailability assessment of lipidic self-emulsifying formulations of halofantrine, *Int. J. Pharmaceut. 167*:155–164 (1998).

81.  L. Z. Benet, C-Y. Wu, M. F. Hebert, V. J. Wacher, Intestinal drug metabolism and antitransport processes: A potential paradigm shift in oral drug delivery, *J. Cont. Rel. 39*:139–143 (1996).

82.  V. J. Wacher, L. Salphati, and L. Z. Benet, Active secretion and enterocytic drug metabolism barriers to drug absorption, *Adv. Drug Deliv. Rev. 20*:99–112 (1996).

83.  K. E. Thummel, K. L. Kunze, and D. D. Shen, Enzyme-catalised processes of first pass hepatic and intestinal drug extraction, *Adv. Drug, Deliv. Rev. 27*:99–127 (1997).

84.  K. E. Thummel, D. O'Shea, M. F. Paine, D. D. Shen, K. L. Kunze, J. D. Perkins, G. R. Wilkinson, Oral first pass elimination of midazolam involves both gastrointestinal and hepatic CYP3A-mediated metabolism, *Clin. Pharmacol. Ther. 59*:491–502 (1996).

85. M. F. Hebert, Contributions of hepatic and intestinal metabolism and P-glycoprotein to cyclosporine and tacrolimus oral drug delivery, *Adv. Drug. Del. Rev.* 27: 201–214 (1997).

86. A. E. M. Vickers, V. Fischer, S. Connors, R. L. Fisher, J. P. Baldeck, G. Maurer, and K. Brendel, Cyclosporin A metabolism in human liver, kidney and intestinal slices, *Drug Metab. Dispos.* 20:802–809 (1992).

87. C-Y. Wu, L. Z. Benet, M. F. Hebert, S. K. Gupta, M. Rowland, D. Y. Gomez, and V. J. Wacher, Differentiation of absorption and first-pass gut and hepatic metabolism in humans: studies with cyclosporine, *Clin. Pharmacol Ther.* 58:492–497 (1995).

88. A. Lampen, U. Christians, F. P. Guengerich, P. B. Watkins, J. C. Kolars, A. Bader, A.-K. Gonschior, H. Dralle, I. Hackbarth, and K. F. Sewing, Metabolism of the immunosuppressant tacrolimus in the small intestine: cytochrome P450, drug interactions, and interindividual variability, *Drug Metab. Dispos.* 23:1315–1324 (1995).

89. P. B. Watkins, The barrier function of CYP3A4 and P-glycoprotein in the small bowel, *Adv. Drug Deliv. Rev.* 27:161–170 (1997).

90. M. E. Fitzsimmons and J. M. Collins, Selective biotransformation of the human immunodeficiency virus protease inhibitor saquinavir by human small intestinal cytochrome P4503A4, *Drug. Metab. Dispos.* 25:256–266 (1997).

91. T. Koudriakova, E. Iatsimirskaia, S. Tulebaev, D. Spetie, I. Utkin, D. Mullet, T. Thompson, P. Vouros, and N. Gerber, In vivo disposition and metabolism by liver and enterocyte microsomes of the antitumour drug rifabutin in rats, *J. Pharmacol. Exp. Therap.* 279:1300–1309 (1996).

92. K. S. Lown, R. Mayo, D. S. Blake, L. Z. Benet, P. B. Watkins, Interpatient variation in intestinal Pgp expression contributes to variable oral kinetics of cyclosporin A, *Clin. Pharmacol. Ther.* 62:1–13 (1997).

93. J. Hunter, M. A. Jepson, T. Tsuruo, N. L. Simmons, B. H. Hirst, Functional expression of p-glycoprotein in apical membranes of human intestinal Caco-2 cell layers: kinetics of vinblastine secretion and interaction with modulators, *J. Biol. Chem.* 268:14991–14997 (1993).

94. H. Saitoh and B. J. Aungst, Possible involvement of multiple P-glycoprotein-mediated efflux systems in the transport of verapamil and other organic cations across the rat intestine, *Pharm Res.* 12:1304–1310 (1995).

95. J. Karlsson, S. M. Kuo, J. Ziemniak, P. Artusson, Transport of celiprolol across human intestinal epithelial (Caco-2) cells: mediation of secretion by multiple transporters including p-glycoprotein, *Br. J. Pharm.* 110:1009–1016 (1993).

96. B. L. Leu and J. Huang, Inhibition of intestinal P-glycoprotein and effects on etoposide absorption, *Cancer Chemother. Pharmacol.* 35:432–436 (1995).

97. V. J. Wacher, C.-Y. Wu, and L. Z. Benet, Overlapping substrate specificities and tissue distribution of Cytochrome P450 3A and P-glycoprotein: Implications for drug delivery and activity in cancer chemotherapy, *Mol. Carcinog.* 13:129–134 (1995).

98. L. Z. Benet, V. J. Wacher, and R. M. Bener, Use of essential oils to increase bioavailability of oral pharmaceutical compounds, United States Patent No 5,665,386, Sep 9 1997.

99. M. M. Nerurkar, P. S. Burton, and R. T. Borchardt, The use of surfactant to enhance

the permeability of peptides through Caco-2 cells by inhibition of an apically polarised efflux system, *Pharm. Res. 13*:528–534 (1996).

100. M. G. Choc and W. T. Robinson, Gastrointestinal variables and water insoluble compounds in normal and disease states, Presentation to the AAPS/CRS/FDA workshop on scientific foundation and applications for the biopharmaceutics classification and in vitro-in vivo correlations, April 14–16, 1997, Arlington VA.

101. C. J. H. Porter, S. A. Charman, and W. N. Charman, Lymphatic transport of halofantrine in the triple-cannulated anaesthetized rat model: Effect of lipid vehicle dispersion, *J. Pharm. Sci. 85*:351–356 (1996).

102. C. J. H. Porter, S. A. Charman, and W. N. Charman, Lymphatic transport of halofantrine in the conscious rat when administered as either the free base or the hydrochloride salt: effect of lipid class and lipid vehicle dispersion, *J. Pharm. Sci. 85*: 357–361 (1996).

103. G. Pantaleo, C. Graziosi, L. Butini, P. A. Pizzo, S. M. Schnittman, D. P. Kotler, and A. S. Fauci, Lymphoid organs as major reservoirs for human immunodeficiency virus, *Proc. Natl. Acad. Sci. USA. 88*:9838–9842 (1991).

104. G. Pantaleo, C. Graziosi, J. F. Demarest, L. Butini, M. Montroni, C. H. Fox, J. M. Orenstein, D. P. Kotler, and A. S. Fauci, HIV infection is active and progressive in lymphoid tissue during the clinically latent stage of disease, *Nature 362*:355–358 (1993).

105. J. R. Casley-Smith, The fine structure and functioning of tissue channels and lymphatics, *Lymphology 12*:177–183 (1980).

106. J. R. Casley-Smith, Are the initial lymphatics normally pulled open by the anchoring filaments? *Lymphology 13*:120–129 (1980).

107. V. V. Yang, P. J. O'Morchoe, and C. C. O'Morchoe, Transport of protein across lymphatic endothelium in the rat kidney, *Microvasc. Res. 21*:75–91 (1981).

108. P. J. O'Morchoe, V. V. Yang, and C. C. O'Morchoe, Lymphatic transport during volume expansion, *Microvasc. Res. 20*:275–294 (1980).

109. C. C. O'Morchoe, K. H. Albertine, and P. J. O'Morchoe, The rate of translymphatic endothelial fluid movement in the canine kidney, *Microvasc. Res. 23*:180–187 (1982).

110. C. J. H. Porter, Drug delivery to the lymphatic system. *Crit. Rev. Drug Carrier Syst. 14*:333–393 (1997).

111. W. N. Charman and V. J. Stella, Estimating the maximal potential for intestinal lymphatic transport of lipophilic drug molecules, *Int. J. Pharmaceut 34*:174–178 (1986).

112. A. Kuksis, Absorption of fat soluble vitamins, *Fat Absorption*, Vol. 2, (A. Kuksis ed.), CRC Press, Boca Raton, 1987, pp. 65–86.

113. K. J. Palin and C. J. Wilson, The effect of different oils on the absorption of probucol in the rat, *J. Pharm. Pharmacol. 36*:641–643 (1984).

114. R. C. Grimus and I. Schuster, The role of the lymphatic transport in the enteral absorption of naftifine by the rat, *Xenobiotica 14*:287–294 (1984).

115. K. Katayama and T Fujita, Studies on lymphatic absorption of 1′, 2′-($^3$H)-Coenzyme $Q_{10}$ in rats, *Chem. Pharm. Bull. 20*:2585–2592 (1972).

116. C. T. Ueda, M. Lemaire, G. Gsell, and K. Nussbaumer, Intestinal lymphatic absorption of cyclosporin A following oral administration in an olive oil solution to rats, *Biopharm. Drug. Dispos. 4*:113–124 (1983).

117. S. M. Sieber, The lymphatic absorption of p,p-DDT and some structurally-related compounds in the rat, *Pharmacology 14*:443–454 (1976).

118. W. N. Charman, T. Noguchi, and V. J. Stella, An experimental system designed to study the in situ intestinal lymphatic transport of drugs in anaesthetized rats, *Int. J. Pharmaceut. 33*:155–164 (1986).

119. W. N. Charman, and V. J. Stella, Effect of lipid class and lipid vehicle volume on the intestinal lymphatic transport of DDT, *Int. J. Pharmaceut. 33*:165–172 (1986).

120. S. M. Sieber, V. H. Cohn, and W. T. Wynn, The entry of foreign compounds into the thoracic duct lymph of the rat, *Xenobiotica 4*:265–284 (1974).

121. J. M. Laher, M. W. Rigler, R. D. Vetter, J. A. Barrowman, and J. S. Patton, Similar bioavailability and lymphatic transport of benzo(a)pyrene when administered to rats in different amounts of dietary fat, *J. Lipid. Res. 25*:1337–1342 (1984).

122. D. L. Busbee, J-S. H. Yoo, J. O. Norman, and C. O. Joe, Polychlorinated biphenyl uptake and transport by lymph and plasma components, *Proc. Soc. Exp. Biol. Med. 179*:116–122 (1985).

123. D. J. Hauss, S. Mehta, G. W. Radebaugh, Targeted lymphatic transport and modified systemic distribution of CI-976, a lipophilic lipid-regulator drug, via a formulation approach, *Int. J. Pharmaceut. 108*:85–93 (1994).

124. R. Nankervis, S. S. Davis, N. H. Day, and P. N. Shaw, Intestinal lymphatic transport of three retinoids in the rat after oral administration: effect of lipophilicity and lipid vehicle, *Int. J. Pharmaceut. 130*:57–64 (1996).

125. G. Y. Kwei, L. B. Novak, L. H. Hettrick, E. R. Reiss, E. K. Fong, T. V. Olah, and A. E. Loper, Lymphatic uptake of MK-386, a sterol 5α-reductase inhibitor, from aqueous and lipid formulations. *Int. J. Pharm. 164*:37–44. (1998).

126. K. Takada, H. Yoshimura, H. Yoshikawa, S. Muranishi, T. Yasumura, and S. Oka, Enhanced selective lymphatic delivery of cyclosporin A by solubilisers and intensified immunosuppressive activity against mice skin allograft, *Pharm. Res. 3*:48–51 (1986).

127. R. Myers and V. J. Stella, Factors affecting the lymphatic transport of penclomedine (NSC-338720), a lipophilic cytotoxic drug: comparison to DDT and hexachlorobenzene. *Int. J. Pharm. 80*:51–62 (1992).

128. W. N. Charman and C. J. H. Porter, Lipophilic prodrugs designed for intestinal lymphatic transport, *Adv. Drug. Deliv. Rev. 19*:149–169 (1996).

129. V. J. Stella and N. L. Pochopin, Lipophilic prodrugs and the promotion of intestinal lymphatic drug transport, *Lymphatic Transport of Drugs* (W. N. Charman and V. J. Stella, eds.), CRC Press, Boca Raton, 1992, pp. 181–210.

130. R. Nishigaki, S. Awazu, M. Hanano, and T. Fuwa, The effect of dosage form on absorption of vitamin A into lymph, *Chem. Pharm. Bull. 24*:3207–3211 (1976).

131. A. Kuksis, Effect of dietary fat on formation and secretion of chylomicrons and other lymph lipoproteins, *Fat Absorption*, Vol 2, (A. Kuksis (ed.), CRC Press, Boca Raton, 1987, Chap. 6.

132. R. Blomhoff P. Helgerud, S. Dueland, T. Berg, J. I. Pederson, K. R. Norum, and C. A. Drevon, Lymphatic absorption and transport of retinol and vitamin D-3 from rat intestine—evidence for different pathways, *Biochim. Biophys. Acta. 772*:109–116 (1984).

133. A. Vost and A. Maclean, Hydrocarbon transport in chylomicrons and high density lipoproteins in rat, *Lipids 19*:423–435 (1984).

134. T. Ichihashi, H. Kinoshita, Y. Takagishi, and H. Yamada, Effect of oily vehicles on absorption of mepitiostane by the lymphatic system in rats, *J. Pharm. Pharmacol. 44*:560–564 (1992).

135. *Geigy Scientific Tables*, 8th Ed. Ciga-Geigy Ltd, Basle, Switzerland, 1984.

136. H. K. Naito, Disorders of lipid metabolism, *Clinical Chemistry Theory Analysis and Correlation* (L. A. Kaplan and A. J. Pesce, eds.), Mosby, St. Louis, Missouri, 1989, pp. 454–483.

137. P. A. Mayes, Lipid transport and storage, *Harper's Biochemistry* (R. K. Murray, D. K. Granner, P. A. Mayes, and V. W. Rodwell, eds.), Prentice Hall, New Jersey, 1996, pp. 254–269.

138. S. Urien, Interaction of drugs with human plasma lipoproteins, *Proceedings of the Symposium on Protein Binding and Drug Transport* (J. P. Tillement and E. Lindenlaub, eds.), F. K. Schattauer Verlag, Stuttgart-New York, 1986, pp. 63–75.

139. K. W. Wasan, S. M. Cassidy, Role of plasma lipoprotiens in modifying the biological activity of hydrophobic drugs, *J. Pharm. Sci. 87*:411–424 (1998).

140. H. P. Shu and A. V. Nichols, Uptake of lipophilic carcinogens by plasma lipoproteins. Structure activity studies, *Biochim. Biophys. Acta 665*:376–384 (1981).

141. S. Urien, E. Albengres, A. Comte, J. R. Kiechel, and J. P. Tillement, Plasma protein binding and erythrocyte partitioning of nicardipine in vitro. *J. Cardiovasc. Pharmacol. 7*:891–898 (1985).

142. S. Glasson, R. Zini, P. D'Athis, J. P. Tillement, J. R. Boissier, The distribution of bound propranolol between the different human serum proteins, *Mol. Pharmacol. 17*:187–191 (1980).

143. O. Chassany, S. Urien, P. Claudepierre, G. Bastian, and J. P. Tillement, Binding of anthracycline derivatives to human serum lipoproteins, *Anticancer Res. 14*: 2353–2355 (1994).

144. J-M. Bard, S. Urien, J-C. Fruchart, and J-P. Tillement, Location of probucol in lipoproteins inferred from compositional analysis of lipoprotein particles. An invitro study, *J. Pharm. Pharmacol. 46*:797–800 (1994).

145. S. Urien, E. Albengres, A. Comte, J. Kiechel, and J. P. Tillement, Plasma protein binding and erythrocyte partitioning of nicardipine in vitro, *J. Cardio. Pharmacol. 7*:891–898 (1985).

146. S. Urien, R. Zini, M. Lemaire, and J. P. Tillement, Assessment of cyclosporine A interactions with human plasma lipoproteins in vitro and in vivo in the rat, *J. Pharm. Exp. Ther. 253*:305–309 (1990).

147. K. M. Wasan, P. H. Pritchard, M. Ramaswamy, W. Wong, E. M. Donnachie, and L. J. Brunner, Differences in lipoprotein lipid concentration and composition modify the plasma distribution of cyclosporine, *Pharm. Res. 14*:1613–1620 (1997).

148. S. Urien, P. Claudepierre, J. Meyer, R. Brandt, and J. P. Tillement, Comparative binding of etretinate and acitretin to plasma proteins and erythrocytes. *Biochem. Pharmacol. 44*:1891–1893 (1992).

149. N. Simon, E. Dailly, P. Jolliet, J.-P. Tillement, and S. Urien, pH dependent binding of ligands to serum lipoproteins, *Pharm. Res. 14*:527–532 (1997).

150. S. Glasson, R. Zini, and J. P. Tillement, Multiple human serum binding of two

thienpyridinic derivatives, ticlopidine and PCR 2362, and their distribution between HSA, a-acid glycoprotein and lipoproteins. *Biochem. Pharmacol 31*:831–835 (1982).

151. D. Sgoutas, W. MacMahon, A. Love, and I. Jerunika, Interaction of cyclosporin A with human lipoproteins, *J. Pharm. Pharmacol. 38*:583–588 (1986).

152. H. L. Verrill, R. E. Girgis, R. E. Easterling, B. S. Malhi, and W. F. Mueller, Distribution of cyclosporine in blood of a renal transplant recipient with type V hyperlipoproteinemia, *Clin. Chem. 33*:423–428 (1987).

153. A. Danon and Z. Chen, Binding of imipramine to plasma proteins: effect of hyperlipoproteinemia, *Clin Pharmacol. Ther. 25*:316–321 (1979).

154. H. A. Eder, The effect of diet on the transport of probucol in monkeys, *Artery 10*: 105–107 (1982).

155. B. Legg, S. K. Gupta, and M. Rowland. A model to account for the variation in cyclosporin binding to plasma lipids in transplant patients. *Ther. Drug Monit. 10*: 20–27 (1988).

156. A. J. Humberstone, C. J. H. Porter, G. A. Edwards, and W. N. Charman, Association of halofantrine with post-prandially derived plasma lipoproteins decreases its clearance relative to administration in the fasted state, *J. Pharm. Sci. 87*:936–942 (1998).

157. S. K. Gupta and L. Z. Benet, High fat meals increase the clearance of cyclosporin. *Pharm. Res. 7*:46–48 (1990).

158. J. Brajtberg, S. Elberg, J. Bolard, G. S. Kobayashi, R. A. Levy, R. E. Ostlund, Jr., D. Schlessinger, and G. Medoff, Interaction of plasma proteins and lipoproteins with amphotericin B, *J. Infect. Dis. 149*:986–997 (1984).

159. K. M. Wasan, K. Vadiei, G. Lopez-Berenstein, and D. R. Luke, Pharmacokinetics, tissue distribution and toxicity of free and liposomal amphotericin B in diabetic rats, *J. Infect. Dis. 161*:562–566 (1990).

160. G. Lopez-Berenstein, Liposomes as carriers of antifungal drugs, *Ann. N.Y. Acad. Sci. 544*:590–597 (1988).

161. D. R. Pontani, D. Sun, J. W. Brown, S. I. Shahied, O. J. Plescia, C. P. Schaffner, G. Lopez-Berestein, and P. S. Sarin, Inhibition of HIV replication by liposomal-encapsulated amphotercin B. *Antiviral Res. 11*:119–125 (1989).

162. G. G. Chabot, R. Pazdur, F. A. Valeriote, and L. H. Baker, Pharmacokinetics and toxicity of a continuous infusion of amphotericin B in cancer patients, *J. Pharm. Sci. 78*:307–310 (1989).

163. K. M. Wasan and J. S. Conklin, Evaluation of renal toxicity and antifungal activity of free and liposomal amphotericin B following a single intravenous dose to diabetic rats with systemic candidiasis, *Antimicrob. Agents Chemother. 40*:1806–1810, (1996).

164. M. H. Koldin, G. S. Kobayashi, J. Brajtburg, and G. Medoff, Effects of elevation of serum cholesterol and administration of amphotericin B complexed to lipoproteins on amphotericin B-induced toxicity to rabbits, *Antimicrob. Agents Chemother. 28*:144–145 (1985).

165. P. Chavanet, V. Joly, D. Rigaud, J. Bolard, C. Carbon, and P. Yeni, Influence of diet on experimental toxicity of amphotericin B deoxycholate, *Antimicrob. Agents Chemother. 38*:963–968 (1994).

166. L. C. Souza, R. C. Maranhao, S. Schrier, and A. Campa, In vitro and in vivo studies

of the decrease of amphotericin B toxicity upon association with a triglyceride rich emulsion, *J. Antimicrob. Agents 32*:123–132 (1993).

167. M. Lemaire and J. P. Tillement, Role of lipoproteins and erythrocytes in the in vitro binding and distribution of cyclosporin in the blood, *J. Pharm. Pharmacol. 34*:715–718 (1982).

168. H. Lithell, B. Odlind, I. Selinus, A. Lindberg, B. Lindstrom, and L. Frodin, Is the plasma lipoprotein pattern of importance for treatment with cyclosporine? *Transplant. Proc. 18*:50–51 (1986).

169. N. De Kippel, J. Sennesael, J. Lamote, G. Ebinger, and J. Keyser, Cyclosporine leukoencephalopathy induced by intravenous lipid solution, *Lancet 339*:1114–1115 (1992).

170. J. Nemunaitis, H. J. Deeg, and G. C. Yee, High cyclosporin levels after bone marrow transplantation associated with hypertriglyceridemia, *Lancet 2*:744–745 (1986).

171. M. Arnadottir, H. Thysell, and P. Nilsson-Ehle, Lipoprotein levels and post heparin lipase activities in kidney transplant recipients: ciclosporin-versus nonciclosporin-treated patients, *Am. J. Kidney Dis. 17*:700–717 (1991).

172. A. M. Gardier, D. Mathe, X. Guedeney, J. Barre, C. Benvenutti, N. Navarro, J. Vernillet, D. Loisance, J. P. Cachera, B. Jacotot, and J. P. Tillement, Effects of plasma lipid evels on blood distribution and pharmacokinetics of cyclosporin A, *Ther. Drug Monit. 15*:274–280 (1993).

173. M. Lemaire, W. M. Partridge, and G. Chaudhuri, Influences of blood components on the tissue uptake indices of cyclosporin in rats, *J. Pharmacol. Exp. Ther. 244*: 740–743 (1988).

174. W. M. Partridge, Carrier mediated transport of thyroid hormones through the rat blood brain barrier. Primary role of albumin bound hormone, *J. Clin. Invest. 64*: 145–154 (1979).

# 5

# Topical Application of Drugs
## Mechanisms Involved in Chemical Enhancement

**Joke A. Bouwstra, B. A. I. van den Bergh, and M. Suhonen**
*Leiden University, Leiden, The Netherlands*

## I. INTRODUCTION

The natural function of the skin is to protect the body against loss of endogenous substances such as water and against undesired influences from exogenous substances in the environment. The main barrier for diffusion of molecules across the skin resides in its outermost layer, the stratum corneum (SC) [1]. The SC consists of keratin-filled dead cells, the corneocytes, which are entirely surrounded by crystalline lamellar lipid regions (Figure 1). The cell boundary of the corneocytes consists of a densely cross-linked protein layer, the corneocyte envelope, that readily protects absorption of drugs into the cells. Consequently, substances mainly diffuse along the lipid regions. This decreases the total skin surface area accessible for diffusion and makes the lipid regions of predominant importance for the skin barrier function [2, 3].

The composition of SC lipids differs greatly from that of cell membranes of living cells. Its major lipid classes are ceramides (CER), cholesterol (CHOL), and free fatty acids (FFA) [4–6]. The length of the CER acyl chains varies between 16 and 33 carbons, whereas the most abundant acyl chain lengths of FFA are $C_{22}$ and $C_{24}$, all significantly longer than those of phospholipids present in plasma membranes. Furthermore, ceramide head groups are small and contain several functional groups that can form lateral hydrogen bonds with adjacent lipids.

**Bouwstra et al.**

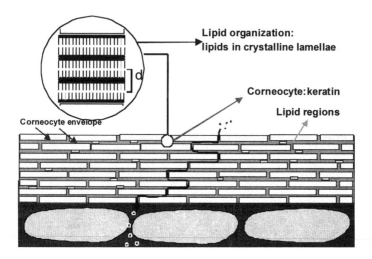

**Figure 1** Schematic presentation of the skin barrier (stratum corneum). The lipids are organized in crystalline lamellar packing. The repeat distances (d) of the two lamellar phases are 6 and 13 nm, respectively.

Because of its important barrier function, the lipid region of the skin has been extensively characterized over the past 20 years. Freeze-fracture electron microscopic studies revealed the presence of lamellae [7, 8] in the intercellular regions, whereas $RuO_4$ postfixation showed an unusual lipid bilayer arrangement of a repeat pattern of electron translucent bands in a sequence of broad-narrow-broad [9–14] oriented along the cell surfaces. By use of the x-ray diffraction technique, it was found that the lipids are arranged in two lamellar phases with periodicities of around 6 and 13 nm, respectively [15–18]. Because the 13-nm phase has a very unusual arrangement and is present in all examined species, this phase is probably most important for the skin barrier. In addition, the lateral packing of the lipids is mainly crystalline [15–20], although liquid phases have also been identified [15, 18]. Because of its predominantly crystalline character, the diffusional resistance is further increased in these lamellar phases. Nuclear magnetic resonance studies revealed a similar packing in model membranes prepared from brain ceramides, CHOL, and palmitic acid [21]. Crystalline hydrated CHOL was also found in pig, human, and mouse SC. Fourier infrared studies revealed that next to these phases a small fraction of lipids co-exist in the liquid phase in human and pig SC [22, 23] that may facilitate mixing and entry of exogenous substances into the skin [18, 20, 24]. This fluid phase was also present in mixtures prepared from SC ceramides, in which next to CHOL, CER, and FFA cholesterol sulfate was incorporated in the lamellae [25].

## II.  VESICLES AFFECT DRUG TRANSPORT ACROSS SKIN

### A.  Vesicles Affect Skin Penetration In Vitro

The first studies that described the beneficial effects of liposomes on the permeation of drugs through skin appeared in 1980 and 1982 [26, 27]. In these publications it was reported that liposomes applied to white rabbit skin *in vivo* favored the disposition of drugs in the epidermis and dermis, whereas the amount of drug found in the various organs was reduced. It was strongly suggested that vesicles penetrated the skin. After the first articles by Mezei and Gulasekharam, a large number of studies were carried out, an assessment of which will be provided in the following.

After the introduction of liposomes as drug delivery systems for the transdermal route, Ganesan and Ho [28, 29] proposed that the drug could be transported either (1) as free solute, (2) as free solute associated with liposomes, or (3) by direct transfer of the drug from the liposomal bilayer to SC without partitioning into the water phase. They assumed that the vesicles neither absorbed intact nor fused with the SC surface. The permeation studies were performed *in vitro* with mouse skin with liposomes prepared from dipalmitoylphosphatidylcholine (DPPC) that forms gel-state bilayers. They encapsulated glucose, progesterone, or hydrocortisone. They observed that (1) no radiolabeled DPPC was detected in the receiver cell, (2) glucose entrapped in liposomes resulted in smaller skin permeability compared with glucose applied in a normal saline solution, and (3) almost no release of the lipophilic drugs from liposomes into the saline solution was found. Very convincingly, the results of these studies revealed that intact liposome penetration, as suggested by Mezei and Gulasekharam [26, 27], was not the active principle.

In a study by Knepp et al. [30, 31] the vesicles composed of egg-phosphatidylcholine (EPC), dioleoylphosphatidylcholine (DOPC), dimirystoylphosphatidylcholine (DMPC), or DPPC were suspended in an agarose gel. Progesterone release from an agarose gel alone was fast compared with the release from liposomes embedded in the agarose gel. As a result, vesicles reduced the progesterone transport rate across hairless mouse skin. Gel-state DPPC resulted in slower skin penetration of progesterone than liquid-crystalline EPC liposomes. Furthermore, intercalation of *cis*-unsaturated fatty acid (oleic acid) in the phospholipid bilayer increased the drug transport rate across the skin by an order of magnitude. In two more recent articles [32, 33] it has been reported that a gel immobilizes the liposomes and, therefore, might affect the interactions between the liposomes and the skin. However, the trend observed in the studies of Knepp et al. [30, 31] in which incorporation of drugs in liquid-crystalline state liposomes resulted in a higher skin permeation rate than in gel-state liposomes was also observed for gel-free formulations (see later). Furthermore, because Knepp et al. used equal concentrations of the drug in the various formulations instead of equal thermodynamic activity, the driving force for the partitioning of the drug from the agarose gel to the skin was most probably greater than in the presence of the liposomes. This might be the reason for the unusually high skin penetration rate of the active substance when applied in the agarose gel compared with in the presence of the liposomes.

Several other studies were carried out to evaluate whether liposome composition affects skin penetration of drugs. In the early nineties the effect of liposomal composition on the disposition of encapsulated cyclosporin A in mouse skin when applied nonocclusively *in vitro* was assessed [34]. Two nonionic surfactant vesicles, bovine brain ceramide liposomes, and liposomes prepared from saturated and unsaturated phosphatidylcholine were included in their study. Importantly, all liposome formulations were saturated with cyclosporin A. In this way an equal thermodynamic activity of cyclosporin A was achieved. After 24 h of application, cyclosporin A in nonionic surfactant vesicles (glyceryl dilaurate/cholesterol/polyoxyethylene-10 stearyl ether) had penetrated into deeper skin strata, and the concentration in the receiver compartment was highest compared with the other formulations. These studies confirmed the results of Knepp et al. [30, 31], although vesicles were applied nonocclusively without a gel. The findings of Knepp et al. [30, 31] were also confirmed by Hofland et al. [35], who found that estradiol incorporated in gel-state nonionic surfactant at saturation concentration resulted in a lower drug transport rate through human SC compared with estradiol in liquid-state bilayers (Figure 2).

In contrast to Knepp et al., however, Hofland et al. [35] demonstrated that a drug applied in liquid-state vesicles resulted in higher penetration rates than when applied in phosphate-buffered saline. This may have been due to differences in study design, because Knepp et al. [30, 31] had incorporated the vesicles in agarose and used *equal drug concentration* in the formulations, whereas Hofland et al. suspended the vesicles in a buffer solution and used equal *estradiol thermodynamic activity*. In the latter case an equal driving force from formulation to stratum corneum has been achieved. Furthermore, in the same study it was demonstrated that drug association with vesicles was more effective than pretreatment with vesicles. When using an equal thermodynamic driving force, these findings strongly suggest that the vesicles do not act as penetration enhancers only [35],

---

**Figure 2**  Penetration-enhancing effect of estradiol by various vesicle formulations. ■, flux obtained after direct application of estradiol saturated vesicles; □, flux from estradiol-saturated phosphate-buffered saline through vesicle treated stratum corneum; +, control, flux from estradiol-saturated phosphate-buffered saline through untreated stratum corneum. (A) decaoxyethyleneoleylether (forming liquid state vesicles); (B) trioxyethylenedodecylether (forming liquid state vesicles), (C) trioxyethyleneocta-decylether (forming gel-state vesicles).

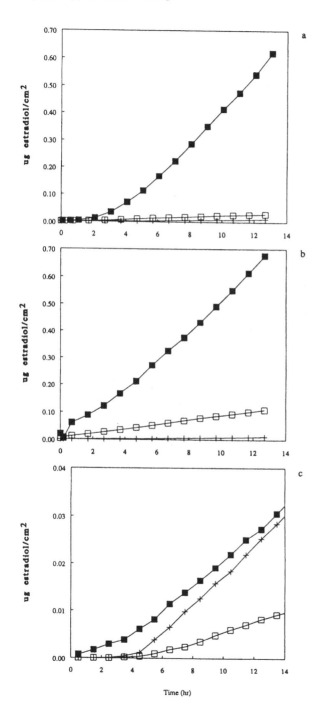

but that, when a proper vesicle composition is chosen, an additional effect can be expected.

In a more recent study, vesicle skin interactions were examined with confocal laser scanning microscopy (CLSM) [36]. A large number of liposome compositions were examined. These studies revealed that the liposome constituents were only present in the outermost layers of the SC and that the constituents only acted as penetration enhancers, which is in contrast to the studies of Hofland et al. Whether these differences in findings are due to another study design or to a difference in vesicle components is not clear.

Hofland [37] studied the vesicles-skin interactions on an ultrastructural level by use of freeze-fracture electron microscopy. When applied occlusively to human skin, the liquid-state vesicle adsorbed on the SC surface, and frequently changes in the lipid organization between the SC superficial cell layers were observed. Occasionally, changes in the deeper layers of the SC, such as the presence of vesicular structures, were observed. The authors asserted that penetration of intact vesicles in the SC would be unlikely to occur and explained the presence of vesicles by the penetration of vesicle constituents that might be able to reorganize to form vesicles in the SC. Furthermore, water pools were observed in the lipid regions between the corneocytes, indicating that phase separation between SC lipids and water occurs. In a more recent study [38], it was found that the vesicles were only occasionally present in the deeper layers of the skin and, therefore, were not expected to affect the diffusion characteristics of drugs across the skin. This is in agreement with studies carried out by Lasch et al. [39]. By use of fluorescence spectroscopy, they found that no intact vesicle penetration occurred.

In a subsequent study, van Hal et al. [40] reported that a decrease in cholesterol content in liquid state bilayers, which increases bilayer fluidity, resulted in an increase in estradiol transport across SC. With confocal laser scanning microscopy, Meuwissen et al. examined the diffusion depth of gel- vs. liquid-state liposomes labeled with fluorescein-dipalmitoylphosphatidylethanolamine (fluorescein-DPPE) with human skin *in vitro* [41] (Figure 3) and rat skin *in vivo* [42] and found that the lipophilic label when applied in liquid-state bilayers onto the skin penetrated deeper into the skin than when applied in gel-state liposomes. Recently, Fresta and Puglisi [43] reported that corticosteroid dermal delivery with skin-lipid liposomes was more effective than delivery with phospholipid vesicles, both with respect to higher drug concentrations in deeper skin layers and therapeutic effectiveness. This is a very surprising result, because skin lipid liposomes are rigid and form stacks of lamellae on the surface of the skin [44]. From the previously mentioned studies it seems clear that the *thermodynamic state of the bilayer* plays a crucial role in the effect of vesicles on drug transport rate across skin *in vitro*.

Korting et al. [45] visualized skin on an ultrastructural level after applica-

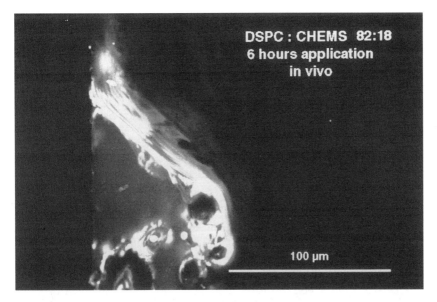

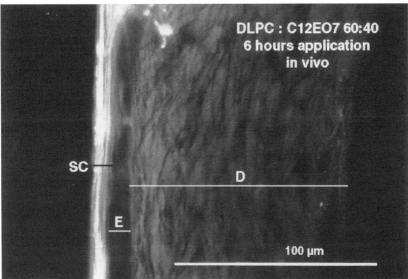

**Figure 3** Gel-state versus liquid-state application of vesicles prepared from dilauryl-phosphatidylcholine and septaoxyethylene alkylethers or distearylphosphatidylcholine and cholesterol hemisuccinate. A cross-section of rat skin visualized with confocal laser scanning microscopy after 6 h application. The dye used was fluorescein-phosphatidyl ethanol amine. The vesicles were applied onto rat skin *in vivo*.

tion of soy lecithin liposomes (liquid-state vesicles) using $OsO_4$ staining and fixation. They observed no intact liposomes in the lower layers of the SC but disposition of lipids derived from liposomes between and within the corneocytes, in agreement with Hofland et al. [37]. Korting et al. disagreed with the interpretations of Foldvari et al. [46], who claimed that intact liposomes penetrate into the skin. In several studies *occlusive* application was compared with *nonocclusive* application. These studies revealed that occlusive application of vesicles suspension was less effective than nonocclusive application [47, 48]. The results were somewhat unexpected, because it has been reported that water is an effective permeation enhancer. However, in case of nonocclusive application, the increased drug transport can be caused by (1) a more profound interaction between the liposomal constituents and the skin and/or (2) the presence of a hydration gradient in the skin. According to Cevc et al. [47] the water gradient is an important driving force for drug diffusion. They claim that even intact vesicles can penetrate through the skin if the vesicles are highly flexible ("transfersomes®") (see later). Size and lamellarity of liposomes may also affect skin penetration rate. du Plessis et al. [49] found no preferential disposition of drug in lower skin strata and the receiver compartment, whether it was applied in small or large vesicles. The authors concluded that intact penetration of liposomes does not occur. Furthermore, it was found that the effect of lamellarity on the disposition of CHOL and cholesterol sulfate in the various skin strata had little effect [49]. It seems that physical parameters such as vesicle size and lamellarity are less important than the application method and the thermodynamic state of the bilayers. These findings again favor *the absence of intact vesicle* penetration across the skin.

Touitou et al. [50] compared penetration enhancers with liposomes. Interestingly, they found that liposomes mainly deposit the drug in the skin, and, therefore, act as an excellent reservoir, whereas the penetration enhancers increased transcutaneous transport.

Liposomes have also been compared with commercial lotions, creams, or penetration enhancers [51–55]. Although these studies provide some insight as to the relative effectiveness of liposomes, no information can be obtained about the underlying mechanisms. Generally, both the thermodynamic activity of the drug in the various formulations differs quite extensively, and the various formulation components themselves interact differently with the skin strata. Although of interest, this issue shall not be discussed further in this chapter.

## B.  Vesicles Affect Skin Penetration *In Vivo*

In 1986, Komatsu et al. [56] applied butylparaben occlusively to the dorsal skin of guinea pigs in a liposome formulation prepared from eggPC, CHOL, and dicetylphosphate. The fate of the liposomes and the active agent was determined with autoradiography. After 48 h, the radioactive DPPC was mainly found on the application site, whereas [14]C-butylparaben radioactivity was found in small intestine, feces, gallbladder, and urinary bladder. These results suggest that not only *in vitro* but also *in vivo* intact penetration of liposomes is unlikely to occur. In a subsequently carried out *in vitro* study [57], relative amounts of lipids and [14]C-butylparaben were varied, and the mechanisms involved in [14]C-butylparaben penetration were examined systematically. The percentage of [14]C-butylparaben transported across the skin decreased as a function of increasing amounts of lipids, whereas the transcutaneous transport rate of radioactive DPPC was much lower than that of [14]C-butylparaben confirming the *in vivo* results. This indicated that the lipids remained in the SC and that co-penetration of butylparaben and phospholipids to deeper layers did not occur [57].

From the finding that the [14]C-butylparaben flux decreased at increasing lipid content, it was concluded that only [14]C-butylparaben dissolved in the water phase contributed to the percutaneous penetration. However, another mechanism may also play a role. An increase in lipid content reduces the thermodynamic activity of butylparaben in the bilayers, which might lead to a decrease in the driving force from the lipid phase to either the water phase or the SC. This may also contribute to a reduced penetration through the skin.

In another, more recent study [58] [14]C-tretinoin was intercalated in soybean lecithin labeled with [3]H-phosphatidylcholine. The [3]H/[14]C ratio in SC remained approximately constant, however, was lower in epidermis, and decreased steeply until a skin depth of approximately 200 μm was reached. The authors concluded that co-penetration of a drug-liposome bilayer is possible in the SC, but that based on the reduced [3]H/[14]C ratio in deeper skin strata, drug and liposomal constituents diffuse separately in these layers.

In a series of studies carried out in the early nineties, it was clearly demonstrated that the transport of protein across the skin was facilitated by a particular liposome formulation, the so-called NAT 106 liposomes [59] and that these liposomes also increased the penetration of [35]S-heparine and [99m]Tc [60]. In a related study it was shown that this liposome suspension also resulted in a decrease in the corneal blood supply as measured by Laser Doppler Flowmetry [61], indicating that *empty* liposomes induce changes in the lower regions of the skin. Interactions between liposomes prepared either from NAT 106, NAT 50, or NAT 89 and skin were examined by freeze-fracture electron microscopy *in vitro* and Fourier transformed infrared spectroscopy (FTIR) [62] *in vivo*. NAT 50 liposomes fused on the surface of the skin, whereas vesicles prepared from NAT 89 penetrated between the upper SC cell layers and quite frequently distinct regions with a rough surface were detected, indicating a mixing between SC lipids and liposomes or the existence of separate domains of phospholipids. Application of vesicles prepared from NAT 106 revealed large changes in the lipid organization in the SC (Figure 4). Almost no intact lipid bilayers were observed between the

**Figure 4** Nat 106 liposomes have a strong effect on the microstructure of the stratum corneum. The corneocytes (C) were swollen considerably, and the smooth ultrastructure of the intercellular lipid lamellae showed flattened spherical structures (see arrows). The linear arranged keratin filaments along the cell boundary of the corneocytes are absent. The scale bar indicates 0.1 μm.

corneocytes. The intercellular regions were characterized by "flattened isolated regions." It seems that liquid-state liposomes prepared from different lipid mixtures interact differently with the SC. Interestingly, the NAT 106 formulation contained the largest fraction PC. The strong interaction with the SC might be explained by the presence of a single-chain lipid, lysophosphatidylcholine, which may act as an edge activator and increase the elasticity of the liposomal bilayer and therefore mimic properties of transfersomes (see later).

In a related study [62], *in vivo* skin hydration was monitored by FTIR after occlusive application of either NAT 106 liposomes prepared in $D_2O$ or pure $D_2O$. Liposomes were superior, compared with pure $D_2O$, in driving $D_2O$ into the skin, and the phospholipid components could be detected in deeper SC layers. The presence of phospholipids deep in the SC confirmed the results observed with freeze fracture, in which strong interactions between phospholipid layers and SC were observed.

Transfersomes® are vesicles prepared from lipids and an "edge activator" that might be a single-chain lipid or surfactant. The edge activator renders the vesicles elastic. As a result of the hydration force in the skin, elastic vesicles can "squeeze" through SC lipid lamellar regions [47]. Transfersomes were much more effective than "conventional" liposomes when applied nonocclusively with respect to mass flow of lipid across the skin. After 8 h of transfersome application in mice, significant amounts were found in blood (approximately 8% of the *applied* dose) and liver (20% of the *recovered* dose). Accordingly, Cevc et al. claimed that 50–90% of the dermally deposited lipids can be transported beyond the level of the SC. Although there is no doubt that transfersomes have advantages with respect to increasing the transport of active material *across the skin*, based on these studies, it was also claimed that the elastic transfersomes penetrated *intact* through SC and viable skin into the blood circulation. The latter met with much skepticism and is still not proven experimentally. It has been particularly questioned whether such large associates are transported across the skin as intact entities, especially because the SC consist of a very tight structure that is designed to act as a barrier (see Introduction).

In the first published study on drugs associated with transfersomes, the transfersome suspension contained relatively high amounts of lidocaine or tetracaine [63]. The transfersomes were tested on rats and on humans, and in both studies they appeared to be more efficient than the conventional liposomes or solutions. The differences found with corticosteroids were somewhat less encouraging [64], which might be due to the lipophilic nature of the drugs. When insulin was associated with transfersomes [65, 66], encouraging results were obtained. Blood glucose levels in mice could be lowered after about 3 h of application, which was significantly different from those achieved with micelles, conventional liposomes. In more recent studies this was confirmed in human trials [66, 67]. Large molecules such as FITC-BSA [68] or [125]I-BSA or large proteins [69] can be transported across the skin when associated with transfersomes. The biodistribution of [125]I-BSA associated with Transfersomes on mouse skin was similar to that of transfersome-associated [125]I-BSA injected subcutaneously. These results show that transfersomes indeed have advantages over vesicles prepared exclusively from double-chain phospholipids and cholesterol. The same group [70–72] intercalated rhodamine-DHPE (1,2-dihexadecanoyl-sn-glycero-3-phosphatidylethanolamine-*N*-lissamine rhodamine B sulfonyl, triethylammonium salt) in the bilayers of transfersomes prepared from soya phosphatidylcholine and sodium cholate. The suspensions were applied nonocclusively on mice skin, and after 4 to 12 h, the skin was examined *ex vivo* using CLSM. They claimed the presence of transport routes in the SC, the intercluster pathway between groups of cells, which should be a preferred route of transport between corneocytes that only partly overlap. However, the skin surface is not flat but contains many wrinkles. In recent experiments, "clefts" with a bright fluorescent appearance were also found that were interpreted as clefts that may correspond to wrinkles in the skin [73]. As to the interpretation of such findings, although Schätzlein and Cevc interpret these regions as being virtual pores between the corneocytes, one should realize that CLSM *cannot* be used to visualize the transport of *vesicles as intact entities*, it can only be used to visualize the transport of the label, which is not necessarily still associated with the vesicle.

An extensive and well-designed study was published by Ogiso et al. [74]. They compared the effect of gel systems in which D-limonene or laurocapram were present as penetration enhancers on the transport of β-histine, in the presence or absence of liposomes (EPC or hydrogenated soy PC) in the gels. In the presence of EPC liposomes, the bioavailability of the drug was greater. Hence, it seems that the fluidity of the membranes is an important factor for penetration enhancement, confirming many earlier studies. Lipid analysis revealed that the amount of ceramides was dramatically reduced in the SC after application of a D-limonene–containing formulation. Hence, extraction of ceramides and replacement with liposomal phospholipid may be the underlying mechanism of the observed increase in penetration of the active agent.

## C. Conclusions

We can generally conclude that drug transport can be adjusted on demand by association of a drug with vesicles. One of the central parameters is the thermodynamic state of the bilayer that can dramatically affect drug permeation across, in addition to interactions with, the SC. It seems that size and lamellarity of vesicles affect drug transport only slightly. Using "traditional" liposomes, intact vesicle transport across stratum corneum does not occur as judged from visualization and permeation studies and from biophysical measurements. Importantly, drug transport increases when single-chain surfactants are intercalated in the vesicles. These single-chain surfactants can act as penetration enhancers, such as in gel-state vesicles, or might result in a decrease in the interfacial tension and render vesicles more deformable, which increases drug transport across the skin even further. However, questions arise as to the underlying mechanism(s). Only a few studies have been carried out until now. Schatzlein and Cevc [70–72] studied the fate of fluorescent labels associated with Transfersomes. However, although information can be obtained about the penetration pathway of the labels, this technique does not provide information about the permeation of the vesicles. A further step has been made by studies of van den Bergh. Vesicles prepared from sucrose ester laurate and PEG-8-laurate appeared to be elastic [75]. The vesicles were applied nonocclusively on hairless mouse skin, after which the skin was visualized with $RuO_4$ fixation in combination with transmission electron microscopy. Domains of stacks in between the lamellae of the SC were clearly visible (Figure 5) [76, 77]. With two-photon excitation spectroscopy, it was found that the penetration pathway of an amphiphilic fluorescent label in these deformable vesicles permeated in microchannels (Figure 6). The localization of these channels is in the lipid regions. However, the shape of the microchannels is different than the clefts proposed by the studies of Cevc. The presence of permeation channels is in contrast to fluorescent label applied in rigid "conventional" vesicles that penetrate homogeneously across the intercellular regions. However, nei-

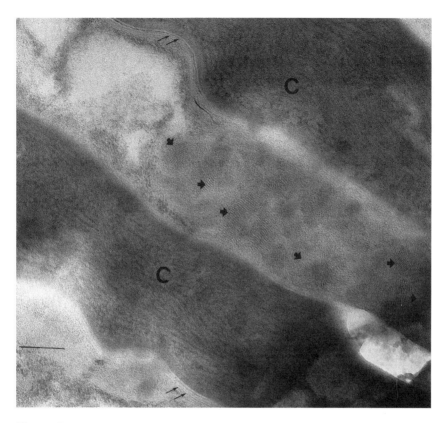

**Figure 5** Transmission electron micrographs of human skin treated with liquid-state PEG-8-L:L-595:CS (70:30:5) vesicles, nonocclusively. Detailed micrographs of skin treated for 16 h. Lamellar stacks are present in the intercellular spaces (see double arrow) only along the cell boundaries. The islands of lamellar stacks are found in the intercellular regions. These stacks most probably originate from vesicle material Arrow, lamellar stack; C, corneocyte. Bar represents 100 nm.

ther morphological changes nor the fluorescent label has been found in the viable epidermis. Therefore, although these vesicles might have favorable properties in increasing drug transport across the skin, targeting to certain cell types in the viable epidermis and dermis or partitioning of intact vesicles in the blood circulation still seems to be a utopia. Targeting will only be possible if vesicles partition into the viable epidermis. However, there is one exception, which is targeting to the hair follicles. A number of publications have now shown that particulate systems, including vesicles [78] and small polymeric beads, favor the partitioning of drugs into the hair follicles.

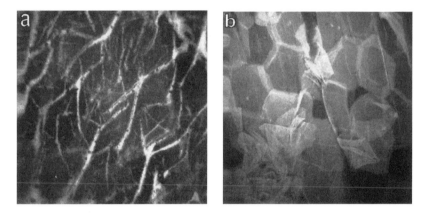

**Figure 6**   TPE images of skin treated with PEG-8-L:L-595:CS (70:30:5) elastic liquid-state vesicles. (**a**) Threadlike channels are formed in the entire stratum corneum. However, these channels were absent in the stratum granulosum or stratum spinosum (not shown). Treatment with rigid vesicles resulted in a homogenous staining of the intercellular regions (**b**).

## III.   PENETRATION ENHANCERS

### A.   Introduction

Increasing drug penetration through the skin requires penetration enhancers that perturb the lipid organization in the SC. Rational development of new penetration enhancers can only be achieved if one understands their action mechanism(s) (i.e., *how* the protective lipid barrier in the skin is perturbed). Depending on their molecular structure, penetration enhancers can modify the lipid organization in various ways: they may (1) intercalate into the lateral sublattices in the lamellae and modify the sublattice, (2) disturb the lamellar packing, (3) form separate small domains in the lateral packing in the lamellae, and (4) form enhancer-rich large separate domains in the intercellular regions. A combination of these effects is also possible. A schematic view of the various interactions is provided in Figure 7.

**Figure 7**   The possible mechanisms involved in the effect of penetration enhancers on the lipid organization of the intercellular domains in the stratum corneum. (**A**) Intercalation of the enhancer in the lipid lamellae. (**B**) Phase separation between enhancer and skin lipids in the lamellae. (**C**) Phase separation between lipid lamellae and an enhancer-rich phase. (**D**) Intercalation of the enhancer in the lipid lamellae and simultaneous phase separation between lipid lamellae and enhancer. (**E**) Phase separation within the lamellae and separation between an enahncer-rich phase and the lamellar phase. (**F**) Disappearance of the lamellar phases.

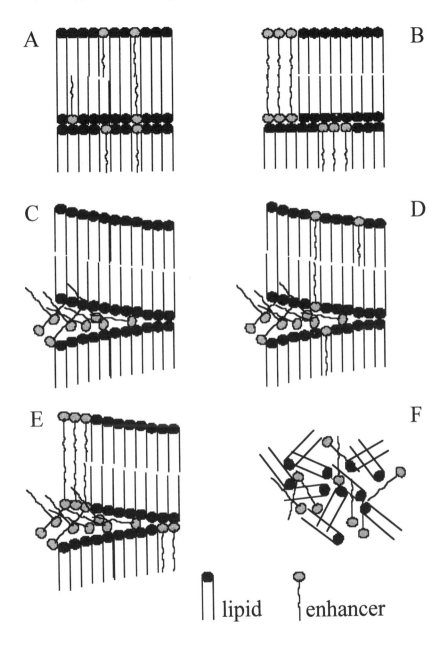

In the following, a summary is given on water and its interactions with skin, as well as on the type of interactions between several classes of amphiphilic molecules and the skin.

## B. The Effect of Water on Skin Penetration

One of the first reports on the interaction between the lipid bilayers and water has been published by van Duzee et al. using thermal analysis [79]. They detected four endothermic transitions that were attributed to either lipid phase transitions or to protein denaturation. Transition temperatures depend on the water content in the SC. Such water-dependent thermal transitions have also been found in phospholipid membranes. However, water reduces the transition temperature of phospholipid membranes to a much greater extent than in the SC. X-ray diffraction revealed that on increasing the water content almost no swelling of the lamellae took place [10, 16–18, 20], indicating that if water is present between the head groups, it will only be present in small quantities. With infrared spectroscopy [80] it was found that water did not affect lipid alkyl chain order at room temperature, whereas electron spin resonance [81] studies revealed that increased water content increased the mobility of the hydrocarbon chains. However, whether this increase in mobility is due to an indirect effect of the swelling of the corneocytes or to a small amount of water located between the head groups is not yet clear. It has been suggested that increased chain mobility might be limited to certain domains in the intercellular regions [82]. This would be in agreement with freeze-fracture observations. With this technique, the lipid lamellae can be visualized as smooth areas at the plane of fracture. However, after extensive treatment with phosphate-buffered saline, next to these smooth regions areas with a rough surface appeared in the intercellular domains, indicating changes in lipid structure. Furthermore, water domains were found not only in the corneocytes but also in the intercellular regions [83] and are often present between the bound lipids of the cornified envelope and the lipid bilayer regions (unpublished observations). The presence of water domains indicates a likely phase separation between the lipid lamellae and the water, because the lipids present in the intercellular domains are rather lipophilic, especially at pH values at which FFAs are not dissociated. Distinct water domains can also change the penetration pathway of the penetrant and increase its permeability. It is unlikely that water domains are continuous channels across the SC, because this would require large interfacial regions, which would be energetically unfavorable.

## C. Amphiphilic Molecules as Penetration Enhancers

The most extensively studied amphiphilic and lipophilic penetration enhancers are oleic acid, oleyl surfactants, terpenes, alcohols, azone, azone analogues [84–86], and FFAs.

For fatty acids and fatty alcohols the enhancement has been related to chain length and degree of saturation. Generally, maximum enhancement occurs at a saturated chain length of $C_{12}$ [87–90] or when the chain possesses one or more double bonds, which makes it more fluid. Studies on the enhancing properties of a series of *cis-* and *trans-*positioned isomers of oleic acid [91–98] have shown that *cis-*oleic acid is a more effective enhancer than the *trans* isomer. Furthermore, incorporation of oleic acid into the SC lipid lamellae resulted in increased fluidity of the alkyl chains at elevated temperatures and in a reduction of the lipid phase transitions at 60 and 70°C (Figure 8). However, the changes in lipid organization could not explain the penetration enhancement at physiological temperatures, in contrast to what had been suggested in previous reports [99]. By use of deuterated oleic acid it appeared that phase separation occurs between oleic-rich domains and SC lipid lamellae at physiological temperatures

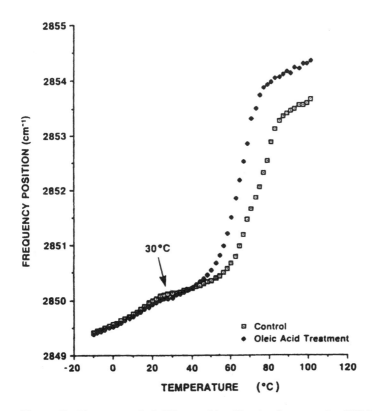

**Figure 8**   The symmetrical $CH_2$ stretching fibration frequency [$v_s$ ($CH_2$)] as a function of temperature for untreated and predeuterated oleic acid–treated porcine stratum corneum. (From Ref. 95.)

that would explain the enhanced permeability at this temperature [95]. Recently, phase separation of oleic acid has been confirmed by freeze-fracture electron microscopy [100]. Furthermore, with fluorescence spectroscopy, it has been found that an overall decrease in order parameters goes in parallel with flux enhancement [101].

In the case of fatty acids, optimal penetration enhancement has been achieved with a chain length of approximately 9–12. In all cases correlation between the effect and chain-length was parabolic [102–106]. Most probably this is due to a balance between the ability of the fatty acid to partition into the SC, which requires a certain degree of lipophilicity, and the ability to change the lipid organization in such a way that increased permeation is achieved when the penetration enhancer itself is in a fluid phase at physiological temperature. For example, long-chain unsaturated fatty acids are not able to create a fluid phase in the SC lipid domains, unless a eutectic mixture is formed with the SC lipids or unless the fatty acid is still "dissolved" in the solvent, in which it has been applied to the skin surface.

Among the $C_{12}$-penetration enhancers is the well-known azone (1-dodecyl-azacycloheptan-2-one), which has been shown to be effective in enhancing the transdermal fluxes of a variety of drugs [107–115]. Because of its molecular structure, azone is expected to attain a "soup-spoon" conformation, in which the seven-membered ring would lie in the plane of the lipid polar head groups [116, 117]. The $C_{12}$ chain seems to be an ideal length to provide sufficient affinity of the molecule for the bilayer, ensuring partitioning into the lipophilic regions. Other 1-alkyl-azacycloheptan-2-one (alkyl-azone) molecules with chain lengths varying between $C_6$ and unsaturated $C_{18}$ have been studied. In recent studies [118] the transport route of $HgCl_2$, in the presence and absence of alkyl-azones, across human skin was visualized at an ultrastructural level. The intercellular route was found dominant, whereas apical corneocytes took up the agent only after longer transport times. Longer alkyl-azones in PG tended to favor the intercellular route even further [119], demonstrating the importance of lipid-enhancer interactions in understanding the mode of action of penetration enhancers on the skin barrier function.

Flux studies revealed that chain lengths of $C_8$ and beyond increased the flux of the hydrophilic drug 9-desglycinamide, 8-arginine vasopressin (DGAVP) and lipophilic drug nitroglycerine [116, 120]. Optimum enhancement for DGAVP was achieved with dodecyl-azone. Thermal analysis revealed that alkyl-azones (dissolved in propylene glycol) modulated the lipid phase transitions of SC [121]. Lipid phase transitions are located at 67 and 85°C. Pretreatment with propylene glycol shifted the two transitions about 10°C downward, whereas the corresponding transition enthalpies did not change. A similar behavior was found with increasing water content. Pretreatment with octyl-azone and longer alkyl-chain azones resulted in a gradual decrease in total transition enthalpy localized and 67 and 85°C. After treatment with dodecyl-azone and those with a longer alkyl chain [122], x-ray diffraction and electron microscopic studies revealed a disturbance of the lamellar ordering in SC [123, 124]. After pretreatment with octyl-azone the SC lipid structure became less ordered. More specifically, a rough, less anisotropic fracture behavior had appeared. Disorder was most pronounced in the center of the intercellular spaces. Close to the corneocyte envelope, much less perturbed lipid lamellae were still visible. An increase of the chain length in the alkyl-azones in PG further aggravated the disordering of the lamellae. However, even in strongly perturbed lipid regions, lamellae were still present in the direct vicinity of the corneocyte envelopes (Figure 9). It is obvious that depending on the alkyl-chain length $N$-alkyl-azones induced perturbation of the lamellae mainly in the center of the intercellular spaces. Furthermore, lipid lamellae were still present, but the stacking of these lamellae completely disappeared, in agreement with the findings obtained by SAXD (see earlier). The exception was found after treatment with oleyl-azone. In this case, even the lamellae disappeared and the appearance of the corneocyte envelope changed entirely. Finally, the lipid lateral packing was examined by wide-angle x-ray diffraction. The pattern of untreated SC displayed two very strong reflections at 0.415 nm and 0.375 nm, respectively, revealing an orthorhombic structure. After treatment with propylene glycol and alkyl-azones in propylene glycol, the position of the reflections did not change. Visualization of $RuO_4$ postfixed thin slices showed similar results. Treatment with $C_{12}$-azone and $C_{18:1}$-azone in PG resulted in a loss of the lamellar stacking in large SC regions (Figure 10). After $C_{18:1}$-azone treatment the lipid, lamellae almost disappeared.

The influence of hexyl-azone in PG on the structure of human SC was similar to that of PG alone. However, when SC was pretreated with octyl-azone in PG and longer alkyl-azones, SAXD revealed a disordering of the lamellar stacking. Both FFEM and FSEM illustrated the nature of the disorder and confirmed the disordering of the lamellae. Furthermore, both techniques revealed that the disordering was more easily induced in the center of the intercellular spaces than close to cell boundaries where lamellae were still present. That a part of the lipid lateral packing remained intact could still be deduced from the WAXD pattern of $N$-alkyl-azone–treated human SC; reflections based on an orthorhombic lateral packing were still present in the corresponding WAXD patterns. When the extent of $C_{12}$-azone loading in human stratum corneum was correlated with its effect on the diffusion coefficient of diazepam, it was clearly demonstrated that the strongest increase in diffusion coefficient of diazepam was obtained at 12% w/w enhancer content of the SC [125]. It is quite surprising that such large quantities of $C_{12}$-azone can be absorbed in the stratum corneum, implying that, on a molar basis, the amount of enhancer in the SC exceeds that

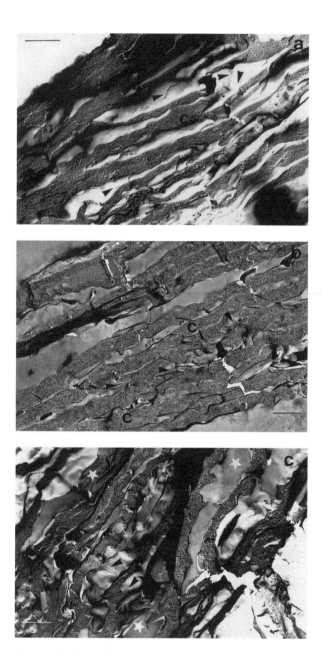

**Figure 9** Freeze-fracture electron microscopy of control (**a**) or after treatment of alkyl-azones (**b–f**). (**b**) Hexyl-azone; (**c**) octyl-azone; (**d**) dodecyl-azone; (**e**) myristyl-azone; and (**f**) oleyl-azone. Bar represents 100 nm. * represents rough structures indicating either separate domains of enhancer or more perturbation of lipid lamellae. Arrows indicate a clear presentation of the intact smooth regions of intact lamellae with steps (fracture across the lamellae). C, Corneocyte; scl, stratum corneum lipid lamellae.

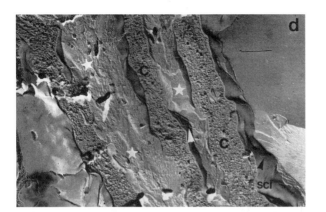

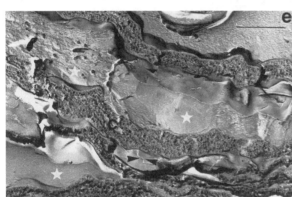

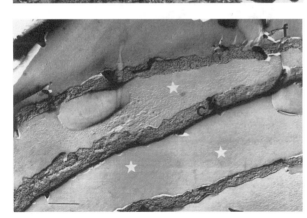

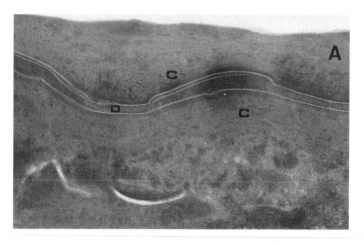

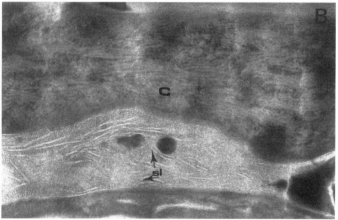

**Figure 10**   Stratum corneum sections after ruthenium tetroxide fixation confirmed these results (see Figure 9(d) for dodecyl-azone [**A**] and Figure 9(**f**) for oleyl-azone [**B**]). Bar represents 100 nm. C, corneocyte; D, desmosome; SL, single lamellae.

of the endogenous lipids. In the same study, an increase in enhancer uptake also resulted in a stronger effect on the lipid thermal transitions around 70°C.

In conclusion, the mechanism by which alkyl-azones interact with the SC and change the barrier properties of the SC is complex but depends on the alkyl chain length.

Cyclic monoterpenes increase the penetration of hydrophilic and lipophilic drugs [125–129]. Cornwell et al. found remarkable penetration enhancement of 5-fluorouracil after pretreatment with the noncyclic terpenes nerolidol, farnesol, and, to a lesser extent, other terpenes [130]. Interestingly, an optimum chain length of the acyclic terpene enhancers was found for nerolidone, which is a $C_8$ chain. The same study showed that terpene enhancement was much stronger with the hydrophilic 5-fluorouracil than the lipophilic estradiol. Penetration enhancement of terpenes when applied neat was different compared with terpenes dissolved in propylene glycol. A remarkable synergistic enhancer effect of propylene glycol and most of the terpenes was observed for transport of 5-fluorouracil, which was less pronounced for estradiol. Subsequent mechanistic studies on D-limonene, 1-8-cineole, and nerolidon [131, 132] showed that terpenes increased the diffusivity of the drugs by increasing bilayer fluidity and that the synergistic effects are more likely caused by increased disruption of the lipid organization than improved drug partitioning into the SC.

## IV. CONCLUSIONS

The mechanisms involved in increasing drug transport into the skin are either due to a facilitated partitioning of the drug from the delivery system into the SC, an increase in the diffusion rate of the drug in the tissue, or a decrease in the penetration pathway. There is no general preference for any particular penetration enhancer or family of penetration enhancers, because their effect depends strongly on interactions with the drug of interest. However, one can conclude that in case of amphiphilic molecules either the dodecyl chain length or the unsaturated $C_{18}$ chain length is most favorable for increasing drug transport across the skin. These penetration enhancers often act by the formation of separate domains in the intercellular lipid regions, possibly in addition to modulating the lipid organization of SC lipids.

## REFERENCES

1. P. W. Wertz and D. T. Downing, Stratum corneum: Biological and biochemical considerations, *Transdermal Drug Delivery* (J. Hadgraft, R. H. Guy, eds.), Marcel Dekker, New York, Basel, 1989, pp. 1–17.
2. H. Boddé, M. Kruithof, J. Brussee, and H. Koerten, Visualisation of normal and enhanced $HgCl_2$ transport through human skin in vitro. *Int. J. Pharm. 253*:13–24 (1989).
3. R. O. Potts and R. H. Guy, Predicting skin permeability, *Pharm. Res. 9*:663–669 (1992).
4. M. Ponec, A. Weerheim, J. Kempenaar, A. M. Mommaas, and D. H. Nugteren, Lipid composition of cultured human keratinocytes in relation to their differentiation. *J. Lipid Res. 29*:949–996 (1988).

5. P. W. Wertz and D. T. Downing, *Physiology, Biochemistry and Molecular Biology in the Skin*, 2nd Ed. (L. A. Goldsmith, ed.), Oxford University Press, 1991, pp. 205–236.

6. N. Y. Schurer and P. M. Elias, The biochemistry and function of stratum corneum, *Adv. in Lipid Res.*, *24*:27–56 (1991).

7. A. S. Breathnach, T Goodman, C. Stolinsky, and M. Gross, Freeze fracture replication of cells of stratum corneum of human cells. *J. Anat.* *114*:65–81 (1973).

8. A. S. Breathnach, Aspects of epidermal structure, *J. Invest. Dermatol.* *65*:2–12 (1975).

9. K. C. Madison, D. C. Swarzendruber, P. W. Wertz, and D. T. Downing, Presence of intact intercellular lipid lamellae in the upper layers of the stratum corneum, *J. Invest. Dermatol.* *88*:714–718 (1987).

10. S. Y. Hou, A. K. Mitra, S. H. White, G. K. Menon, R. Ghadially, P. Elias, Membranes structure in normal and essential fatty acid-deficient stratum corneum; characterization of ruthenium tetroxide staining and X-ray diffraction, *J. Invest. Dermatol.* *96*:215–223 (1991).

11. J. A. Bouwstra, O. Sibon, M. A. Salomons-de Vries, F. Spies. Interactions between nanodispersions and human skin, Proceedings of the Controlled Rel. Society Meeting, Orlando, 1992, pp. 481–482.

12. D. T. Downing, Lipid and protein structures in the permeability barrier of mammalian epidermis, *J. Lipid Res.* *33*:301–314, 1992.

13. M. Fartasch, I. D. Bassakas, and T. L. Diepgen. Disturbed extrusion mechanism of lamellar bodies in dry non-eczematous skin atopics, *Br. J. Dermatol.* *127*:221–227 (1992).

14. D. C. Swartzendruber, Studies of epidermal lipids using electron microscopy, *Semin. Dermatol.* *11*:157–161 (1992).

15. S. H. White, D. Mirejovsky, and G. I. King, Structure of lamellar lipid domains and corneocyte envelopes of murine stratum corneum. An X-ray diffraction study, *Biochemistry* *27*:3725–3732 (1988).

16. J. A. Bouwstra, G. S. Gooris, J. A. van der Spek, and W. Bras, Structural investigations of human stratum corneum by small-angle X-ray scattering, *J. Invest. Dermatol.* *97*:1005–1012 (1991).

17. J. A. Bouwstra, G. S. Gooris, W. Bras, and D. T. Downing, Lipid organization in pig stratum corneum, *J. Lipid Res.* *36*:685–695 (1995).

18. J. A. Bouwstra, G. S. Gooris, J. A. van der Spek, S. Lavrijsen, and W. Bras, The lipid and protein structure of mouse stratum corneum: A wide and small angle diffraction study, *Biochim. Biophys. Acta* *1212*:183–192 (1994).

19. J.-C. Garson, J. Doucet, J.-L. Lévêque, and G. Tsoucaris, Oriented structure in human stratum corneum revealed by X-ray diffraction, *J. Invest. Dermatol.* *96*:43–49 (1991).

20. J. A. Bouwstra, G. S. Gooris, M. A. Salomons-de Vries, J. A. van der Spek, and W. Bras, Structure of human stratum corneum as a function of temperature and hydration: A wide-angle X-ray diffraction study, *Int. J. Pharm.* *84*:205–216 (1992).

21. D. B. Fenske, J. L. Thewalt, M. Bloom, and N. Kitson, Models of stratumcorneum intercellular membranes: 3H NMR of microscopically oriented multilayers, *Biophys. J.* *67*:1562–1573 (1994).

22. C. L. Gay, R. H. Guy, G. M. Golden, V. H. W. Mak, and M. L. Francoeur, Characterization of low-temperature (i.e. < 65 °C) lipid transitionsin human stratum corneum, *J. Invest. Dermatol. 103*:233–239 (1994).

23. B. Ongpipattanakul, M. L. Francoeur, and R. O. Potts, Polymorphism in stratum corneum lipids, *Biochim. Biophys. Acta 1190*:115–122 (1994).

24. J. Engblom, S. Engström, and K. Fontell, The effect of the skin penetration enhancer Azone® on fatty acid-sodium soap-water mixtures, *J. Control. Rel. 33*:299–305 (1995).

25. J. A. Bouwstra, G. S. Gooris, F. E. R. Dubbelaar, A. M. Weerheim, M. Ponec, Cholesterol sulfate, and fatty acids affect the skin stratumcorneum lipids organisation, *J. Invest. Dermatol. Proc. 3*:69–74 (1998).

26. M. Mezei and V. Gulasekharam, Liposomes-a selective drug delivery system for the topical route of administration, I. Lotion dosage forms, *Life Sci. 26*:1473–1477 (1980).

27. M. Mezei, and V. Gulasekharam, Liposomes—a selective drug delivery system for the topical route of administration. II. Gel dosage form, *J. Pharm. Pharmacol. 34*: 473–474 (1980).

28. N. F. H. Ho, M. G. Ganesan, and G. L. Flynn, Mechanism of topical delivery of liposomally entrapped drugs, *J. Contr. Rel. 2*:61–65 (1985).

29. M. G. Ganesan, N. D. Weiner, G. L. Flynn, and N. F. H. Ho, Influence of liposomal drug entrapment on percutaneous absorption, *Int. J. Pharm. 20*:139–154 (1984).

30. V. M. Knepp, R. S. Hinz, F. C. Szoka, and R. H. Guy, Controlled drug release from a novel liposomal delivery system, i. Investigations of transdermal potential, *J. Contr. Rel. 5*:211–221 (1988).

31. V. M. Knepp, F. C. Szoka, and R. H. Guy, Controlled drug release from a novel liposome delivery system. II transdermal delivery characteristics, *J. Contr. Rel. 12*: 25–30 (1990)

32. J. Lasch and J. A. Bouwstra, Interactions of external lipids (lipid vesicles) with the skin. *J. Liposome Res. 5*:543–569 (1995).

33. M. Foldvari, B. Jarvis, and C. J. N. Ogueijofor, Topical dosage form of liposomal tetraciane: effect of additives on the in vitro release and in vivo efficacy, *J. Contr. Rel. 27*:193–205 (1993).

34. S. M. Dowton, Z., Hu, C. Ramachandran, D. F. H. Wallach, and N. Weiner, Influence of liposomal composition on topical delivery of encapsulated cyclosporin A, I. An in vitro study using hairless mouse skin, *STP Pharma Sciences 3*:404–407 (1993).

35. H. E. J. Hofland, R. van der Geest, H. E. Boddé, H. E. Junginger, and J. A. Bouwstra. Estradiol permeation from nonionic surfactant vesicles through human stratum corneum in vitro, *Pharma. Res. 11*:659–664 (1994).

36. M. Kirjaivainen, A. Urtti, I. Jääskelänen, T. M. Suhonen, P. Paronen, R. Valjakka-Koskela, J. Kiesvaara, and J. Mönkkönen, Interactions of liposomes with human skin in vitro—the influence of lipid composition and structure, *Biochim. Biophys. Acta 1304*:179–189 (1996).

37. H. E. J. Hofland, J. A. Bouwstra, F. Spies, H. E. Bodde, J. Nagelkerke, C. Cullander, H. E. Junginger, Interactions between non-ionic surfactant vesicles and human stratum corneum in vitro, *J. Liposome Res. 5*:241–263 (1995).

38. D. A. van Hal, Nonionic surfactant vesicles for dermal and transdermal drug delivery, thesis, Leiden University, 1994, pp. 149–176.

39. J. Lasch, R. Laub, and W. Wohlrab. How deep do intact liposomes penetrate into human skin? *J. Control. Release 18*:55–58 (1991).

40. D. van Hal, A. van Rensen, T. de Vringer, H. Junginger, and J. Bouwstra, Diffusion of estradiol from non-ionic surfactant vesicles through human stratum corneum in vitro, *J. Invest. Dermatol. 6*:72–78 (1996).

41. M. M. Meuwissen, L. Mougin, H. E. Junginger, and J. A. Bouwstra, Transport of model drugs through skin in vitro and in vivo by means of vesicles, *Proceed. Intern. Symp. Control. Rel. Bioact. Mater. 23*:303–304 (1996).

42. M. E. M. J. van Kuyk-Meuwissen, L. Mougin, H. E. Junginger, J. A. Bouwstra, Application of vesicles to rat skin in vivo: a confocal laser scanning microscopy study. *J. Contr. Rel. 56*:189–196 (1998).

43. M. Fresta and G. Puglisi, Corticosteroid dermal delivery with skin-lipid liposomes. *J. Contr. Rel. 44*:141–151 (1997).

44. B. A. I. van den Bergh, M. A. Salomons-de Vries and J. A. Bouwstra, Interactions between liposomes and stratum corneum studied by freeze substitution electron microscopy, *Int. J. Pharm. 167*:57–67 (1998).

45. H. C. Korting, W. Stolz, M. H. Schmid, G. Maierhofen. Interactions of liposomes with human epidermis reconstructed in vitro. *Br. J. Dermatol. 132*:571–579 (1995).

46. M. Foldvari, A. Gesztes, and M. Mezei, Dermal drug delivery by liposome encapsulation, clinical and electron microscopic studies. *J. Microsc. Encapsulation 7*:479–489 (1990).

47. G. Cevc and G. Blume, Lipid vesicles penetrate into intact skin owing to the transdermal osmotic gradients and hydration force, *Biochim. Biophys. Acta. 1104*:226–232 (1992).

48. M. E. M. J. van Kuyk-Meuwissen, H. E. Junginger, and J. A. Bouwstra, Interactions between liposomes and human skin in vitro, a confocal laser scanning microscopy study, *Biochim. Biophys. Acta 1371*:13–39 (1998).

49. J. du Plessis, C. Ramanchandran, N. Weiner, and D. G. Muller, The influence of particle size of liposomes on the disposition of drugs into the skin, *Int. J. Pharm. 103*:277–282 (1994).

50. E. Touitou, F. Levi-Schaffer, N. Dayan, F. Alhaique, and F. Riccieri, Modulation of caffeine skin delivery by carrier design: Liposomes versus permeation enhancers, *Int. J. Pharm. 153*:131–136 (1994).

51. J. Lasch and W. Wohlrab, Liposome-bound cortisol: A new approach to cutaneous therapy, *Biomed. Biochim. Acta 45*:1295–1299 (1986).

52. W. Wohlrab, and J. Lasch, Penetration kinetics of liposomal hydrocortisone in human skin, *Dermatologica 174*:18–22 (1987).

53. C. Michel, T. Purmann, E. Mentrup, E. Seiller, and J. Kreuter. Effect of liposomes on percutaneous penetration of lipophilic materials. *Int. J. Pharm. 84*:93–105 (1992).

54. R. Singhy and S. P. Vyass, Topical liposomal system for localized and controlled drug delivery, *J. Dermatol. Sci. 13*:107–111 (1996).

55. M.-K Kim, S.-J. Chung, M.-H. Lee, A.-R. Cho, and C.-K. Shim, Targeted and sustained delivery of hydrocortisone to normal and stratum corneum-removed skin

without enhanced skin absorption using a liposome gel, *J. Contr. Rel. 46*:243–251 (1997).

56.  H. Kumatsu, K. Higazi, H. Okamoto, K. Miyakwa, M. Hashida, and H. Sezaki, Preservative activity and in vivo percutaneous penetration of butylparaben entrapped in liposomes, *Chem. Pharm. Bull. 34*:3415–3422 (1986).

57.  H. Komatsu, H. Okamoto, K. Miyagawa, M. Hashida, and H. Sezaki. Percutaneous absorption of butylparaben from liposomes in vitro, *Chem. Pharm. Bull. 34*:3423–3430 (1986).

58.  V. Masini, F. Bonte, A. Meybeck, and J. Wepierre, Cutaneous bioavailability in hairless rats of tretinoin in liposomes or gel, *J. Pharm. Sciences 82*:17–21 (1993).

59.  C. Artmann, J. Roding, M. Ghyczy, and H. G. Pratzel, Liposomes from soya phospholipids as percutaneous drug carriers, *Artzneim.-Forsch/Drug Res. 40(II), 12*: 1363–1365 (1990).

60.  C. Artmann, J. Roding, M. Ghyczy, and H. G. Pratzel. Liposomes from soya phospholipids as percutaneous drug carriers, *Artzneim.-Forsch/Drug Res. 40(II), 12*: 1365–1368 (1990).

61.  W. Gehring, M. Ghyzhy, M. Gloor, Ch. Heitzler, and J. Roding. Significance of empty liposomes Alone and as drug carriers in dermatology, *Artzneim.-Forsch/ Drug Res. 40(II). 12*:1368–1371 (1990).

62.  H. E. Bodde, L. A. R. M. Pechtold, M. T. A. Subnel, and F. H. N. de Haan, Monitoring in vivo skin hydration by liposomes using infrared spectroscopy in conjunction with tape stripping, *Liposome Dermatics* (O. Braun Falco, H. C. Korting, eds.), Springer Verlag 1992, pp. 137–149.

63.  M. E. Planas, P. Gonzales, L. Rodriguez, S. Sanches, G. Cevc, Noninvasive percutaneous induction of topical analgesic by a new type of drug carrier, a prolongation of local pain insensitivity by anesthetic liposomes. *Anesth. Analg. 75*:615–621 (1992).

64.  G. Cevc, G. Blume, and A. Schatzlein, Transfersomes-mediated transepidermal delivery improves the regio-specificity and biological activity of corticosteroids in vivo, *J. Cont. Rel. 45*:211–226 (1997).

65.  G. Cevc, Dermal insulin, *Frontiers in Insulin Pharmacology* (M. Berger, A. Gries, eds.), Georg Thieme, Stuttgart, 1993, pp. 61–74.

66.  G. Cevc, D. Gebauer, J. Stieber, A. Satzlein, and G. Blume, Ultraflexible vesicles, Transfersomes, have an extremely low pore penetration resistance and transport therapeutic amounts of insulin across the intact mammalian skin, *Biochim. Biophys. Acta 1368*:201–215 (1998).

67.  G. Cevc, Transfersomes, liposomes and other lipid suspensions on the skin: permeation enhancement, vesicle penetration, and transdermal drug delivery, *Crit. Rev. Therap. Drug Carrier Systems, 13*:257–388 (1996).

68.  A. Paul and G. Cevc, Non-invasive administration of protein antigens: transdermal immunization with the bovine serum albumin in transfersomes, *Vaccine Res. 4*:145 (1995).

69.  A. Paul, G. Cevc, and B. K. Bachhawat. Transdermal immunization with large proteins by means of ultradeformable drug carriers. *Eur. J. Immunol. 25*:3521–5324 (1995).

70.  A. Schätzlein and G. Cevc, Skin penetration by phospholipid vesicles, trans-

fersomes, as visualized by means of confocal laser scanning microscopy, *Phospholipids: Characterization, Metabolism, and Novel Biological Applications* (G. Cevc and F. Paltauf, eds.), AOCS Press, Champaign, IL, 1993, pp. 189–207.

71. A. Schatzlein and G. Cevc. Non-uniform cellular packing of the stratum corneum and permeability barrier function of intact skin: a high-resolution confocal laser scanning microscopy study using highly deformable vesicles (Transfersomes). *Br. J. Dermatol. 138*:583–592 (1998).

72. G. Cevc, Transfersomes, liposomes and other lipid suspensions on the skin: permeation enhancement, vesicle penetration, and transdermal drug delivery, *Crit. Rev. Therapeutic Drug Carrier Systems 13*:257–388 (1996).

73. M. M. E. J. van Kuyk-Meuwissen, H. E. Junginger, and J. A. Bouwstra, Application of vesicles in human skin in vitro: a confocal laser scanning microscopy study, *Biochim. Biophys. Acta 1371*:31–39 (1998).

74. T. Ogiso, N. Niinaka, and M. Iwaki. Mechanism for enhancement effect of lipid disperse system on percutaneous absorption, *J. Pharm. Sci. 85*:57–64 (1996).

75. B. A. I. van den Bergh, Thesis, Leiden University, The Netherlands, 1999.

76. B. A. I. van den Bergh, J. Bouwstra, H. E. Junginger, and P. W. Wertz, Elasticity of vesicles affects hairless mouse skin structure and permeability. *J. Contr. Rel. 62*:367–397 (1999).

77. B. A. I. van den Bergh, J. Vroom, H. Gerritsen, H. E. Junginger, and J. Bouwstra. Interactions of elastic vesicles with human skin in vitro: electron microscopy and two-photon excitation microscopy, *Biochim. Biophys. Acta 1461*:155–173 (1999).

78. L. Li, V. K. Lishko, and R. M. Hoffman, Liposomes can specifically target entrapped malanin to the hair follicles in histocultured skin, *In Vitro Cell. Dev. Biol. 29A*:192–194 (1993).

79. B. F. van Duzee, Thermal analysis of human stratum corneum, *J. Invest. Dermatol. 65*:404–408 (1975).

80. V. M. Mak, R. O. Potts, and R. H. Guy, Does hydration affect intercellular lipid organisation in the stratum corneum? *Pharm. Res. 8*:1064–1065 (1991).

81. A. Alonso, N. Meirelles, and M. Tabak, Effect of hydration upon the fluidity of intercellular membranes of stratum corneum: an EPR study, *Biochim. Biophys. Acta 1237*:6–15 (1995).

82. C. L. Gay, R. H. Guy, G. M. Golden, V. M. Mak, and M. L. Francoeur, Characterization of low-temperature (i.e., < 65 degrees C) lipid transitions in human stratum corneum, *J. Invest. Dermatol. 103*:233–239 (1994).

83. D. A. van Hal, E. E. Jeremiasse, H. E. Junginger, F. Spies, and J. A. Bouwstra, Structure of fully hydrated human stratum corneum: a freeze fracture electron microscopy study, *J. Invest. Dermatol. 106*:89–95 (1996).

84. B. B. Michniak, M. R. Player, J. M. Chapman, Jr, and J. W. Sowell, Sr, In vitro evaluation of a series of Azone analogs as dermal penetration enhancers. I, *Int. J. Pharm. 91*:85–93 (1993).

85. B. B. Michniak, M. R. Player, L. C. Fuhrman, J. M. Christensen, J. M. Chapman, and J. W. Sowell, In vitro evaluation of a series of Azone analogs as dermal penetration enhancers. II (Thio)amides, *Int. J. Pharm. 91*: (1993).

86. J. Hadgraft, Skin penetration enhancement, *Prediction of Percutaneous Penetra-*

*tion*, Vol. 3 (K. R. Brain, V. J. James, and K. A. Walters, eds.), STS Publishing, Cardiff, UK, 1993, pp. 138–148.

87. B. J. Aungst, N. J. Rogers, and E. Shefter, Enhancement of naloxone penetration through human skin in vitro using fatty acids, fatty alcohols, surfactants, sulfoxides and amides. *Int. J. Pharm. 33*:225–234 (1995).

88. B. J. Aungst, Structure/effect studies of fatty acid isomers as skin penetration enhancers and skin irritants, *Pharm. Res. 6*:244–247 (1989).

89. E. R. Cooper, Increased skin permeability for lipophilic molecules, *J. Pharm. Sci. 73*:1153–1156 (1984).

90. N. Tsuzuki, O. Wong, and T. Higuchi, Effect of primary alcohols on percutaneous absorption. *Int. J. Pharm. 46*:19–23 (1988).

91. M. Yamada, Y. Uda, and Y. Tanigawara, Mechanism of enhancement of percutaneous absorption of molsidomine by oleic acid, *Chem. Pharm. Bull. 35*:3399–3406 (1987).

92. P. Green and J. Hadgraft, Facilitated transfer of cationic drugs across a lipoidal membrane by oleic acid and lauric acid, *Int. J. Pharm. 37*:251–255 (1987).

93. G. M. Golden, J. E. McKie, and R. O. Potts, Role of stratum corneum lipids fluidity in transdermal drug flux, *J. Pharm. Sci. 76*:25–28 (1987).

94. V. H. W. Mak, R. O. Potts, and R. H. Guy, Oleic acid concentration and effect in human stratum corneum: noninvasive determination by attenuated total reflectance infrared spectroscopy in vivo, *J. Control. Rel. 31*:263–269 (1994).

95. M. L. Francoeur, G. M. Golden, and R. O. Potts, Oleic acid: its effects on stratum corneum in relation to transdermal drug delivery, *Pharm. Res. 7*:621–626 (1990).

96. B. Ongpipattanakul, R. Burnette, and R. O. Potts, Evidence that oleic acid exists as a separate phase within stratum corneum, *Pharm. Res. 8*:350–354 (1991).

97. A. Naik, L. Pechtold, R. O. Potts, and R. H. Guy, Mechanism of skin penetration enhancement, in vivo, in man, *Prediction of Percutaneous Penetration*, Vol. 3 (K. R. Brain, V. J. James, and K. A. Walters, eds.), STS Publishing, Cardiff, UK, 1993, pp. 161–165.

98. H. Tanojo, H. E. Junginger, and H. E. Bodde, Effects of oleic acid on human transepidermal water loss using ethanol or propylene glycol as vehicles, *Prediction of Percutaneous Penetration*, Vol. 3 (K. R. Brain, V. J. James, K. A. Walters, eds.), STS Publishing, Cardiff, UK, 1993, pp. 319–324.

99. B. W. Barry, Mode of action of penetration enhancers in human skin, *J. Contr. Rel. 6*:85–97 (1987).

100. H. Tanojo, A. van Bos-van Geest, J. A. Bouwstra, H. E. Junginger, and H. E. Bodde, In vitro human skin barrier perturbation by oleic acid: thermal analysis and freeze fracture electron microscopy, *Thermochimica Acta 293*:77–85 (1997).

101. M. D. Garrison, L. M. Doh, R. O. Potts, and R. H. Guy, Effect of oleic acid on human epidermis: fluorescence spectroscopic investigation. *J. Cont. Rel. 31*:263–269 (1994).

102. T. Ogiso and M. Shintani, Mechanism for the enhancement effect of fatty acids on the percutaneous absorption of propranolol, *J. Pharm. Sci. 79*:774–779 (1990).

103. B. J. Aungst, N. J. Rogers, and E. Shefter, Enhancement of naloxone penetration through human skin in vitro using fatty acids, fatty alcohols, surfactants, sulphoxides and amides, *Int. J. Pharm. 33*:225–234 (1986).

104. C. K. Lee, T. Uchida, N. S. Kim, and S. Goto, Skin permeation enhancement of tegafur by ethanol/panasate 800 or ethanol/water binary vehicle and combined effect of fatty acids and fatty alcohols, *J. Pharm. Sci. 82*:1155–1159 (1993).

105. Y. Takeuchi, H. Yasukawa, Y. Yamaoka, Y. Kato, Y. Morimoto, Y. Fukomori, and T. Fukada, Effect of fatty acids, fatty amines and propylene glycol on rat stratum corneum lipids and proteins in vitro measured by Fourier Transformed Infrared/Attenuated Total Reflectance spectroscopy. *Chem. Pharm. Bull. 40*:1887–1892 (1992).

106. H. Tanojo, J. A. Bouwstra, H. E. Junginger, and H. E. Bodde, In vitro human skin barrier modulation by fatty acids: skin permeation and thermal analysis studies, *Pharm Res. 14*:42–49 (1997).

107. R. Stoughtton and W. McClure, Azone—a new non-toxic enhancer for cutaneous penetration, *Drug. Dev. Ind. Pharm. 9*:725–744 (1983).

108. P. S. Bannerjee, Transdermal penetration of vasopressin. II: The influence of Azone in in vitro and in vivo penetration, *Int. J. Pharm. 49*:199–204 (1989).

109. J. W. Wiechers, B. F. H. Drenth, F. A. W. Adolfsen, G. M. M. Groothuis, and R. A. de Zeeuw, Disposition and metabolic profiling of the penetration enhancer Azone, I. In vivo studies urinary profiles of hamster, rat, monkey, and man, *Pharm. Res. 5*: 496–499 (1990).

110. P. W. Swart, F. A. M. Toulouse, and R. A. De Zeeuw, The influence of azone on the transdermal penetration of the dopamine $D_2$ agonist MN-0923 in freely moving rats, *Int. J. Pharm. 88*:165–170 (1992).

111. A. Ruland and J. Kreuter, Influence of various penetration enhancers on the in vitro permeation of amino acids across hairless mouse skin, *Int. J. Pharm. 85*:7–17 (1992).

112. K. Sugibayashi, K. Hosaya, Y. Moromoto, and W. Higuchi, Effect of the absorption enhancer azone on the transport of 5 fluorouracil across hairless rat skin, *J. Pharm. Pharmacol. 37*:578–580 (1985).

113. P. Wotton, B. Mollgaard, J. Hadgraft, and A. Hoelgaard, Vehicle effect on topical drug delivery. III. Effect of azon on the cutaneous permeation of metronidazole and propylene glycol, *Int. J. Pharm. 24*:19–26 (1985).

114. N. Sheth, D. Freeman, W. Higuchi, and S. Spruance, The influence of azone, propylene glycol and polyethylene glycol on in vitro skin penetration of trifluoro-thymidine, *Int. J. Pharm. 28*:201–209 (1986).

115. J. Hadgraft, Skin penetration enhancement, *Prediction of Percutaneous Penetration*, Vol. 3 (K. R. Brian, V. J. James, and K. A. Walters, eds.), STS Publishing, Cardiff, UK, 1993, pp. 138–148.

116. A. J. Hoogstraate, J. Verhoef, J. Brussee, A. P. IJzerman, F. Spies, and H. E. Bodde, Kinetics, ultrastructural aspects and molecular modelling of transdermal peptide flux enhancement by N-alkylazacycloheptanones. *Int. J. Pharm. 76*:37–47 (1991).

117. D. Lewis and J. Hadgraft, Mixed monolayers of dipalmitoylphosphatidylcholine with Azone or oleic acid at the air-water interface. *Int. J. Pharm. 65*:211–218 (1990).

118. H. E. Bodde, I. van den Brink, H. K. Koerten, and F. H. N. de Haan, Visualization of in vitro percutaneous penetration of $HgCl_2$; transport through intercellular space versus cellular uptake through desmosomes, *J. Control Rel. 15*:227–236 (1991).

119. H. E. Boddé, M. A. M. Kruithof, J. Brussee, and H. K. Koerten, Visualisation of normal and enhanced HgCl$_2$ transport through human skin, *Int. J. Pharm. 53*:12–34 (1989).

120. H. L. G. M. Tiemessen, H. E. Bodde, M. Van Koppen, W. C. Bauer, and H. E. Junginger, A two-chambered diffusion cell with improved flow through characteristics through for studying the drug permeation of biological membranes, *Acta Pharm. Techn. 34*:99–101 (1988).

121. J. A. Bouwstra, L. Peschier, J. Brussee, and H. Bodde, Effect of N-alkyl-azocyclo-heptan-2-ones including azone on the thermal behaviour of human stratum corneum, *Int. J. Pharm. 52*:249–260 (1989).

122. J. A. Bouwstra, G. S. Gooris, M. A. Salomons-de Vries, and W. Bras, The influence of N-alkyl-azones on the ordering of the lamellae in human stratum corneum, as determined by small angle x-ray diffraction, *Int. J. Pharm. 79*:141–148 (1992).

123. J. A. Bouwstra, M. A. Salomons-de Vries, B. A. I. van den Bergh, and G. S. Gooris, Changes in lipid organisation of the skin barrier by N-alkyl-azocycloheptanones: a visualisation and X-ray diffraction study, *Int. J. Pharm. 144*:81–89 (1996).

124. F. R. Bezema, E. Marttin, P. E. H. Roemele, M. A. Salomons, J. Brussee, H. J. M. de Groot, F. Spies, and H. E. Bodde, $^2$H NMR and freeze fracture electron microscopy reveal rapid isotropic motion and lipid perturbation by azone in human stratum corneum, *Prediction of Percutaneous Penetration*, Vol. 3 (K. R. Brain, V. J. James, K. A. Walters, eds.), STS Publishing, Cardiff, UK, 1993, pp. 8–17.

125. M. Hori, S. Satoh, H. I. Maibach, and R. H. Guy, Enhancement of propranolol hydrochloride and diazepam skin absorption in vitro: effect of enhancer lipophilicity. *J. Pharm. Sci. 80*:32–35 (1991).

126. A. C. Williams and B. W. Barry, The enhancement index concept applied to terpene penetration enhancers for human skin and model lipophilic (oestradiol) and hydrophylic (5-fluorouracil) drugs, *Int. J. Pharm. 76*:157–161 (1991).

127. A. C. Williams and B. W. Barry, Terpenes and the lipid-protein partitioning theory of skin penetration enhancement, *Pharm. Res. 8*:17–24 (1991).

128. H. Okabe, K. Takayama, A. Ogura, and T. Nagai, Effect of limonene and related compounds on the percutaneous absorption of indomethacine. *Drug Design Delivery, 4*:313–321 (1989).

129. H. Okamoto, M. Ohyabu, M. Hashida, and H. Sezaki, Enhanced penetration of mytomicin C through hairless mouse and rat skin by enhancers with terpene moieties, *J. Pharm. Pharmacol. 39*:531–534 (1987).

130. P. A. Cornwell, Mechanism of action of terpene penetration enhancers in human skin, Bradford, Thesis, 1993, pp. 111–162.

131. P. A. Cornwell, B. W. Barry, J. A. Bouwstra, and G. S. Gooris, Small angle X-ray diffraction investigations on the lipid barrier in human skin, *Prediction of Percutaneous Penetration*, Vol. 3 (K. R. Brain, V. J. James, and K. A. Walters, eds.), STS Publishing, Cardiff, UK, 1993, pp. 18–26.

132. P. A. Cornwell, C. P. Stoddart, J. A. Bouwstra, and B. W. Barry, Wide-angle x-ray diffraction of human stratum corneum: Effect of hydration and terpene enhancer treatment, *J. Pharm. Pharmacol. 46*:938–950 (1994).

# 6
# Drug Targeting by Retrometabolic Design
## Soft Drugs and Chemical Delivery Systems

**Nicholas Bodor and Peter Buchwald**
*University of Florida, Gainesville, Florida*

## I. INTRODUCTION

One of the most important goals of pharmaceutical research and development is targeted drug delivery, defined as optimization of the therapeutic index by localizing the pharmacological activity of the drug to the site of action. It is important to distinguish this broad definition as the ability to achieve a desired pharmacological response at a selected site without undesired interaction at other sites from a narrower definition of the basic targeting concept. Within this narrower definition, a specific drug receptor is considered as target, and the objective is to improve fit, affinity, and binding to this receptor that ultimately will trigger the pharmacological activity.

Ever since the development of the receptor theory, attempts have been directed toward developing new therapeutic agents that have a singular target, that is, agents that bind only to a specific receptor. It was hoped that this way any aberrant toxicity would be avoided, and only the desired therapeutic gain would be produced. Unfortunately, the situation is not so simple. Most highly active new therapeutic agents designed to bind to a specific receptor ultimately had to be discarded when unacceptable toxicity or unavoidable side effects were encountered in later stages of the development. There are a number of reasons for this. First, side effects are usually related to the intrinsic receptor affinity responsible for the desired activity. Second, although in most cases the desired response should be localized to some organ or cell, various receptors are often distributed throughout the whole body. Third, for most drugs, metabolism generates multiple metabolites that can have an enhanced or a different type of biological activity or can be toxic.

Beyond receptor targeting, something additional has to be done: one needs to localize drugs at the desired site of action. Successful targeting, meaning preferential delivery, would lead to reduced drug dosage, decreased toxicity, and increased treatment efficacy. With reasonable biological activity at hand, targeting to the site of action should be superior to molecular manipulations aimed at refining receptor-substrate interactions. However, successful drug targeting is a complicated problem, because any drug introduced into the body encounters or must bypass various organs, cells, membranes, enzymes, and receptors before reaching its designated target. Nevertheless, during the past two decades, significant efforts have been focused on the field of site-specific, targeted drug delivery systems. Most of these efforts were directed as improving the delivery of already known or marketed drugs. Although improvements—even some significant ones—were achieved, we have to acknowledge the fact that drugs developed in the past usually reached the clinical stage without consideration of their targeting. Hence, more often than not a currently accepted and approved drug is not particularly suitable for targeting manipulations. This recognition led to the idea, already argued in many of our previous publications, that something additional has to be done to enhance the therapeutic index: whenever possible, *targeting and metabolism considerations should be included in the drug design process from the beginning.* It is hoped that future drugs will be designed with a preferred metabolic route and targeting in mind, and the actual new chemical entity will have *site specificity and selectivity built into its molecular structure.*

During the past years, significant development and transformation took place in the large field that is now considered site-specific drug delivery. Various attempts were made to classify all these efforts. This review uses a mechanism-based classification that differentiates between physical, biological, and chemical targeting, a classification that should be most useful for medicinal chemists. This chapter concentrates on chemical drug targeting and describes advanced chemical-enzymatic–based drug targeting systems obtained with strategies that are part of an approach designated now as *retrometabolic drug design.*

Other general classifications are also possible. For example, one can differentiate among first-, second-, and third-order targeting [1]. First-order targeting refers to restricted drug distribution to the site of action (organ or tissue). Second-order targeting refers to selective drug delivery to specific cells (e.g., tumor cells), and third-order targeting refers to directed drug release at predetermined intracellular sites. A number of reviews describe different aspects of drug targeting dealing mainly with issues related to second- and third-order targeting [2–8]. The same classification, in a renamed form, was used in an extensive review on drug delivery systems [9]. Another more general approach was presented in a comprehensive review by Tomlinson [10].

## II. PRINCIPLES OF RETROMETABOLIC DRUG DESIGN

As we have often argued, metabolic considerations should be an integral part of any drug design process. Rational drug design can be accomplished only by incorporating metabolic considerations into the design process from the very beginning. Retrometabolic approaches represent a novel, systematic method to accomplish this goal. By combining structure-activity relationships (SAR) with structure-metabolism relationships (SMR), they allow the design of safe, localized compounds.

As illustrated by the *retrometabolic drug design* loop in Figure 1, retrometabolic drug design approaches include two distinct methods to improve the therapeutic index of a drug (D). One is *soft drug* (SD) design. Soft drugs are active isosteric–isoelectronic analogues of a lead compound, but they are deactivated

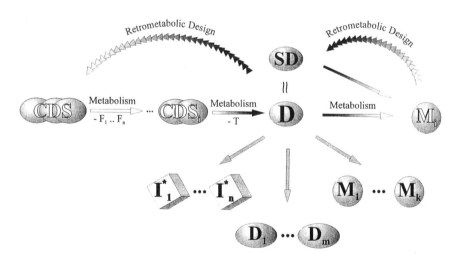

**Figure 1**  The retrometabolic drug design loop, including chemical delivery system (CDS) and soft drug (SD) design. Straight arrows represent metabolic changes; curved arrows represent retrometabolic design approaches. The general metabolic conversion of a drug is also illustrated. D, drug; $M_i$, inactive metabolite(s); $D_1 \ldots D_m$, analogue metabolites; $M_1 \ldots M_k$, other metabolites; $I_1^* \ldots I_n^*$, potential reactive intermediates; $CDS_i$, metabolites of the original CDS formed during the sequential metabolic conversion that removes the modifier functions ($F_n$) and the targetor (T) moiety.

in a predictable and controllable way after achieving their therapeutic role. The other method is *chemical delivery system* (CDS) design. A CDS is defined as a biologically inert molecule that requires several steps in its conversion to the active drug and that enhances drug delivery to a particular organ or site. Although both approaches involve chemical modifications to obtain an improved therapeutic index and both require enzymatic reactions to fulfill drug targeting, there is not much in common between the principles of SD design and those of CDS-based retrometabolic drug design. Although the CDS is inactive by definition, and sequential enzymatic reactions provide the differential distribution and drug activation, SDs are active therapeutic agents designed to be rapidly metabolized into inactive species.

Owing to the considerable flexibility of retrometabolic drug design, for certain lead compounds a large number of possible analogue structures can be designed, and finding the best drug candidate among them may prove tedious and difficult. Fortunately, computer methods developed to calculate various molecular properties, such as molecular volume, surface area, charge distribution, polarizability, aqueous solubility, and partition coefficient [11–18] allow more quantitative design. The capabilities of quantitative design have been further advanced by developing expert systems that combine the various structure-generating rules and predictive software to provide an analogy-based ranking order [19–23]. The approach is general in nature and can be used starting with essentially any lead.

## III. SOFT DRUGS

Soft drugs are active isosteric–isoelectronic analogues of a lead compound, but they are deactivated in a predictable and controllable way after achieving their therapeutic role [21, 24–29]. They are designed to be rapidly metabolized into inactive species and, hence, to simplify the transformation-distribution-activity profile of the lead. Consequently, soft drugs are new therapeutic agents obtained by building in the molecule, in addition to the activity, the most desired way in which the molecule is to be deactivated and detoxified subsequent to exerting its biological effects. The desired activity is generally local, and the soft drug is applied or administered near the site of action. Therefore, in most cases, they produce pharmacological activity locally, but their distribution away from the site results in a prompt metabolic deactivation that prevents any kind of undesired pharmacological activity or toxicity. The resulting differential distribution is not the result of a classical drug targeting but can be regarded as a *reversed targeting*.

The soft drug concept was introduced in 1976 [24] and reiterated in 1980 [30–32]. Since then, five distinct types have been identified [21, 25, 27–29, 33]: (1) soft analogues, (2) soft drugs based on the inactive metabolite approach, (3) controlled-release endogenous agents, (4) activated soft compounds, and (5) active metabolite-based drugs. General design principles have been reviewed in a number of articles, and examples for practical use of each of these classes are provided in the literature [21, 27–29, 34, 35]. The *inactive metabolite* and the *soft analogue* approaches have been the most useful and successful strategies for designing safe and selective drugs. In agreement with the principles set forth in the previous chapters, they provide compounds that are structural analogues of known active drugs but have a metabolically, preferentially hydrolytically, sensitive spot built into their structure. This allows a one-step controllable decomposition into inactive, nontoxic moieties as soon as possible after the desired effect is achieved and avoids other types of metabolic routes.

During evolution, living organisms developed not only fine-tuned metabolic mechanisms for endogenous chemicals but also several defensive mechanisms to detoxify xenobiotics. Most metabolic processes used in the attempt to eliminate invading foreign chemicals by transforming them into more hydrophilic or more easily conjugated compounds are oxidative in nature. Unfortunately, many of these mechanisms are indiscriminate, and detoxifying enzymes, such as cytochrome P-450 or *N*-acetyltransferase, can generate toxic reactive intermediates (e.g., epoxides, radicals) from otherwise nontoxic compounds [36]. Chemicals and xenobiotics are, therefore, not always metabolized only into more hydrophilic and less toxic substances but also into highly reactive chemical species that then can react with various macromolecules and cause tissue damage or elicit antigen production. In addition, oxygenases that mediate most of these critical metabolic pathways exhibit not only interspecies but also interindividual variability and are subject to inhibition and induction [36]. In different individuals, half-lives of various foreign compounds may vary as much as 10- to 50-fold [36]. Furthermore, the rates of hepatic mono-oxygenase reactions are at least two orders of magnitude lower than the slowest of the other enzymatic reactions [37]. These mono-oxygenase reactions are slow, because they only have very few substrate molecules to react with. The substrate for NADPH-cytochrome P-450 reductase is not the exogeneous substrate *per se*, but the ferricytochrome P-450–substrate complex, which is present in considerably lower concentration [37]. Therefore, it is usually desirable to avoid oxidative pathways and slow, easily saturable oxidases and to design soft drugs that are inactivated by hydrolytic enzymes. In addition, because diseases can alter organs responsible for metabolism of blood-borne substances, rapid metabolism can be more reliably carried out by ubiquitous esterases. In critically ill patients, it is better not to rely on metabolism or clearance by organs such as liver or kidney, because blood flow and enzyme activity in these organs can be seriously impaired. Many structures susceptible for rapid enzymatic hydrolysis are chemically sufficiently stable to provide the required shelf life. To illustrate the concept, a more detailed description of one of the most successful soft drugs developed is included.

## A. Loteprednol Etabonate

Topical corticosteroids represent an important class of drugs used to treat *ocular* inflammations and allergies, but a number of contraindications severely limit their usefulness. In addition to the general systemic corticosteroid side effects, they can also produce a number of ocular complications such as elevation of the intraocular pressure (IOP) and resultant steroid-induced glaucoma, posterior subcapsular cataract formation, secondary ocular infection, retardation of wound healing, uveitis, mydriasis, transient ocular discomfort. A soft drug approach proved useful and resulted in an active corticosteroid that is void of these serious side effects. Loteprednol etabonate (**4**) (Figure 2), a soft steroid developed in our laboratories [38–58], has recently received Food and Drug Administration (FDA) approval as the active ingredient of two ophthalmic preparations, Lotemax™ and Alrex™. With this, it became the only corticosteroid that has FDA approval for use in all inflammatory and allergy-related ophthalmic disorders, including inflammation after cataract surgery, uveitis, allergic conjunctivitis, giant papillary conjunctivitis (GPC, an inflammatory condition most commonly associated with the use of contact lenses), etc. It is also being developed for treatment of asthma, rhinitis, colitis, and dermatological problems.

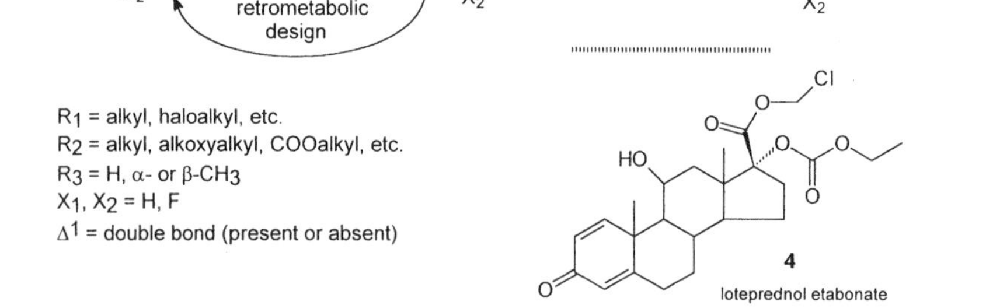
1
soft steroid
2
inactive metabolite
3
retrometabolic design
R1 = alkyl, haloalkyl, etc.
R2 = alkyl, alkoxyalkyl, COOalkyl, etc.
R3 = H, α- or β-CH3
X1, X2 = H, F
Δ1 = double bond (present or absent)
4
loteprednol etabonate

**Figure 2** Design and metabolism of soft corticosteroids (1) based on the inactive metabolite approach. The acid metabolites (2, 3) are inactive, but suitable substitution at the 17α-hydroxy and 17β-carboxy functions ($R_1$, $R_2$) can restore corticosteroid activity and also allow facile one-step deactivation. Loteprednol etabonate (4), a soft steroid, is an active anti-inflammatory compound that lacks the IOP-elevating side effect of the other steroids used ophthalmically.

As most other corticosteroids, hydrocortisone undergoes a variety of oxidative and reductive metabolic conversions. One of its major metabolic routes is oxidation of the dihydroxyacetone side chain, which ultimately leads to formation of cortienic acid (**3**: $R_3$, $X_1$, $X_2$ = H, no $\Delta^1$). Cortienic acid is a major metabolite excreted in human urine, and it lacks corticosteroid activity; therefore, it is an ideal lead for the *inactive metabolite* approach [27, 59, 60]. The design process (Figure 2) can directly involve the important pharmacophores found in the 17α and 17β side chains. Suitable isosteric/isoelectronic substitutions of the α-hydroxy and β-carboxy substituents with esters or other types of functions should restore the original corticosteroid activity without restoring the potential to produce adverse effects. Modifications of the 17β carboxyl function, besides the modifies in the 17α and the customary activity enhancing structural modifications (introduction of $\Delta^1$, fluorination at 6α and/or 9α, methyl introduction at 16α or 16β), led to a host of more or less active analogues represented by the general structure **1**. More than 120 of these soft steroids have been synthesized. Critical functions for activity are clearly the haloester in the 17β and the novel carbonate [39] and ether [61] substitutions in 17α-positions that provided the best activity.

We concentrated on 17α carbonates instead of 17α esters to enhance stability and, hence, to prevent formation of mixed anhydrides from 17α esters after hydrolysis of the 17β esters. Such mixed anhydrides were assumed toxic and probably cataractogenic. The mechanism of steroid-induced cataract is somewhat obscure [62], but the most prominent hypothesis involves the formation of Schiff bases between the steroid C-20 ketone group and nucleophilic groups such as ε-amino groups of lysine residues of proteins followed by a Heyns rearrangement [63] involving the adjacent C-21 hydroxyl group that results in stable ketoimine products (Figure 3) [64–66]. This covalent binding results in destabilization of the protein structure allowing further modification, (i.e., oxidation), leading to

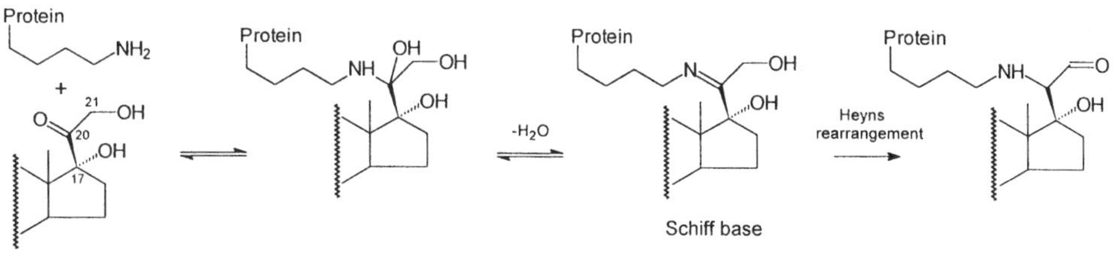

**Figure 3** Mechanism of steroid-induced cataract according to the most prominent hypothesis. It involves first the formation of Schiff bases between the steroid C-20 ketone group and nucleophilic groups such as ε-amino groups of lysine residues of proteins and then a Heyns rearrangement involving the adjacent C-21 hydroxyl group that results in stable amine-linked adducts.

cataract. The carbonates were expected to be less reactive than the corresponding esters owing to the lower electrophilicity of the carbonyl carbon. The $17\alpha$ carbonates were a new class of corticosteroids, which turned out to be difficult to obtain on normal corticosteroid derivatives. However, after oxidative removal of the C-21 carbon, their synthesis proved relatively easy [39]. Initial activities were determined by classical cotton pellet granuloma tests and by human vasoconstrictor studies [25, 27, 33, 67, 68]. A variety of $17\beta$ esters were synthesized, and they showed very different activities. Because this position is an important pharmacophore that is quite sensitive to small modifications, the freedom of choice was relatively limited. For example, although chloromethyl or fluoromethyl esters showed very good activity, the chloroethyl or $\alpha$-chloroethylidene derivatives were very weak. Simple alkyl esters also proved virtually inactive. Consequently, chloromethyl esters of various $17\alpha$ carbonates with different substituents on the steroid skeleton were selected for further investigation. For a number of derivatives, the therapeutic index was determined as the ratio between the anti-inflammatory activity and the thymus involution activity. As illustrated in Table 1, classical steroids, regardless of their intrinsic activity, have similar therapeutic indices, but loteprednol etabonate, the soft steroid selected for final development, provides a significant improvement. Many of the other soft steroids also showed a dramatic improvement in the therapeutic index [59, 69], and even their intrinsic activities were quite remarkable. Recent studies on binding to rat lung cytosolic corticosteroid receptors showed that the binding affinity of some of these compounds approaches or even exceeds those of the most potent corticosteroids known.

Selection of the final candidate for development was based on various properties. In addition to the therapeutic index, availability, synthesis, and "softness" (the rate and easiness of metabolic deactivation) also had to be considered. Loteprednol etabonate (**4**, chloromethyl $17\alpha$-ethoxycarbonyloxy-$11\beta$-hydroxy-3-oxoandrosta-1,4-diene, $17\beta$-carboxylate; **1**: $\Delta^1$, $R_1 = CH_2Cl$, $R_2 = COOC_2H_5$,

**Table 1**   Comparison of Loteprednol Etabonate with Other Steroids

| Treatment | N | $ED_{50}$[a] | Rel. pot. | $TED_{50}$[b] | Rel. pot. | $TI$[c] |
|---|---|---|---|---|---|---|
| Loteprednol etabonate (0.1%) | 8 | 178.0 | 0.48 | 10,000 | 0.02 | 24.0 |
| Hydrocortisone $17\alpha$-butyrate (0.1%) | 8 | 121.0 | 0.70 | 369 | 0.57 | 1.3 |
| Betamethasone $17\alpha$-valerate (0.12%) | 8 | 84.8 | 1.00 | 212 | 1.00 | 1.0 |
| Clobetasone $17\alpha$-propionate (0.1%) | 8 | 2.9 | 29.70 | 11 | 19.30 | 1.5 |

[a] Anti-inflammatory activity in the cotton pellet granuloma test ($\mu$g/pellet).
[b] Thymolysis potency ($\mu$g/pellet).
[c] Therapeutic index: the ratio of the relative potency for the $ED_{50}$ to the relative potency for the $TED_{50}$; betamethasone $17\alpha$-valerate has been arbitrarily assigned a value of 1.

$R_3$ = H, $X_1$, $X_2$ = H), a soft steroid derived from prednisolone, was selected for clinical development, and it was successfully developed into a unique ophthalmic anti-inflammatory/antiallergic compound. Early studies in rabbits [40, 43] and rats [44] demonstrated that, consistent with its design, **4** is indeed active, is metabolized into its predicted metabolites (PJ-91, **2**; PJ-90, **3**; $\Delta^1$, $R_2$ = $COOC_2H_5$, $R_3$ = H, $X_1$, $X_2$ = H), and these metabolites are inactive [39]. Loteprednol etabonate had a terminal half-life ($t_{1/2}$) of 2.8 h in dogs after IV administration of a 5-mg/kg dose [45]. It did not affect the IOP in rabbits (Figure 4) [43], an observation confirmed later in various human studies [51, 58]. A long-term ($\geq$28 days) use study showed that IOP elevation greater than 10 mmHg, a dreaded side effect of steroid therapy, occurred in 1.7%, 0.5%, and 6.7% of patients taking loteprednol etabonate, vehicle, and prednisolone acetate, respectively [58]. For patients who did not wear contact lenses, the same numbers were 0.6%, 1.0%, and 6.7%. Loteprednol etabonate has, therefore, a lower propensity to cause clinically significant elevations in IOP than prednisolone acetate, and, in patients not wearing contact lenses, this propensity is similar to that found in subjects receiving vehicle. Clinical studies also proved that it is a safe and effective treatment for GPC [50, 55], seasonal allergic conjunctivitis [54, 56], postoperative inflammation [70], or uveitis. It provides ophthalmologists with well-tolerated and effective means of

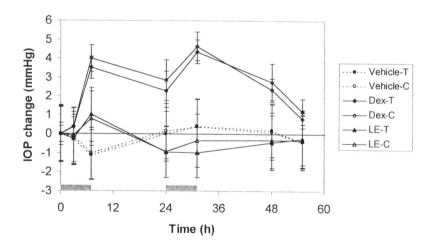

**Figure 4** Change in IOP of normotensive rabbits treated with 0.1% dexamethasone (Dex), 0.1% loteprednol etabonate (LE), and vehicle (50% w/w 2-hydroxypropyl-β-cyclodextrin water solution). Twelve rabbits were investigated in crossover experiments; drugs were administered in one eye (100 μL) every hour during the 8-h periods marked on the graph. (Data from Ref. 43; represent mean ± SE both for treated (T) and control (C) eyes.)

treating GPC that can be used without the need for patients to stop wearing their contact lenses. Two multicenter, randomized, double-masked, placebo-controlled, parallel-group studies were also conducted on a total of 424 patients with a mean age of 70 years [70, 71]. In one study, 64% of patients had complete resolution of anterior chamber inflammation compared with 29% of those receiving placebo. In the other study, the percentages were 55% and 28%, respectively ($p < 0.001$). On the basis of such study results, it was concluded recently that "loteprednol is truly a designer drug of the 90's and the drug of choice for postoperative anterior chamber inflammation" [71].

Proving effective reversed targeting, which results from the soft nature of this steroid, systemic levels or effects cannot be detected even after chronic ocular administration [57]. Plasma levels of loteprednol etabonate and its primary metabolite (PJ-91) were less than the 1 µg/L detection limit in 10 healthy volunteers who received the drug in both eyes eight times daily for 2 days and four times daily for a further 41 days [57]. In addition to its already approved uses, loteprednol etabonate is also being developed for the treatment of colitis, atopic dermatitis, and asthma based on promising results from animal studies [47, 48].

## IV. CHEMICAL DELIVERY SYSTEMS

Chemical delivery systems (CDSs), positioned on the other side of the retrometabolic drug design loop, represent novel and systematic ways of targeting active biological molecules to specific target sites or organs on the basis of predictable enzymatic activation. In principle, chemical drug delivery systems should include any drug targeting system that requires a chemical reaction to produce it. Consequently, they should include those systems where there is a covalent link between the drug and the so-called carrier, and, accordingly, at least one chemical bond needs to be broken to release the active component. However, in a stricter sense used here, chemical drug delivery systems refer to inactive chemical derivatives of a drug obtained by one or more chemical modifications so that the newly attached moieties are monomolecular units (generally comparable in size to the original molecule) and provide a site-specific or site-enhanced delivery of the drug through multistep enzymatic and/or chemical transformations [26, 29, 72–75].

During the chemical manipulations, two types of bioremovable moieties are introduced to convert the drug into an inactive precursor form. A *targetor* (T) moiety is responsible for targeting, site-specificity, and lock-in, whereas *modifier functions* ($F_1 \ldots F_n$) serve as lipophilizers, protect certain functions, or fine-tune the necessary molecular properties to prevent premature, unwanted metabolic conversions. The CDS is designed to undergo sequential metabolic conversions, disengaging the modifier functions and finally the targetor, after this moiety fulfills its site- or organ-targeting role. The CDS concept evolved from the prodrug concept [76–78] but became essentially different by the introduction of multistep activation and targetor moieties. Within the present formalism, one can say that prodrugs contain one or more F moieties for protected or enhanced overall delivery, but they do not contain T. Thus, they generally fail to achieve true drug targeting, which is the major pathway to improve the therapeutic index.

With a CDS, targeting is achieved by design: recognizing specific enzymes found primarily, exclusively, or at higher activity at the site of action, or exploiting site-specific transport properties such as, for example, those of the blood–brain barrier (BBB). The strategically predicted multienzymatic transformations result in a differential distribution of the drug. The CDS concept has been applied in a variety of drug-targeting problems, and successful deliveries to the brain, to the eye, and to other organs have been achieved [29, 75, 79–87]. CDSs can be divided into several distinct types: (1) enzymatic physical-chemical-based CDSs, (2) site-specific enzyme-activated CDSs, and (3) receptor-based (transient anchor-type) CDSs.

## A. Brain-targeting Chemical Delivery Systems

Brain-targeting chemical delivery systems represent just one class of CDSs. However, they represent the most developed class and can be classified as enzymatic physical-chemical–based CDSs. Within this approach, the drug is chemically modified to introduce the protective function(s) and the targetor moiety. On administration, the resulting CDS is distributed throughout the body. Predictable enzymatic reactions convert the original CDS by removing some of the protective functions and modifying the T moiety, leading to a precursor form, shown here as $T^+$-D, which is still inactive, but has significantly different physicochemical properties (Figure 5). These intermediates are continuously eliminated from the "rest of the body." Because of the presence of a specific membrane or other distributional barrier, efflux-influx processes at the targeted site are not the same, and they will provide a specific concentration here, ultimately allowing release of the active drug only at the site of action.

For example, the BBB can be regarded as a biological membrane that is permeable to most lipophilic compounds but not to hydrophilic molecules, and in most cases, these transport criteria apply to both sides of the barrier. Thus, if a lipophilic compound that can enter the brain is converted there to a hydrophilic molecule, one can assume that it will be "locked-in": it will no longer be able to come out. Targeting is assisted because the same conversion taking place in the rest of the body accelerates peripheral elimination and further contributes to brain targeting.

In principle, many targetor moieties are possible for a general system of this kind [72, 88–91], but the one based on the 1,4-dihydrotrigonelline ↔ trigonelline

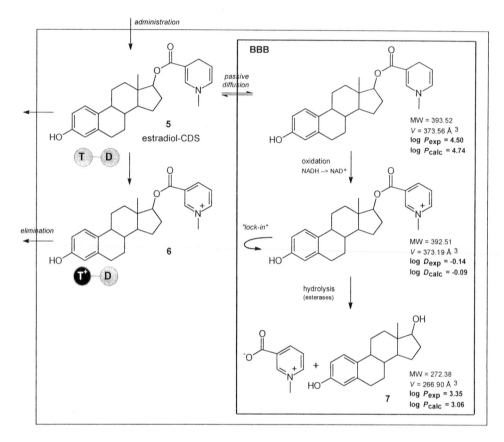

**Figure 5** Illustration of the "lock-in" mechanism for estradiol-CDS. Octanol-water log partition (log $P$) and distribution coefficients (log $D$) are shown to illustrate the significant changes in partition properties. The lipophilic CDS (5) (log $P > 4$) can easily cross the blood–brain barrier (BBB), but the hydrophilic intermediate (6) (log $D < 0$) is no longer able to come out providing a sustained release of the active estradiol (7).

(coffearine) system, in which the lipophilic 1,4-dihydro form (T) is converted *in vivo* to the hydrophilic quaternary form ($T^+$), proved the most useful. This conversion takes place easily everywhere in the body, because it is closely related to the ubiquitous NADH $\leftrightarrow$ NAD$^+$ coenzyme system associated with numerous oxidoreductases and cellular respiration [92,93]. Because oxidation takes place with direct hydride transfer [94] and without generating highly active or reactive radical intermediates, it provides a nontoxic targetor system [95]. Furthermore, because for small quaternary pyridinium ions rapid elimination from the brain, probably caused by involvement of an active transport mechanism that eliminates small organic ions, has been shown [96, 97], the $T^+$ moiety formed during the final release of the active drug D from the charged $T^+$-D form will not accumulate within the brain.

Although the charged $T^+$-D form is locked behind the BBB into the brain, it is easily eliminated from the body as a result of the acquired positive charge, which enhances water solubility. After a relatively short time, the delivered drug D (as the inactive, locked-in $T^+$-D) is present essentially only in the brain, providing sustained and brain-specific release of the acting drug. It has to be emphasized again that the system not only achieves delivery to the brain, but it provides preferential delivery, which means brain targeting. Ultimately, this should allow smaller doses and reduce peripheral side effects. Furthermore, because the "lock-in" mechanism works against the concentration gradient, it provides more prolonged effects. Consequently, these CDSs can be used not only to deliver compounds that otherwise have no access to the brain but also to retain lipophilic compounds within the brain, as it has indeed been achieved, for example, with a variety of steroid hormones. During the last decade, the system has been explored with a wide variety of drug classes (e.g., anti-infective agents, anticancer agents, anticonvulsants, antioxidants, antivirals, cholinesterase inhibitors, monoamine oxidase (MAO) inhibitors, neurotransmitters, nonsteroidal anti-inflammatory drugs (NSAIDs), steroid hormones) [29, 75]. To illustrate the concept, a more detailed description of estradiol-CDS will be included here.

Recently, the approach has been extended to achieve successful brain deliveries of neuropeptides, such as enkephalin, thyrotropin-releasing hormone (TRH), and kyotorphin analogues as well [83, 86, 98–100]. Successful brain-targeted delivery of peptides is an even more difficult task than delivery of other drugs, and three issues have to be solved simultaneously: enhance passive transport by increasing the lipophilicity, ensure enzymatic stability to prevent premature degradation, and exploit the "lock-in" mechanism to provide targeting. The solution we suggested is a complex *molecular packaging* strategy, in which the peptide unit is part of a bulky molecule, dominated by lipophilic modifying groups that direct BBB penetration and prevent recognition by peptidases [85]. Such a brain-targeted packaged peptide delivery system contains the following major components: the redox *targetor* (T); a *spacer* function (S), consisting of strategically used amino acids to ensure timely removal of the charged targetor from the peptide; the *peptide* itself (P); and a bulky *lipophilic moiety* (L) attached through an ester bond or sometimes through a C-terminal *adjuster* (A) at the carboxyl terminal to enhance lipid solubility and to disguise the peptide nature of the molecule (Figure 6). To achieve delivery and sustained activity with such complex systems, it is very important that the designated enzymatic reactions take place in a specific sequence. On delivery, the first step must be the conversion of the targetor to allow for "lock-in." This must be followed by removal of the

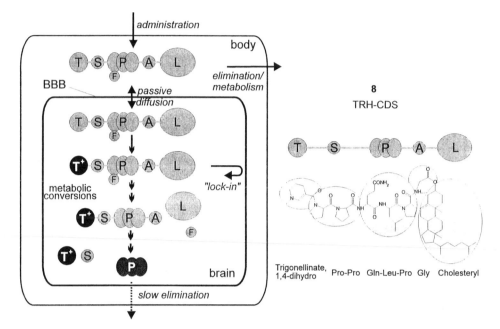

**Figure 6**  Schematic representation of the molecular packaging and sequential metabolism used for brain targeting of neuropeptides. TRH-CDS (8) is included to provide a concrete illustration for the targetor (T), spacer (S), peptide (P), adjuster (A), and lipophilic (L) moieties.

L function to form a direct precursor of the peptide that is still attached to the charged targetor. Subsequent cleavage of the T-S moiety finally leads to the active peptide (Figure 6).

Because, for these CDSs, adequate changes in physicochemical (i.e., partition) properties during the sequential metabolism are crucial for successful targeting, it is of considerable importance to acceptably predict such properties starting from molecular structure. A distinctively simple molecular size-based model (QLogP) [15, 16, 18] recently developed by us to predict log $n$-octanol-water partition coefficients proved useful here. On the basis of this model that works on a large variety of molecules and on the basis of experimental distribution data of quaternary pyridinium-type compounds including a number of CDSs, we concluded that oxidation of the 1,4-dihydrotrigonelline T moiety causes a change of 4–5 log units in partition properties [22, 23]. Such a significant change explains the success of the ''lock-in'' mechanism. The mechanism, together with structures, metabolic conversions, calculated and measured physicochemical properties, is summarized for brain-targeting estradiol-CDS (5) in Figure 5.

Unfortunately, the same physicochemical characteristics that allow for successful chemical delivery also complicate the development of acceptable pharmaceutical formulations. The increased lipophilicity allows partition into deep brain compartments but also confers poor aqueous solubility. The oxidative lability, which is needed for the ''lock-in'' mechanism, and the hydrolytic instability, which releases the modifier functions or the active drug, combine to limit the shelf life of the CDS. Cyclodextrins may provide a possible solution. During the last decade, the usefulness of cyclodextrin inclusion- complexes in improving the pharmaceutical characteristics of various drugs became well established [101]. Indeed, the corresponding inclusion complex with 2-hydroxypropyl-$\beta$-cyclodextrin (HP$\beta$CD) solved essentially all problems with estradiol-CDS [102]. Its aqueous solubility was enhanced about 250,000-fold in a 40% (w/v) HP$\beta$CD solution (from 65.8 ng/mL–16.36 mg/mL), and its stability was also significantly increased allowing formulation in acceptable form. The rate of ferricyanide-mediated oxidation, a good indicator of oxidative stability, was decreased about 10-fold, and shelf life was increased about fourfold, as indicated by $t_{90}$ and $t_{50}$ values in a temperature range of 23–80°C [102]. The cyclodextrin complex even provided better distribution by preventing retention of the solid material precipitated in the lung. Promising results were obtained for testosterone-CDS [103], lomustine-CDS [104], and for benzylpenicillin-CDS [105] as well.

## B. Estradiol-CDS

Among all CDSs, estradiol-CDS is in the most advanced investigation stage, and it is currently undergoing phase I and II clinical trials. Estrogens are lipophilic steroids that are not impeded in their entry to the central nervous system (CNS). They can readily penetrate the BBB and achieve high central levels after peripheral administration, but, unfortunately, estrogens are poorly retained within the brain. Therefore, to maintain therapeutically significant concentrations, frequent doses have to be administered. Constant peripheral exposure to estrogens has been related, however, to a number of pathological conditions including cancer, hypertension, and altered metabolism [106–109]. Because the CNS is the target site for many estrogenic actions, brain-targeted delivery may provide safer and more effective agents. Estrogen CDSs could be useful in reducing the secretion of luteinizing hormone-releasing hormone (LHRH) and, hence, in reducing the secretion of luteinizing hormone (LH) and gonadal steroids. As such, they could be used to achieve contraception and to reduce the growth of peripheral steroid-dependent tumors, such as those of the breast, uterus, and prostate, and to treat endometriosis. They also could be useful in stimulating male and female sexual behavior, and in the treatment of menopausal vasomotor symptoms ("hot flushes") [110]. Other potential uses are in neuroprotection, in the reduction of body weight, or in the treatment of depression and various types of dementia, including Alzheimer's disease [108, 111, 112]. Alzheimer's disease, which still has no specific cure, results in progressively worsening symptoms that range from memory loss to declining cognitive ability. It affects an estimated 10% of the population older than 65 years of age and almost 50% of those older than 85 years of age [113].

Estradiol ($E_2$) (7, Figure 5) is the most potent natural steroid. It contains two hydroxy functions: one in the phenolic 3 position and one in the 17 position. With these synthetic handles, three possible CDSs can be designed attaching the targetor at the 17-, at the 3-, or at both positions. Attachment at either position, but especially at the 17 position, should greatly decrease the pharmacological activity of $E_2$, because these esters are known not to interact with estrogen receptors [114].

Since its first synthesis in 1986 [115], $E_2$-CDS (5) has been investigated in several models [116–130]. *In vitro* studies with rat organ homogenates as the test matrix indicated half-lives of 156.6 min, 29.9 min, and 29.2 min (T at the 17 position) in plasma, liver, and brain homogenates, respectively [115]. Thus, $E_2$-CDS is converted to the corresponding quaternary form ($T^+$-$E_2$) (6) faster in the tissue homogenates than in plasma. This is consistent with the hypothesis of a membrane-bound enzyme, such as the members of the NADH transhydrogenase family, acting as oxidative catalyst. These studies also indicated a very slow production of $E_2$ from $T^+$-$E_2$, suggesting a possible slow and sustained release of estradiol from brain deposits of $T^+$-$E_2$.

To detect doses of $E_2$-CDS (5), $T^+$-$E_2$ (6), and $E_2$ (7) of physiological significance, a selective and sensitive method was needed. This problem was solved using a precolumn-enriching high-performance liquid chromatography system [122] that allowed accurate detection in plasma samples and organ homogenates with limits of 10, 20, and 50 ng/mL or ng/g for $T^+$-$E_2$, $E_2$-CDS, and $E_2$, respectively. This study proved that in rats, $E_2$ released from the $T^+$-$E_2$ intermediate formed after IV $E_2$-CDS administration has an elimination half-life of more than 200 h (Figure 7) and brain $E_2$-levels are elevated four to five times longer after administration than after simple estradiol treatment [122]. Proving effective targeting, another study also found that steroid levels between 1 and 16 days after $E_2$-CDS treatment were more than 12-fold greater in brain samples than in plasma samples [125]. Studies in orchidectomized rats proved that a single IV injection of $E_2$-CDS (3 mg/kg) suppressed LH secretion by 88%, 86%, and 66% relative to dimethyl sulfoxide (DMSO) controls at 12, 18, and 24 days, respectively, and that $E_2$ levels were not elevated relative to the DMSO control at any sampling time [118]. A single IV administration of doses as low as 0.5 mg/kg to ovariectomized rats induced prolonged (3–6 weeks) pharmacological effects as measured

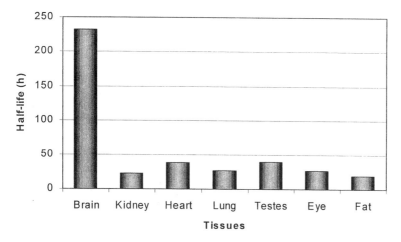

**Figure 7** Elimination half-lives in various tissues for the $T^+-E_2$ (6) formed after IV administration of 38.1 μmol/kg $E_2$-CDS (5) in rats (Data are from Ref. 122.)

by LH suppression [116, 118, 125], reduced rate of weight gain [119, 123–125], or, in castrated male rats, re-establishment of copulatory behavior [117]. A large number of other encouraging results have been obtained in various animal models and phase I/II clinical trials; most of them have been reviewed previously [74, 126, 128]. Clinical evaluations suggest a potent central effect with only marginal elevations in systemic estrogen levels; therefore, $E_2$-CDS may become a useful and safe therapy for menopausal symptoms or for estrogen-dependent cognitive effects.

Recently, $E_2$-CDS also was shown to provide encouraging neuroprotective effects. In ovariectomized rats, pretreatment with $E_2$-CDS decreased the mortality caused by middle cerebral artery (MCA) occlusion from 65–16% [129]. Even when administered 40 or 90 min after MCA occlusion, $E_2$-CDS reduced the area of ischemia by 45–90% or 31%, respectively. Another recent study provided evidence that treatment with $E_2$-CDS can protect cholinergic neurons in the medial septum from lesion-induced degeneration [130].

## V. CONCLUSIONS

Retrometabolic approaches are novel systematic methods aimed to improve the therapeutic index by a thorough integration of structure-activity and structure-metabolism relationships in the drug design process. The particular advantage of these approaches is to enhance, sometimes very significantly, drug targeting to the site of action. They include two distinct approaches to design soft drugs and chemical delivery systems, respectively. Soft drugs and chemical delivery systems are opposite in terms of how they achieve their drug-targeting role, but they have in common the basic concept of designed metabolism to control drug action and targeting. For CDSs, the molecule is designed to be inactive and to undergo strategic enzymatic activation to release the active agent only at the site of action. Delivery of this kind was successfully achieved to the brain, to the eye, and to other organs such as lungs. On the other hand, soft drugs are intrinsically potent new drugs that are strategically deactivated after they achieve their therapeutic role. These approaches are general in nature and can be applied to essentially all drug classes. To illustrate the concepts, two of the more successful examples, loteprednol etabonate, a soft steroid, and estradiol-CDS, were reviewed in more detail.

## REFERENCES

1. K. J. Widder, A. E. Senyei, and D. F. Rannes, Magnetically responsive microspheres and other carriers for the biophysical targeting of antitumor agents, *Adv. Pharmacol. Chemother. 16*:213 (1979).
2. G. Poste, R. Kirsh, and T. Koestler, The challenge of liposome targeting in vivo, *Liposome Technology*, Vol. 3 (G. Gregoriadis, ed.), CRC Press, Boca Raton, 1984, p. 1.
3. M. S. Poznansky and M. S. Juliano, Biological approaches to the controlled delivery of drugs: a critical review, *Pharmacol. Rev. 36*:277 (1984).
4. C. R. Gardner, Potential and limitations of drug targeting: an overview, *Biomaterials 6*:153 (1985).
5. E. E. Tomlinson and S. S. E. Davis, *Site-Specific Drug Delivery: Cell Biology, Medical, and Pharmaceutical Aspects*, John Wiley and Sons, New York, 1987.
6. V. V. Ranade, Drug delivery systems. 1. Site-specific drug delivery using liposomes as carriers, *J. Clin. Pharmacol. 29*:685 (1989).
7. D. K. F. Meijer, R. W. Jansen, and G. Molema, Drug targeting systems for antiviral agents: options and limitations, *Antiviral Res. 18*:215 (1992).
8. S. M. Moghimi, (ed.), Themed issue. Targeting of drugs and delivery systems, *Adv. Drug Deliv. Rev. 17*:1 (1995).
9. D. R. Friend and S. Pangburn, Site-specific drug delivery, *Med. Res. Rev. 7*:53 (1987).
10. E. Tomlinson, Theory and practice of site-specific drug delivery, *Adv. Drug. Del. Rev. 1*:87 (1987).
11. N. Bodor, Z. Gabanyi, and C.-K. Wong, A new method for the estimation of partition coefficient, *J. Am. Chem. Soc. 111*:3783 (1989).
12. N. Bodor, A. Harget, and M.-J. Huang, Neural network studies. 1. Estimation of the aqueous solubility of organic compounds, *J. Am. Chem. Soc. 113*:9480 (1991).

13. N. Bodor and M.-J. Huang, An extended version of a novel method for the estimation of partition coefficients, *J. Pharm. Sci. 81*:272 (1992).

14. N. Bodor and M.-J. Huang, A new method for the estimation of the aqueous solubility of organic compounds, *J. Pharm. Sci. 81*:954 (1992).

15. N. Bodor and P. Buchwald, Molecular size based approach to estimate partition properties for organic solutes, *J. Phys. Chem. B 101*:3404 (1997).

16. P. Buchwald and N. Bodor, Octanol-water partition of nonzwitterionic peptides: Predictive power of a molecular size based model, *Proteins 30*:86 (1998).

17. P. Buchwald and N. Bodor, Molecular size-based model to describe simple organic liquids, *J. Phys. Chem. B 102*:5715 (1998).

18. P. Buchwald and N. Bodor, Octanol-water partition: Searching for predictive models, *Curr. Med. Chem. 5*:353 (1998).

19. N. Bodor and M.-J. Huang, Computer-aided design of new drugs based on retrometabolic concepts, *Computer-Aided Molecular Design. Applications in Agrochemicals, Materials, and Pharmaceuticals* (C. H. Reynolds, M. K. Holloway, and H. K. Cox, eds.), American Chemical Society, Washington, DC, 1994, p. 98.

20. N. Bodor and M.-J. Huang, Computational approaches to the design of safer drugs and their molecular properties, *Computational Chemistry: Reviews of Current Trends*, Vol. 1 (J. Leszczynski, ed.); World Scientific, Singapore, 1996, p. 219.

21. N. Bodor, Design of biologically safer chemicals, *Chemtech 25(10)*:22 (1995).

22. N. Bodor, P. Buchwald, and M.-J. Huang, Computer assisted design of new drugs based on retrometabolic concepts, *SAR QSAR Environ. Res. 8*:41 (1998).

23. N. Bodor, P. Buchwald, and M.-J. Huang, The role of computational techniques in retrometabolic drug design strategies, *Computational Molecular Biology* (J. Leszczynski, ed.), Elsevier, Amsterdam, 1999, p. 569.

24. N. Bodor, Novel approaches for the design of membrane transport properties of drugs, *Design of Biopharmaceutical Properties through Prodrugs and Analogs* (E. B. Roche, ed.), Academy of Pharmaceutical Sciences, Washington, D.C, 1977, p. 98.

25. N. Bodor, Soft drugs: strategies for design of safer drugs, *Strategy in Drug Research. Proceedings of the 2nd IUPAC-IUPHAR Symposium on Research, Noordwijkerhout, The Netherlands* (J. A. K. Buisman, ed.); Elsevier, Amsterdam, 1982, p. 137.

26. N. Bodor, Novel approaches to the design of safer drugs: Soft drugs and site-specific chemical delivery systems, *Adv. Drug Res. 13*:255 (1984).

27. N. Bodor, The soft drug approach, *Chemtech 14*:28 (1984).

28. N. Bodor, Soft drugs, *Encyclopedia of Human Biology*, Vol. 7 (R. Dulbecco, ed.), Academic Press, San Diego, 1991, p. 101.

29. N. Bodor and P. Buchwald, Drug targeting via retrometabolic approaches, *Pharmacol. Ther. 76*:1 (1997).

30. N. Bodor, J. J. Kaminski, and S. Selk, Soft drugs. 1. Labile quaternary ammonium salts as soft antimicrobials, *J. Med. Chem. 23*:469 (1980).

31. N. Bodor and J. J. Kaminski, Soft drugs. 2. Soft alkylating compounds as potential antitumor agents, *J. Med. Chem. 23*:566 (1980).

32. N. Bodor, R. Woods, C. Raper, P. Kearney, and J. Kaminski, Soft drugs. 3. A new class of anticholinergic agents, *J. Med. Chem. 23*:474 (1980).

33. N. Bodor, Designing safer drugs based on the soft drug approach, *Trends Pharmacol. Sci. 3*:53 (1982).
34. A. Korolkovas, *Essentials of Medicinal Chemistry*, 2nd ed, Wiley & Sons, New York, 1988.
35. T. Nogrady, *Medicinal Chemistry. A Biochemical Approach*, 2nd ed, Oxford University Press, New York, 1988.
36. J. R. Gillette, Effects of induction of cytochrome P-450 enzymes on the concentration of foreign compounds and their metabolites and on the toxicological effects of these compounds. *Drug Metab. Rev. 10*:59 (1979).
37. G. J. Mannering, Hepatic cytochrome P-450-linked drug-metabolizing systems, *Concepts in Drug Metabolism part B* (B. Testa and P. Jenner, eds.); Marcel Dekker, Inc, New York, 1981, p. 53.
38. N. Bodor and M. Varga, Effect of a novel soft steroid on the wound healing of rabbit cornea, *Exp. Eye Res. 50*:183 (1990).
39. P. Druzgala, G. Hochhaus, and N. Bodor, Soft drugs. 10. Blanching activity and receptor binding affinity of a new type of glucocorticoid: loteprednol etabonate, *J. Steroid Biochem. 38*:149 (1991).
40. P. Druzgala, W.-M. Wu, and N. Bodor, Ocular absorption and distribution of loteprednol etabonate, a soft steroid, in rabbit eyes, *Curr. Eye Res. 10*:933 (1991).
41. M. Alberth, W.-M. Wu, D. Winwood, and N. Bodor, Lipophilicity, solubility and permeability of loteprednol etabonate: a novel, soft anti-inflammatory corticosteroid, *J. Biopharm. Sci. 2*:115 (1991).
42. N. S. Bodor, S. T. Kiss-Buris, and L. Buris, Novel soft steroids: effects on cell growth in vitro and on wound healing in the mouse, *Steroids 56*:434 (1991).
43. N. Bodor, N. Bodor, and W.-M. Wu, A comparison of intraocular pressure elevating activity of loteprednol etabonate and dexamethasone in rabbits, *Curr. Eye Res. 11*: 525 (1992).
44. N. Bodor, T. Loftsson, and W.-M. Wu, Metabolism, distribution, and transdermal permeability of a soft corticosteroid, loteprednol etabonate, *Pharm. Res. 9*:1275 (1992).
45. G. Hochhaus, L.-S. Chen, A. Ratka, P. Druzgala, J. Howes, N. Bodor, and H. Derendorf, Pharmacokinetic characterization and tissue distribution of the new glucocorticoid soft drug loteprednol etabonate in rats and dogs, *J. Pharm. Sci. 81*:1210 (1992).
46. T. Loftsson and N. Bodor, The pharmacokinetics and transdermal delivery of loteprednol etabonate and related soft steroids, *Adv. Drug Del. Rev. 14*:293 (1994).
47. N. Bodor, T. Murakami, and W.-M. Wu, Soft drugs. 18. Oral and rectal delivery of loteprednol etabonate, a novel soft corticosteroid, in rats-for safer treatment of gastrointestinal inflammation, *Pharm. Res. 12*:869 (1995).
48. N. Bodor, W.-M. Wu, T. Murakami, and S. Engel, Soft drugs. 19. Pharmacokinetics, metabolism and excretion of a novel soft corticosteroid, loteprednol etabonate, in rats, *Pharm. Res. 12*:875 (1995).
49. I. K. Reddy, M. A. Khan, W.-M. Wu, and N. S. Bodor, Permeability of a soft steroid, loteprednol etabonate, through an excised rabbit cornea. *J. Ocul. Pharmacol. Ther. 12*:159 (1996).
50. J. D. Bartlett, J. F. Howes, N. R. Ghormley, J. F. Amos, R. Laibovitz, and B.

Horwitz, Safety and efficacy of loteprednol etabonate for treatment of papillae in contact lens-associated giant papillary conjunctivitis, *Curr. Eye Res. 12*:313 (1993).

51. J. D. Bartlett, B. Horwitz, R. Laibovitz, and J. F. Howes, Intraocular pressure response to loteprednol etabonate in known steroid responders, *J. Ocul. Pharmacol.* 9:157 (1993).

52. J. F. Howes, H. Baru, M. Vered, and R. Neumann, Loteprednol etabonate: comparison with other steroids in two models of intraocular inflammation, *J. Ocul. Pharmacol. 10*:289 (1994).

53. P. Asbell and J. Howes, A double-masked, placebo-controlled evaluation of the efficacy and safety of loteprednol etabonate in the treatment of giant papillary conjunctivitis, *CLAO J. 23*:31 (1997).

54. S. J. Dell, D. G. Shulman, G. M. Lowry, and J. Howes, A controlled evaluation of the efficacy and safety of loteprednol etabonate in the prophylactic treatment of seasonal allergic conjunctivitis, *Am. J. Ophthalmol. 123*:791 (1997).

55. M. H. Friedlaender and J. Howes, A double-masked, placebo-controlled evaluation of the efficacy and safety of loteprednol etabonate in the treatment of giant papillary conjunctivitis, *Am. J. Ophthalmol. 123*:455 (1997).

56. S. J. Dell, G. M. Lowry, J. A. Northcutt, J. Howes, G. D. Novack, and K. Hart, A randomized, double-masked, placebo-controlled parallel study of 0.2% loteprednol etabonate in patients with seasonal allergic conjunctivitis, *J. Allergy Clin. Immunol. 102*:251 (1998).

57. J. Howes and G. D. Novack, Failure to detect systemic levels and effects of loteprednol etabonate and its metabolite, PJ-91, following chronic ocular administration, *J. Ocul. Pharmacol. Ther 14*:153 (1998).

58. G. D. Novack, J. Howes, R. S. Crockett, and M. B. Sherwood, Change in intraocular pressure during long-term use of loteprednol etabonate, *J. Glaucoma 7*:266 (1998).

59. N. Bodor, The application of soft drug approaches to the design of safer corticosteroids, *Topical Corticosteroid Therapy: A Novel Approach to Safer Drugs* (E. Christophers, A. M. Kligman, E. Schöpf, and R. B. Stoughton, eds.); Raven Press Ltd, New York, 1988, p. 13.

60. N. Bodor, Design of novel soft corticosteroids, *Topical Glucocorticoids with Increased Benefit-Risk Ratio*, Vol. 21 (H. Korting, ed.); Karger AG, Basel, 1993, p. 11.

61. P. Druzgala and N. Bodor, Regioselective O-alkylation of cortienic acid and synthesis of a new class of glucocorticoids containing a 17α-alkoxy, a 17α-(1'-alkoxyethyloxy), a 17α-alkoxymethyloxy, or a 17α-methylthiomethyloxy function, *Steroids 56*:490 (1991).

62. J. E. Dickerson, Jr., E. Dotzel, and A. F. Clark, Steroid-induced cataract: New perspectives from in vitro and lens culture studies, *Exp. Eye Res. 65*:507 (1997).

63. K. Heyns and W. Koch, Über die Bildung eines Aminozuckers aus d-Fruktose und Ammoniak, *Z. Naturforsch. B 7B*:486 (1952).

64. R. Bucala, J. Fishman, and A. Cerami, Formation of covalent adducts between cortisol and 16α-hydroxyestrone and protein: Possible role in the pathogenesis of cortisol toxicity and systemic lupus erythematosus, *Proc. Natl. Acad. Sci. USA 79*: 3320 (1982).

65. R. Bucala, M. Gallati, S. Manabe, E. Cotlier, and A. Cerami, Glucocorticoid-lens

protein adducts in experimentally induced steroid cataracts, *Exp. Eye Res. 40*:853 (1985).

66. R. C. Urban, Jr. and E. Cotlier, Corticosteroid-induced cataracts, *Surv. Ophthalmol. 31*:102 (1986).

67. N. Bodor, Soft drugs: principles and methods for the design of safe drugs, *Med. Res. Rev. 3*:449 (1984).

68. N. Bodor, Prodrugs versus soft drugs, *Design of Prodrugs* (H. Bundgaard, ed.); Elsevier, Amsterdam, 1985, p. 333.

69. N. Bodor, Designing safer ophthalmic drugs, *Trends in Medicinal Chemistry '88 Proceeding of the Xth International Symposium on Medicinal Chemistry* (H. van der Goot, G. Domany, L. Pallos, and H. Timmerman, eds.); Elsevier, Amsterdam, 1989, p. 145.

70. The Loteprednol Etabonate Postoperative Inflammation Study Group 2, A double-masked, placebo-controlled evaluation of 0.5% loteprednol etabonate in the treatment of postoperative inflammation, *Ophthalmology 105*:1780 (1998).

71. Advanstar Communications Inc., *Ophthalmology Times*, September (1998).

72. N. Bodor and M. E. Brewster, Problems of delivery of drugs to the brain, *Pharmacol. Ther. 19*:337 (1983).

73. N. Bodor, Redox drug delivery systems for targeting drugs to the brain, *Ann. NY Acad. Sci. 507*:289 (1987).

74. N. Bodor and M. E. Brewster, Chemical delivery systems, *Targeted Drug Delivery*, Vol. 100 (R. L. Juliano, ed.), Springer-Verlag, Berlin, 1991, p. 231.

75. N. Bodor and P. Buchwald, Recent advances in the brain targeting of neuropharmaceuticals by chemical delivery systems, *Adv. Drug Delivery Rev. 36*:229 (1999).

76. H. Bundgaard (ed.), *Design of Prodrugs*, Elsevier Science, Amsterdam, 1985.

77. N. Bodor and J. J. Kaminski, Prodrugs and site-specific chemical delivery systems, *Annu. Rep. Med. Chem. 22*:303 (1987).

78. V. J. Stella (ed.), Themed issue. Low molecular weight prodrugs, *Adv. Drug Deliv. Rev. 19*:111 (1996).

79. N. Bodor, New methods of drug targeting, *Trends in Medicinal Chemistry '90* (S. Sarel, R. Mechoulam, and I. Agranat, eds.), Blackwell Scientific Publications, Oxford, 1992, p. 35.

80. N. Bodor, Drug targeting and retrometabolic drug design approaches, *Adv. Drug Delivery Rev. 14*:157 (1994).

81. N. Bodor, Retrometabolic drug design concepts in ophthalmic target-specific drug delivery, *Adv. Drug Delivery Rev. 16*:21 (1995).

82. N. Bodor, Targeting drugs to the brain by sequential metabolism, *Discovery of Novel Opioid Medications*, Vol. 147 (R. Rapaka and H. Sorer, eds.); NIDA, Rockville, 1995, p. 1.

83. N. Bodor, L. Prokai, W.-M. Wu, H. H. Farag, S. Jonnalagadda, M. Kawamura, and J. Simpkins, A strategy for delivering peptides into the central nervous system by sequential metabolism, *Science 257*:1698 (1992).

84. N. Bodor, H. H. Farag, G. Somogyi, W.-M. Wu, M. D. C. Barros, and L. Prokai, Ocular-specific delivery of timolol by sequential bioactivation of its oxime and methoxime analogs, *J. Ocul. Pharmacol. 13*:389 (1997).

85. N. Bodor and L. Prokai, Molecular packaging: peptide delivery to the central ner-

vous system by sequential metabolism, *Peptide-Based Drug Design: Controlling Transport and Metabolism* (M. Taylor and G. Amidon, eds.), American Chemical Society, Washington, DC, 1995, p. 317.

86. K. Prokai-Tatrai, L. Prokai, and N. Bodor, Brain-targeted delivery of a leucine-enkephalin analogue by retrometabolic design, *J. Med. Chem. 39*:4775 (1996).

87. M. Saah, W.-M. Wu, K. Eberst, E. Marvanyos, and N. Bodor, Design, synthesis, and pharmacokinetic evaluation of a chemical delivery system for drug targeting to lung tissue, *J. Pharm. Sci. 85*:496 (1996).

88. T. Ishikura, T. Senou, H. Ishihara, T. Kato, and T. Ito, Drug delivery to the brain. DOPA prodrugs based on a ring-closure reaction to quaternary thiazolium compounds, *Int. J. Pharm. 116*:51 (1995).

89. E. Pop, Optimization of the properties of brain specific chemical delivery systems by structural modifications, *Curr. Med. Chem. 4*:279 (1997).

90. G. Somogyi, S. Nishitani, D. Nomi, P. Buchwald, L. Prokai, and N. Bodor, Targeted drug delivery to the brain via phosphonate derivatives. I. Design, synthesis, and evaluation of an anionic chemical delivery system for testosterone, *Int. J. Pharm. 166*:15 (1998).

91. G. Somogyi, P. Buchwald, D. Nomi, L. Prokai, and N. Bodor, Targeted drug delivery to the brain via phosphonate derivatives. II. Anionic chemical delivery system for zidovudine (AZT), *Int. J. Pharm. 166*:27 (1998).

92. J. Rydström, J. B. Hoek, and L. Ernster, The nicotinamide nucleotide transhydrogenases, *The Enzymes*, Vol. 13 (P. D. Boyer, ed.), Academic Press, New York, 1976.

93. J. B. Hoek and J. Rydström, Physiological roles of nicotinamide nucleotide transhydrogenase, *Biochem. J. 254*:1 (1988).

94. N. Bodor, M. E. Brewster, and J. J. Kaminski, Reactivity of biologically important reduced pyridines. Part III. Energetics and mechanism of hydride transfer between 1-methyl-1,4-dihydronicotinamide and the 1-methylnicotinamide cation, a theoretical study, *J. Mol. Struct. (Theochem.) 206*:315 (1990).

95. M. E. Brewster, K. S. Estes, R. Perchalski, and N. Bodor, A dihydropyridine conjugate which generates high and sustained levels of the corresponding pyridinium salt in the brain does not exhibit neurotoxicity in cynomolgus monkeys, *Neurosci. Lett. 87*:277 (1988).

96. N. Bodor, R. G. Roller, and S. J. Selk, Elimination of a quaternary pyridinium salt delivered as its dihydropyridine derivative from brain of mice, *J. Pharm. Sci. 67*: 685 (1978).

97. E. Palomino, D. Kessel, and J. P. Horwitz, A dihydropyridine carrier system for sustained delivery of 1′, 3′-dideoxynucleosides to the brain, *J. Med. Chem. 32*:622 (1989).

98. L. Prokai, X.-D. Ouyang, W.-M. Wu, and N. Bodor, Chemical delivery system to transport a pyroglutamyl peptide to the central nervous system, *J. Am. Chem. Soc. 116*:2643 (1994).

99. N. Bodor and P. Buchwald, All in the mind, *Chem. Br. 34(1)*:36 (1998).

100. P. Chen, N. Bodor, W.-M. Wu, and L. Prokai, Strategies to target kyotorphin analogues to the brain, *J. Med. Chem. 41*:3773 (1998).

101. J. Szejtli, Medicinal applications of cyclodextrins, *Med. Res. Rev. 14*:353 (1994).

102. M. E. Brewster, K. E. Estes, T. Loftsson, R. Perchalski, H. Derendorf, G. Mullersman, and N. Bodor, Improved delivery through biological membranes. XXXI. Solubilization and stabilization of an estradiol chemical delivery system by modified β-cyclodextrins, *J. Pharm. Sci.* 77:981 (1988).

103. W. R. Anderson, J. W. Simpkins, M. E. Brewster, and N. Bodor, Brain-enhanced delivery of testosterone using a chemical delivery system complexed with 2-hydroxypropyl-β-cyclodextrin, *Drug Des. Del.* 2:287 (1988).

104. K. Raghavan, T. Loftsson, M. E. Brewster, and N. Bodor, Improved delivery through biological membranes. XLV. Synthesis, physical-chemical evaluation, and brain uptake studies of 2-chloroethyl nitrosourea delivery system, *Pharm. Res. 9*: 743 (1992).

105. E. Pop, T. Loftsson, and N. Bodor, Solubilization and stabilization of a benzylpenicillin chemical delivery system by 2-hydroxypropyl-β-cyclodextrin, *Pharm. Res.* 8:1044 (1991).

106. N. M. Kaplan, Cardiovascular complications of oral contraceptives, *Annu. Rev. Med. 29*:31 (1978).

107. K. Fotherby, Oral contraceptives, lipids and cardiovascular disease, *Contraception 31*:367 (1985).

108. R. A. Lobo, Benefits and risks of estrogen replacement therapy, *Am. J. Obstet. Gynecol. 173*:982 (1995).

109. J. D. Yager and J. G. Liehr, Molecular mechanism of estrogen carcinogenesis, *Annu. Rev. Pharmacol. Toxicol. 36*:203 (1996).

110. G. V. Upton, Therapeutic considerations in the management of the climacteric, *J. Reprod. Med. 29*:71 (1984).

111. A. Maggi and J. Perez, Role of female gonadal hormones in the CNS: clinical and experimental aspects, *Life Sci. 37*:893 (1985).

112. K. Yaffe, G. Sawaya, I. Lieberburg, and D. Grady, Estrogen therapy in postmenopausal women—Effects on cognitive function and dementia, *JAMA—J. Am. Med. Assoc. 279*:688 (1998).

113. L. Gopinath, Outsmarting Alzheimer's disease, *Chem. Br. 34(5)*:38 (1998).

114. L. Janocko, J. M. Lamer, and R. B. Hochberg, The interaction of C-17 esters of estradiol with the estrogen receptor, *Endocrinology 114*:1180 (1984).

115. N. Bodor, J. McCornack, and M. E. Brewster, Improved delivery through biological membranes. XXII. Synthesis and distribution of brain-selective estrogen delivery systems, *Int. J. Pharm. 35*:47 (1987).

116. J. W. Simpkins, J. McCornack, K. S. Estes, M. E. Brewster, E. Shek, and N. Bodor, Sustained brain-specific delivery of estradiol causes long-term suppression of luteinizing hormone secretion, *J. Med. Chem. 29*:1809 (1986).

117. W. R. Anderson, J. W. Simpkins, M. E. Brewster, and N. Bodor, Evidence for the reestablishment of copulatory behavior in castrated male rats with a brain-enhanced estradiol-chemical delivery system, *Pharmacol. Biochem. Behav. 27*:265 (1987).

118. K. S. Estes, M. E. Brewster, J. W. Simpkins, and N. Bodor, A novel redox system for CNS-directed delivery of estradiol causes sustained LH suppression in castrate rats, *Life Sci. 40*:1327 (1987).

119. K. S. Estes, M. E. Brewster, and N. S. Bodor, A redox system for brain targeted estrogen delivery causes chronic body weight decrease in rats, *Life Sci. 42*:1077 (1988).

120. W. R. Anderson, J. W. Simpkins, M. E. Brewster, and N. Bodor, Effects of a brain-enhanced chemical delivery system for estradiol on body weight and serum hormones in middle-aged male rats, *Endocr. Res. 14*:131 (1988).

121. M. E. Brewster, K. S. Estes, and N. Bodor, Improved delivery through biological membranes. XXXII. Synthesis and biological activity of brain-targeted delivery systems for various estradiol derivatives, *J. Med. Chem. 31*:244 (1988).

122. G. Mullersman, H. Derendorf, M. E. Brewster, K. S. Estes, and N. Bodor, High performance liquid chromatographic assay of a central nervous system (CNS)-directed estradiol chemical delivery system and its application after intravenous administration in rats, *Pharm. Res. 5*:172 (1988).

123. J. W. Simpkins, W. R. Anderson, R. Dawson, Jr., E. Seth, M. Brewster, K. S. Estes, and N. Bodor, Chronic weight loss in lean and obese rats with a brain-enhanced chemical delivery system for estradiol, *Physiol. Behav. 44*:573 (1988).

124. J. W. Simpkins, W. R. Anderson, R. Dawson, Jr., and N. Bodor, Effects of a brain-enhanced chemical delivery system for estradiol on body weight and food intake in intact and ovariectomized rats, *Pharm. Res. 6*:592 (1989).

125. D. K. Sarkar, S. J. Friedman, S. S. C. Yen, and S. A. Frautschy, Chronic inhibition of hypothalamic-pituitary-ovarian axis and body weight gain by brain-directed delivery of estradiol-17β in female rats, *Neuroendocr. 50*:204 (1989).

126. M. E. Brewster, J. W. Simpkins, and N. Bodor, Brain-targeted delivery of estrogens, *Rev. Neurosci. 2*:241 (1990).

127. M. E. Brewster, M. S. M. Bartruff, W. R. Anderson, P. J. Druzgala, N. Bodor, and E. Pop. Effect of molecular manipulation on the estrogenic activity of a brain-targeting estradiol chemical delivery system, *J. Med. Chem. 37*:4237 (1994).

128. K. S. Estes, M. E. Brewster, and N. Bodor, Evaluation of an estradiol chemical delivery system (CDS) designed to provide enhanced and sustained hormone levels in the brain, *Adv. Drug Del. Rev. 14*:167 (1994).

129. J. W. Simpkins, G. Rajakumar, Y.-Q. Zhang, C. E. Simpkins, D. Greenwald, C. J. Yu, N. Bodor, and A. L. Day, Estrogens may reduce mortality and ischemic damage caused by middle cerebral artery occlusion in the female rat, *J. Neurosurg. 87*:724 (1997).

130. O. Rabbani, K. S. Panickar, G. Rajakumar, M. A. King, N. Bodor, E. M. Meyer, and J. W. Simpkins, 17β-estradiol attenuates fimbrial lesion-induced decline of ChAT-immunoreactive neurons in the rat medial septum, *Exp. Neurol. 146*:179 (1997).

# 7

# Neoglyco- and Neopeptide Albumins for Cell-Specific Delivery of Drugs to Chronically Diseased Livers

**Leonie Beljaars, Barbro N. Melgert, Dirk K. F. Meijer, Grietje Molema, and Klaas Poelstra**
*Groningen University Institute for Drug Exploration, Groningen, The Netherlands*

## I. INTRODUCTION

Patients with chronic liver diseases, such as viral hepatitis, liver cirrhosis, and hepatocellular carcinoma, are often exposed to prolonged treatment with drugs. Although the desired effects of the applied drugs are frequently observed, often the adverse effects of these drugs limit their chronic application. Such drugs are suitable candidates to be coupled to drug carriers for a target-cell selective targeting. Drug targeting can be applied for therapeutic and for diagnostic purposes. The primary aim of drug targeting for therapeutic use is to manipulate the whole body distribution of drugs, that is, to prevent distribution to nontarget cells and concomitantly increase the drug concentration in target cells [1–5]. Carrier molecules are designed for selective cellular uptake, taking advantage of specific receptors or binding sites present on the surface membrane of the target cell. In addition to cellular specificity, the extracellular release of drugs from carriers, the internalization and intracellular routing of the carrier, and the intracellular release rate of the coupled drug are crucial factors determining the success of targeting.

Efficient delivery of the drug-targeting preparations to the target cell depends on the chemical and physical properties of the conjugate and on the make up of various cell types in the body [6, 7]. The physicochemical factors may involve size, charge, hydrophobicity, and conformational aspects both of the drug and the drug carrier. Physiological factors that play a role are the injection site and administration route, the (micro)circulation at the injection site and in the target tissue, the cell types present in the target tissue, and the arrangement and the accessibility of the target cells within the tissue. Importantly, the pathological state of the tissue often determines the density of the target receptors. In the liver, the uptake rate in cells depends on the endocytotic capacity of the different hepatic cells (parenchymal cells, Kupffer cells, endothelial cells, and hepatic stellate cells) in relation to the administered dose and on the structural characteristics of the endothelial cell lining in relation to the size of the carrier.

An obvious but relevant question is: What is the benefit of targeting drugs into the liver realizing that most drugs already achieve high hepatic concentrations? Indeed, the liver is the major organ in the body equipped for uptake, detoxification, and excretion of xenobiotics into bile by means of carrier-mediated mechanisms or a versatile apparatus for metabolic transformations. As a consequence, many drugs are rapidly cleared from the blood and display high first-pass clearances by the liver. It should be realized, however, that the total hepatic uptake is predominantly caused by hepatocytes, whereas Kupffer cells can largely contribute to hepatic uptake of particulate material (phagocytosis). Therefore, drugs that enter the liver as such or in the form of covalent carrier conjugates will not necessarily reach the required cell type. Moreover, if drugs are accumulated in the liver, their residence time in the organ and within the favored cell type is influenced by the excretion, metabolism, and reflux to the bloodstream. Therefore, the challenge is to obtain selective accumulation of drugs in one specific cell type and to sustain intracellular (therapeutic) levels for longer periods. In addition, the advantage of the relatively good accessibility of the liver allows many opportunities to evaluate the concept of drug targeting. This review focuses on liver cell-specific drug-targeting preparations in relation to chronic liver diseases with emphasis on the albumin type of carrier.

## II. THE LIVER

The liver (about 2–5% of total body weight) is the organ in the body that is involved in metabolic homeostasis, which includes processing of vitamins, amino acids, carbohydrates and lipids, the synthesis of serum proteins, biotransformation of circulating metabolites, detoxification and excretion of endogenous substrates and xenobiotics into bile, and phagocytosis of foreign proteins and particles, such as viruses and bacteria or bacteria products such as endotoxin. The resident cell types present in the liver, the parenchymal cells, the nonparenchymal cells (endothelial cells, Kupffer cells, hepatic stellate cells, and pit cells), and the bile duct epithelial cells, all have a specific localization within the liver lobule (Figure 1), which is directly related to their functions.

The rest of the liver volume (about 15%) consists of intravascular space, the space of Disse, lymphatic vessels, and extracellular matrix molecules [8]. These matrix proteins, located predominantly in the space of Disse and around blood vessels, consist mainly of basement membrane molecules (collagen type IV, laminin, and fibronectin) and fibronectin) and small amounts of collagen type I, III, VI, undulin, tenascin, and proteoglycans. The matrix proteins determine the specific phenotype and functions of many resident hepatic cells [9–11].

## A. Parenchymal Cells

Parenchymal cells (PC), or hepatocytes, originate from epithelial cells and represent most of the total number of liver cells (65%). Because of their relatively large size, hepatocytes are microscopically clearly visible after staining liver sections with hematoxylin and eosin (HE). Also, the hepatocytes store glycogen, which can be identified histochemically with periodic acid–Schiff (PAS) reagent. The hepatocytes are primarily responsible for the uptake of endogenous products and xenobiotics at the sinusoidal membrane of the cell and their subsequent metabolism and excretion into bile by means of the canalicular membrane.

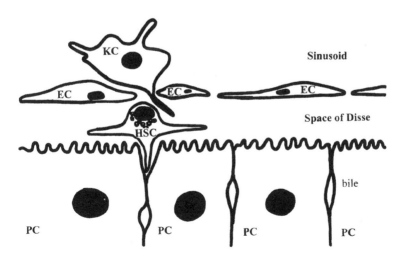

**Figure 1** Schematic representation of the architecture of the liver. Endothelial cells (EC) and Kupffer cells (KC) are located in close contact with the bloodstream, whereas hepatic stellate cells (HSC) and parenchymal cells (PC) reside in or behind the space of Disse.

The microvillous sinusoidal membrane accounts for 72% of the parenchymal cell surface. This membrane contains many specialized carrier proteins involved in the internalization of relatively small molecules (MW 200–1500), including cationic, uncharged, and anionic substrates (as reviewed in [12–14]. Furthermore, the sinusoidal membrane contains receptors that are involved in the endocytosis of glycoproteins and lipoproteins, hormones, and growth factors (for example epidermal growth factor). Of note, in addition to the low density lipoprotein (LDL) or high density lipoprotein (HDL) receptors, the uptake of lipoproteins by hepatocytes is mediated by scavenger receptors (type B1). This scavenger receptor specifically binds native and modified lipoproteins and anionic phospholipids but fails to bind the polyanions (like polyinosinic acid) that represent the classical ligands for class A scavenger receptor [15, 16]. A summary of receptors present on the sinusoidal membrane of hepatocytes is displayed in Table 1. The asialoglycoprotein receptor (ASGPR) on hepatocytes has been widely used to selectively target compounds to this cell type exploiting galactosylated proteins [1, 17, 18], see section VI.

## B. Kupffer Cells

Fifteen percent of all liver cells are Kupffer cells (KC). They constitute 80–90% of the fixed tissue macrophages of the reticuloendothelial system. Kupffer cells are predominantly found at the periportal region of the liver (zone 1), which is an ideal location to monitor the composition of blood entering the liver. They are the major site of phagocytosis of endotoxin, micro-organisms (parasites, bacteria, viruses), old blood cells, fibrin degradation products, immune complexes, particulate substances, and tumor cells, mostly by receptor-mediated endocytosis [19–23]. The receptors present on the cell membrane of Kupffer cells are summarized in Table 1. Kupffer cells are well equipped for their phagocytotic functions, because they contain many lysosomes and pinocytotic vesicles, a broad variety of lysosomal enzymes, together with ultrastructural features like microvilli, lamellapodia, filopodia [24, 25]. Furthermore, Kupffer cells are the major antigen-presenting cells in the liver [26, 27], a property that has to be considered in the design of Kupffer cell–targeted therapies (see Section IV). Another function of these cells is the ability to produce a variety of cytokines, growth factors, and other mediators. The production and secretion of biologically active compounds is strongly enhanced after activation of this cell type in pathological conditions (for reviews see [24, 28–31]). During endotoxemia the most important cytokines produced by Kupffer cells are TNF-$\alpha$, IL-1, and IL-6, whereas during fibrosis TGF-$\beta$1, TGF-$\alpha$, PDGF, TNF-$\alpha$, and IL-1 are prominent factors released by Kupffer cells. In addition, Kupffer cells generate reactive oxygen species and eicosanoids (65% of the total liver production) in response to an inflammatory stimulus.

Table 1 Receptors (R) Present on the Cell Surface of the Hepatic Cells That Are Involved in Protein Endocytosis and May Be Targets for Drug-targeting Preparations

| Hepatocytes | Kupffer cells | Endothelial cells | Hepatic stellate cells |
|---|---|---|---|
| Asialoglycoprotein-R (ASGPR, galactose recognition, size <10 nM) | Mannose/$N$-acetyl glucosamine-R | Mannose/$N$-acetyl glucosamine-R | M6P/IGF II-R |
| HDL-R | Galactose particle-R (galactose recognition, >10 nM) | Scavenger-R (Class AI and AII) | $\alpha_2$-Macroglobulin-R |
| LDL-R | Galactose specific-R (a.o. fucose recognition) | Fc-R (immune complexes) | Ferritin-R |
| IgA-R | Fc-R (immune complexes, opsonized material) | Matrix compounds (hyaluronan, fibronectin, denatured collagens, PIIINP) | Uroplasminogen-R |
| Scavenger-R (class BI) | Scavenger-R (class AI, AII, B1, MARCO, CD36, and macrosialin) | | Thrombin-R |
| Transferrin-R | LDL-R matrix compounds (fibronectin) | | RBP-R matrix compounds (integrins, collagen type VI, fibronectin, CD$_{44}$) |
| Insulin-R | Complement-R (C3b and C1q) LPS-R $\alpha_2$-macroglobulin-R | | |

By light microscop, Kupffer cells cannot be easily identified after histo-chemical staining with HE. However, incubation of liver sections with a buffered diaminobenzidin solution and hydrogen peroxide reveals the high peroxidase activity characteristic these cells [32]. By Immunohistochemistry, Kupffer cells can be distinguished from other hepatic cells or from infiltrating monocytes with the monoclonal antibody ED2 in rat tissue [33] or with the monoclonal antibody CD68 in human livers [34].

## C.  Endothelial Cells

Endothelial cells (EC) are from mesenchymal origin and represent about 20% of the total number of liver cells. They can be divided into two types: vascular endothelial cells and sinusoidal endothelial cells. In liver sections, vascular endo-thelial cells can be readily identified after HE staining in contrast to the sinusoidal endothelial cells that may be visualized with the monoclonal antibody HIS52 (anti-rat endothelial cell antigen-1) or in case of human livers with for instance anti-von Willebrand factor or anti-gp96 [35, 36]. Unlike the cells of the vascular endothelium, the sinusoidal liver endothelial cells lack an underlying fibrous basement membrane. In addition, the sinusoidal endothelial cell lining contains pores, called fenestrae, that are grouped in so-called sieve plates. These features allow direct contact between the cells located in the space of Disse and the plasma, which implicates an undisturbed exchange of molecules between blood and liver cells. Loss of the fenestrae and the formation of a collagenous basal membrane occurs in liver cirrhosis [31, 37, 38]. Whereas the basal membrane usually contains a thin layer of collagen type IV, in pathological conditions this is replaced by a matrix consisting mainly of collagen type I and III [39]. This may not only have consequences for the exchange of nutrients but also for drug-targeting preparations that are directed at hepatic cells. Similar to Kupffer cells, endothelial cells are activated in acute and chronic inflammatory conditions and contribute to the total hepatic production of eicosanoids, reactive oxygen species, cytokines, and growth factors. In case of liver fibrosis, IL-1, IL-6, PDGF, and TGF-$\beta$ are important mediators produced by endothelial cells [24, 29, 40]. Fur-thermore, endothelial cells are actively involved in the recruitment of inflamma-tory cells into the inflamed liver. Adhesion of circulating immune cells to endo-thelial cells is mediated by the expression of adhesion molecules, which are up-regulated by endotoxin and various cytokines [41–43]. During the fibrotic process, endothelial cells may also contribute to the fibrogenesis by the produc-tion of extracellular matrix compounds like collagen type IV, I, III [44, 45] and fibronectin [46].

Endothelial cells have a well-developed endocytotic capacity [24, 25, 47]. Several receptors at the endothelial cell membrane rapidly clear specific sub-stances from the blood. These endocytotic receptors include the scavenger receptors type AI and AII that recognize molecules with a net negative charge. These class A receptors are sensitive to polyinosinic acid in contrast to the class B scavenger receptors [21, 48, 49]. Other receptors present on the cell membrane of endothelial cells are mannose/$N$-acetyl glucosamine receptor [47, 50], receptors for matrix compounds (hyaluronan receptor, fibronectin receptor, and receptors recognizing denatured collagens) [51–53], and Fc receptors (involved in the clearance of IgG complexes) [54, 55]. Insulin, transferrin, and ceruloplasmin bind to specific receptors, enabling the transcytosis of these substances through endothelial cells [56–58]. In addition, receptors for growth factors and cytokines are present on the cell membrane of endothelial cells. Receptors expressed by hepatic endothelial cells are summarized in Table 1.

## D. Hepatic Stellate Cells

The hepatic stellate cell (HSC) has gained increasing interest for its participation in liver diseases, in particular liver cirrhosis [59–68]. HSC, formerly known as fat-storing cells, lipocytes, Ito cells, perisinusoidal or parasinusoidal cells, represent only 5–8% of the total number of liver cells in normal livers. These mesenchymal cells are situated in the space of Disse between the sinusoidal endothelial cells and the basolateral membranes of hepatocytes and maintain close contact with these cells through contractile cellular branches. This intimate association between stellate cells and the other hepatic cells facilitates intercellular transport of compounds and paracrine stimulation by soluble mediators. The major function of this cell type is the storage of vitamin A as reflected by their characteristic vitamin A–lipid droplets that can be visualized by fluorescence microscopy (328 nm). The liver contains 90% of the total body content of vitamin A, of which 75% is stored in HSC. Besides vitamin A storage, HSC are also involved in the synthesis of extracellular matrix proteins and matrix degrading enzymes and the regulation of the sinusoidal blood flow.

In normal healthy livers, HSC express a quiescent phenotype. They maintain a compact cell shape and contain many vitamin A–lipid droplets, and these cells display a low proliferative activity. Quiescent HSC are not readily discernable in tissue sections of normal liver stained with HE, but they can be visualized after gold/silver impregnation or after detection of the characteristic fat droplets by oil-red–O staining [61]. Immunohistochemical markers for rat HSC are desmin for the periportally localized cells [61, 69, 70] and glial fibrillary acidic protein (GFAP) for the pericentral cells [71, 72]. Another marker for stellate cells is vimentin, but this antibody stains other mesenchymal cells like Kupffer and endothelial cells as well [73]. In human livers, α-smooth muscle actin is the best immunohistochemical marker for identification of HSC [61, 74]. Besides zonal differences for the immunohistochemical markers, HSC also display zonal heterogeneity in normal livers with regard to the size, the amount of vitamin A–lipid droplets, and the amount or length of the cellular branches [70]. Near the portal area (zone 1), the HSC are small with short cellular branching, and they contain minute vitamin A–lipid droplets, whereas HSC located in zone 2 display extended cellular branching and abundantly store vitamin A–lipid droplets. Near the central vein in zone 3, HSC become more elongated, assuming a dendritic appearance, whereas vitamin A storage and desmin immunoreactivity are reduced or absent.

After liver injury, HSC acquire an activated phenotype. This activation is caused by continuous stimulation by reactive oxygen species, cytokines, and growth factors produced mainly by Kupffer cells, but also by endothelial cells, infiltrated inflammatory cells, and hepatocytes [75, 76]. This process is depicted in Figure 2. In addition to this paracrine activation, HSC are subjected to many autocrine loops that may lead to a perpetuating process [40, 62, 77]. The key mediators in these processes are TGF-β [39, 78, 79] and PDGF, the latter being the most potent mitogen for HSC identified so far [39, 80, 81]. Constant stimulation of HSC elicits chronic liver diseases like cirrhosis. The activation of hepatic stellate cells generally consists of two phases. In the initiation phase, the cell starts to proliferate and migrates toward regions of injury. Early changes in gene expression and phenotype make the HSC susceptible to cytokines and growth factors produced by other cells. As mentioned in the previous paragraph, oxidative stress is able to mediate the activation and proliferation of HSC. In the cascade of molecular events occurring in the cell, reactive oxygen species induce the activation of NFκB and the induction of c-myb nuclear expression [82]. In addition to these nuclear transcription factors, Lalazar et al. found *in vivo* an induction of the genes corresponding to zinc finger transcriptional regulatory protein (BTEB2) during early HSC activation in a rat model of liver fibrosis. Other genes induced at early HSC activation encode for type II TGF-β receptor, glutathione peroxidase, transferrin, and cellular retrotransposons [83]. These genes may be the targets for new therapeutical agents or may serve as markers to evaluate the effect of (targeted) antifibrotic drugs. In the second phase, the perpetuation phase, HSC become more activated and produce large amounts of extracellular matrix proteins. The transformed HSC, a myofibroblast-like cell, can be identified by a spindle-shaped smooth muscle cell appearance, loss of vitamin A–lipid droplets [84], an enlarged rough endoplasmatic reticulum, a high proliferative state, the production of many matrix proteins including collagen type I and III, and the expression of the cytoskeleton filament α-smooth muscle actin [85]. The most prominent marker to identify activated HSC immunohistochemically is therefore α-smooth muscle actin [61]. During activation, the main function of the hepatic stellate cells shifts from vitamin A storage to an active participation in the remodeling of the extracellular matrix. This includes the synthesis and secretion of various matrix proteins [86], the synthesis and secretion of matrix-degrading metalloproteinases (MMP) but also inhibitors of these proteases, the tissue inhibitors

inciting stimulus

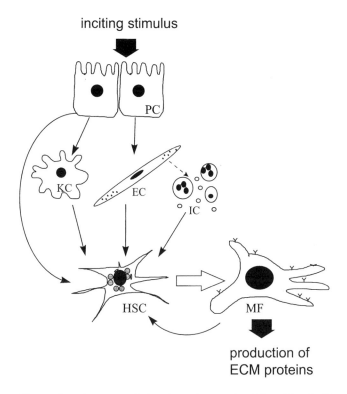

production of
ECM proteins

**Figure 2** Diagram outlining the pathogenesis of liver fibrosis. Damage to parenchymal cells (PC) results in the activation of Kupffer cells (KC) and endothelial cells (EC) and the influx of inflammatory cells (IC). These cells release growth factors, cytokines, and ROS that induce the activation and proliferation of hepatic stellate cells (HSC). HSC gradually transform into a myofibroblasts (MF), the major producers of extracellular matrix proteins (ECM).

of metalloproteinases (in particular TIMP-1 and -2) [86]. In addition, the production of several cytokines and growth factors [40] and the regulation of the sinusoidal lumen by endothelin [67] contribute to the pathogenesis of liver fibrosis (summarized in Table 2).

Hepatic stellate cell activation is also associated with the induction of other liver diseases. In livers of patients with chronic hepatitis, an increased number of activated HSC were detected [87]. Recently, an interaction of hepatitis C virus (HCV) with HSC was reported [88]. Furthermore, HSC activation is associated with the development of liver tumors, for instance, hepatocellular carcinomas (HCC). Mediators like TGF-α and TGF-β derived from dysplastic hepatocytes

**Table 2**  Compounds Produced by Activated HSC in Fibrotic Livers

| Category | Compounds | Examples |
|---|---|---|
| Mediators and growth factors | Prostanoids | Prostaglandins (PGF2$\alpha$, PGD2, PGI2, PGE2) and leukotriens (LTC$_4$, LTB$_4$) |
| | Vasoactive mediators | Endothelin, nitric oxide |
| | Cytokines | IL-6, $\alpha_2$-macroglobulin, macrophage–colony stimulating factor (M-CSF), monocyte chemotactic peptide-1 (MCP-1), platelet activating factor (PAF) |
| | Growth factors | TGF-$\beta$, PDGF, HGF, EGF, IGF-I, IGF-II, aFGF, stem cell factor, |
| Extracellular matrix proteins | Collagens | Types I, III, IV, V, VI, XIV |
| | Proteoglycans | Heparan-, dermatan- and chondroitin sulfates, perlecan, syndecan-1, biglycan, decorin |
| | Glycoproteins | Fibronectin, laminin, merosin, tenascin, nidogen, undulin, hyaluronic acid |
| Matrix proteases and inhibitors | Matrix proteases | MMP-1, MMP-2, stromelysin, MT-MMP |
| | Protease inhibitors | TIMP-1, TIMP-2, plasminogen activator inhibitor-1, C-1 esterase inhibitor |

or from monocytes and macrophages near the tumor probably elicit the proliferation and activation of HSC within the dysplastic nodules. Extracellular matrix protein production by HSC may be beneficial to tumor survival and growth by maintaining its structure, the capsule formation, and the microcirculation within HCC. The proliferation and differentiation of the dysplastic hepatocytes within HCC may also be influenced by their matrix interactions [61, 89, 90].

On the basis of their role in the progression of many liver diseases, the hepatic stellate cells are an important target for pharmacological interventions. Low drug accumulation in this cell type requires drug targeting approaches to improve the therapeutic outcome. This need for targeting of drugs to HSC during liver cirrhosis is acknowledged by several groups [68, 91–94]. To obtain a cell-specific drug delivery to this cell type, relevant target receptors should be identified on the cell membrane of HSC. It has been demonstrated that the stellate cells, just like any other hepatic cell type, are able to internalize compounds. Some of these internalization receptors are present on quiescent HSC, such as the retinol-binding protein (RBP) receptor [95], whereas other receptors are induced after activation of this cell type. Examples of the latter group of receptors are several growth factor and cytokine receptors, the mannose 6-phosphate/insulin-like growth factor II (M6P/IGFII) receptor [96–98], $\alpha_2$-macroglobulin receptor [99, 100], ferritin receptor [101], and matrix receptors [102]. For a summary of the receptors expressed by HSC see Table 1.

## III. CHRONIC LIVER DISEASES OF INTEREST FOR DRUG TARGETING

As mentioned before, drug targeting to the liver may be a promising therapeutic approach for hepatic diseases with a chronic character. Examples of such diseases are liver cirrhosis, viral hepatitis and other infectious liver diseases, liver carcinomas or metastases of tumors, and hepatic autoimmune diseases (hemochromatosis, Wilson's disease, and $\alpha_1$-antitrypsine deficiency). The problem with the available pharmacotherapy in these diseases is that most drugs are not liver-specific and often exhibit undesirable toxicity. In the next paragraphs, we describe the pathosis of chronic liver diseases that are the subject of experimental therapies based on the application of drug delivery systems. This knowledge is important for the development of specific carriers and for the identification of molecular regulatory pathways that may serve as targets for therapeutical interventions.

### A. Liver Cirrhosis or Fibrosis

Liver cirrhosis is among the top 10 causes of death in the Western world. The disease occurs after chronic damage to hepatic cells, mainly hepatocytes, which can be caused by viral hepatitis, chronic alcohol abuse or toxic injury, biliary disease, and metabolic liver disorders [64]. Liver cirrhosis is characterized by an abnormal deposition of connective tissue in the liver, which hampers the normal functions of the liver. Other features of the disease are general tissue damage, chronic inflammation, and the conversion of normal liver architecture into structurally abnormal nodules. Secondary to these anatomical changes are disturbances in the liver function and in the hemodynamics leading to portal hypertension and intrahepatic shunting [39, 64, 103].

Cirrhosis results from the inability of the liver to restore liver homeostasis [39, 64, 104]. After a single damaging event or disturbance, the liver restores the normal situation by the production of cytokines, growth factors, and extracellular matrix constituents. This process can be envisioned as physiological wound repair. After chronic activation of liver cells, however, the local regulatory repair mechanisms are not able to keep up with the disturbances. The concerted action of many cell types will create several autocrine and paracrine activation loops leading to a perpetuating process as depicted in Figure 3.

Although the cellular and molecular mechanisms underlying fibrosis are not fully explained, it is assumed that an inflammatory reaction is the initiating factor in the early stage of fibrosis and that this inflammatory process continues during the fibrotic process [77, 105, 106]. Kupffer and endothelial cells are considered to be the most important resident cells involved in the local production of inflammatory mediators [24, 28, 29, 76]. Besides causing the activation of HSC, the inflammatory mediators induce the expression of adhesion molecules, such as ICAM-1 and VCAM on endothelial cells, that direct neutrophils and monocytes into the inflamed liver tissue [41–43]. Expression of adhesion molecules is also shown for KC and stellate cells [42, 107]. Furthermore, chemotactic compounds are released by endothelial and KC to attract immune competent cells

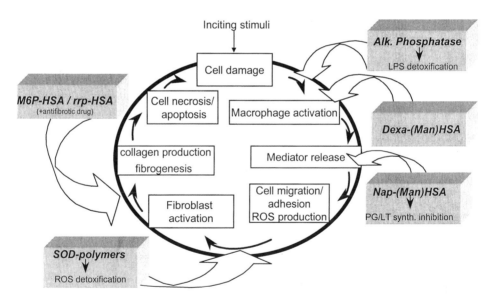

**Figure 3**  The process of fibrosis envisioned as a vicious circle. Different stages of the disease provide the opportunity for pharmacotherapeutical interventions. The targeted therapy used in our laboratory, carriers, and coupled drugs is shown in this diagram. Alkaline phosphatase, modified forms of superoxide dismutase (SOD), dexamethasone conjugated to HSA (dexa-HSA) or to mannoseHSA (dexa-ManHSA), and naproxen conjugated to HSA are aimed at different stages of the inflammatory process, whereas M6P-HSA and albumins modified receptor-recognizing peptides (rrp-HSA) are aimed at the myofibroblasts to target antifibrotic drugs.

from the bloodstream [108]. Infiltrated neutrophils release reactive oxygen species and cytokines that further promote the hepatocellular damage, the activation of stellate cells, and the fibrogenesis [109, 110] Therefore, anti-inflammatory drugs like glucocorticosteroids are putative therapeutics during fibrogenesis with KC and endothelial cells as their main target [111].

The central event in the development of liver fibrosis is the enhanced sinusoidal deposition of extracellular matrix proteins that are mainly produced by activated HSC [86, 112, 113] and to a minor extent by endothelial cells [44–46] and hepatocytes [114, 115]. So far, no evidence has been found that KC are directly involved in the production of extracellular matrix proteins [39]. The accumulation of extracellular matrix proteins is caused by a disturbed balance between the synthesis and the degradation of the matrix proteins. This imbalance leads to a 5 to 10-fold increase in the total amount of matrix molecules and to an altered composition of the extracellular matrix. In contrast to normal livers, the sinusoids in fibrotic livers are stuffed with the fibrillar collagens type I and III. This collagenization of the sinusoids, referred to as sinusoidal capillarization, causes severe disturbances of the blood flow and an impaired exchange of proteins between the liver cells and blood. Furthermore, this capillarization is accompanied by a loss of fenestration of the sinusoidal endothelial lining, which further hampers the diffusion of proteins between plasma and hepatic cells.

At present, experimental treatments of liver fibrosis are mostly directed at the interference with the matrix deposition in the basis of its prominent role in the loss of liver architecture and function [91, 93, 116]. Several levels in the process of matrix deposition are potential targets for pharmacological interventions. These include the inhibition of HSC activation and proliferation, the interference with collagen synthesis, and the enhancement of matrix degradation (summarized in Table 3). Another approach makes up the induction of apoptosis in HSC, because apoptosis represents an important mechanism terminating the proliferation of activated HSC [117–120]. To date, however, most antifibrotic drugs are not effective in attenuating the fibrotic process, leaving a liver transplantation as the only adequate therapeutic alternative at end-stage fibrosis. Cell-specific drug delivery is a relevant option for improving the effectiveness and safety profile of antifibrotic pharmacotherapy. To obtain effective pharmacotherapy, the proper antifibrotic drug should be directed at the proper hepatic cell type. This means in general that anti-inflammatory drugs should be targeted to Ke and endothelial cells, whereas drugs that interfere with matrix deposition should be targeted to stellate cells.

## B. Infectious Liver Diseases

Infectious liver diseases are caused by infiltration of viruses, bacteria, or parasites into liver cells. The most common chronic viral infections are caused by the hepatitis B (HBV) and hepatitis C virus (HCV). These viruses all have a specific

**Table 3**  Potential Therapeutic Intervention of Liver Cirrhosis, Which Can Be Directed to Different Stages of the Disease

| Level | Antifibrotic drugs |
|---|---|
| Removal of inciting stimulus | |
| Reduction of hepatic inflammation and cell injury | Corticosteroids |
| | NFκB inhibitors |
| | Free radical scavengers |
| | Ursodeoxycholic acid |
| Inhibition of HSC activation and proliferation | NFκB inhibitors |
| | Antibodies to TGFβ/PDGF |
| | Interferon-γ |
| Interference with matrix production | Proline analogues |
| | Prolyl hydroxylase inhibitors |
| | Interferon-γ |
| | Retinoids |
| Enhancement of matrix degradation | Polyunsaturated lecithins |
| | Colchicine |
| | Adenosine |
| | Collagenase activity |

affinity for the liver, in particular for hepatocytes, in common. Worldwide, more than 300 million people are chronic carriers of HBV. HBV is an enveloped virus with a double-stranded circular DNA genome. The uptake of viruses into cells is thought to be receptor mediated. Although HBV is capable of infecting several tissues, its replication occurs almost exclusively within the hepatocyte. Chronic HBV infection is a major cause of mortality throughout the world, because it is often associated with cirrhosis and HCC [121]. The most effective experimental treatment currently available for HBV infection is interferon-α (IFN-α) [122, 123]. However, the clinical effects of IFN-α therapy are disappointing, because only a subset of patients respond favorably to IFN-α, and a lot of adverse effects occur. Moreover, treatment with IFN-α is quite expensive. Therefore, the search for effective and safe inhibitors of viral replication continues. Targeting of antiviral nucleotide analogues to the hepatocyte may be one of the potential approaches for future therapy [124].

Other viruses with affinity for the liver are human immunodeficiency virus (HIV) and cytomegalovirus (CMV). CMV-DNA has been found in hepatocytes, bile ducts, and vascular cells. CMV is a dormant virus but may be reactivated, for instance, on immunosuppressive therapy after organ transplantation. It may cause vanishing bile duct syndrome and (allograft) rejection [125, 126]. The main target for HIV infection are CD4-positive cells, such as T4 lymphocytes, monocytes, and macrophages [127]. Kupffer cells have also been shown to become HIV infected, and therefore they may be a target for antiviral therapy. The negatively charged albumin carriers (sucHSA and acoHSA) and mannosylated albumins are taken up in KC and can hence be exploited to target antiviral drugs to these cells to improve the clinical efficacy of these drugs [128, 129].

The liver is also the target organ for many parasites. These parasites may either reside in the liver to replicate, or they may be phagocytosed and subsequently degraded by liver cells. For instance, malaria parasites have a short developmental stage in hepatocytes after the infection without damaging the organ [130]. The target cell for the Leishmaniasis parasites (*Leishmania donovani*), causing visceral leishmaniasis or kala azar, are KC [131, 132]. Another example of an infectious disease is schistosomiasis. The eggs of the schistosomes (*Schistosoma mansoni*) are carried via the portal system into the liver, where they are trapped in the small capillaries. When the egg load is high, they elicit a granulomatous inflammation, and finally liver cirrhosis may occur [133]. Also in the case of parasitic infections, an increase in the pharmacotherapeutic effect of antiparasitic agents can be achieved by directing the proper drug to the proper hepatic cell type.

## C. Liver Cancer

Primary liver cancer, or HCC, is a rare type of cancer in Western countries, but occurs frequently in Africa and Asia. HCC is often the sequel to chronic viral hepatitis, cirrhosis, nutritional deficiencies, or specific toxins. More common types of cancer occurring in the liver are metastatic diseases, which originate mainly from primary gastrointestinal tumors. For the growth of these and other solid tumors, sprouting of the vascular system, called angiogenesis, is essential to provide an adequate blood supply to the tumor cells. Nutrients and oxygen are needed for the proliferation of tumor cells [134–136].

The prognosis for both HCC and metastases is poor. Treatment of either type of cancer can consist of debulking of the tumor load by surgery followed by chemotherapy or radiotherapy. In the case of multiple metastases, surgical debulking of the tumor is not feasible. Furthermore, a number of factors hamper the use of chemotherapeutic agents, such as intolerable toxicity of the drugs, limited accessibility of tumor tissue, occurrence of multidrug resistance, and heterogeneity of tumor tissue. Drug targeting approaches may be applied to increase the effectiveness and safety profile of cytostatic drugs. The drugs may be selectively delivered to tumor cells, taking advantage of the expression of tumor-specific markers against which antibodies can be developed. However, most carrier molecules lack the specificity for tumor tissue, whereas their accessibility to the tumor tissue also constitutes a major problem [1]. In addition to these tumor-directed strategies, recent explanation of processes involved in angiogenesis, the sprouting of the vascular system to provide new blood vessels to the solid tumors, have led to the development of numerous antiangiogenic compounds now tested in clinical trials [137]. Furthermore, blockade of the tumor blood flow may be an effective strategy to lower the tumor burden in animal models, rendering the endothelial cell a putative target cell for antitumor therapy [138]. These approaches have been extensively reviewed recently [139–141] and will not be discussed here in more detail.

## IV. CELL-SPECIFIC CARRIER SYSTEMS

As mentioned in the Introduction, drug targeting alters the distribution of drugs in the body by directing its accumulation to a specific pathological site. Different forms of targeting can be distinguished [1, 142]. Passive targeting means that side effects are avoided by preventing uptake at nontarget sites, whereas selective accumulation and activation of drugs in the target cells is called active targeting. If the carrier itself has a therapeutic effect, called intrinsic activity, which is superimposed on the effect of the drug coupled to this carrier, this is referred to as dual targeting [143]. Examples of carrier systems with potential intrinsic activities are negatively charged albumins [144], lactoferrin [145], alkaline phosphatase [146], and superoxide dismutase (SOD) [147]. Others, such as mannose 6-phosphate modified albumin and monoclonal antibodies also can possibly be considered carriers with intrinsic activities, because they can occupy specific receptors involved in the progression of the disease. For example, mannose 6-phosphate modified albumin binds to the M6P/IGF II receptor, which is involved in the activation of the fibrogenic mediator TGFβ [96, 98]. A carrier molecule is composed of a core molecule and a homing device. The core protein used in our studies is the plasma protein albumin [1] but other macromolecules are also exploited for this purpose [3, 148]. Homing devices such as sugar molecules, charged molecules, or (cyclic) peptides containing a receptor-binding domain are attached to the core protein to obtain a macromolecule with selectivity for a certain receptor expressed on the cell membrane of a specific cell type. Covalent attachment of drugs to such a carrier results in the formation of a drug-carrier conjugate. Cellular handling and intracellular release of the drug from the carrier may warrant the use of spacer molecules (e.g., enzymatically degradable or acid-sensitive) between drug and carrier, enabling proper release of therapeutically active drugs from the conjugate within the cell [149–151]. A number of important features have to be taken into account in the design of drug-carrier conjugates. In addition to the target specificity of the carrier, biocompatibility, toxicity, immunogenicity of the conjugate, and the number of functional groups in the carrier for chemical attachment of drugs are also relevant factors that determine the overall success of the conjugate [1, 2].

For selective delivery of drugs to the liver, both soluble and particle carrier systems can be used. Examples of the particle type of carrier are liposomes [152–154], HDL and LDL particles [3, 155, 156], microspheres and nanoparticles (including albumin particles) [157], whereas monoclonal antibodies [158], immunotoxins [159], bile acids [160], polymers [148, 161, 162], neo(glyco)proteins [1, 4, 163], and enzymes like SOD and alkaline phosphatase [147] represent the soluble type of carrier. An advantage of the particle carriers such as liposomes is that drugs can easily be incorporated, either dissolved in the aqueous phase or in the lipid phase without the requirement of a covalent linkage between drug and carrier as is necessary for the soluble carriers. A major disadvantage of most particulate carriers (>100–150 nm) is that their extravasation from the blood into the liver tissue and other organs is limited because of the existence of the endothelial barrier, with the exception of small liposomes (<100 nm) and that most particle carriers are susceptible for uptake by the reticuloendothelial system. Avoidance of this reticuloendothelial system makes progress by using smaller sized particles [152] or by masking the particle surface with polyethylene glycol [164]. Targeting with particle carriers often represents a passive targeting strategy. Active targeting using liposomes is achieved by the application of site-directed ligands to the liposome surface [165, 166]. An example of this active targeting is demonstrated by Kamps et al. [167], who prepared negatively charged liposomes that are selectively taken up in endothelial cells by scavenger receptors [167]. This approach may create novel therapeutic opportunities for liposome-based drug targeting. Another drawback for the application of particle carrier systems for KC targeting is that chronic administration of these carriers may block the functions of KC if the carrier is slowly degraded and accumulates in the cell. This may cause chronic toxicity of the carrier [168, 169].

Soluble carriers differ from the particle carriers mainly by their smaller size. A major advantage of the soluble molecules is that they easily pass the liver endothelial lining. Consequently, cells that are not directly in contact with the blood may also be a target. A disadvantage of this type of carrier may be that drugs need to be covalently attached to the carriers. This may influence the physicochemical properties of both the drug and the carrier or may lead to the release of drugs in pharmacological inactive forms. To overcome inadequate drug release, small spacer molecules can be applied when engineering protein-drug conjugates (see earlier) [149]. Because of their smaller size and the covalent attachment of drugs, soluble carriers will deliver smaller amounts of drugs per carrier molecule than the particle carriers. A serious drawback of the soluble carriers such as antibodies and modified (glyco)proteins is the immunogenic response observed in particular after repeated administration. Soluble polymers are less immunogenic, but these carriers often are less biodegradable compared with the other soluble carriers [170]. A decrease in the occurrence of antigen induction will probably occur after a rapid uptake in the target cells, thereby preventing the exposure to antigen-presenting cells. Furthermore, the use of species-specific (homologous) proteins has also been shown to attenuate the immunogenicity [171–173].

## V.   CRITICAL ASPECTS OF CONJUGATES TARGETED TO THE LIVER

From the moment of injection to the moment of pharmacological action, several aspects are of critical relevance for drug-targeting outcome [6, 174, 175]. Most carriers use specific receptors on the cell membrane for their cell entry. These receptors are generally not exclusively expressed on one cell type only, and therefore the difference in receptor density between target and nontarget cells is important with respect to target site–specific accumulation. This selectivity of drug action may be further increased by exploiting a spacer between the carrier and the drug, which is only cleaved in specific (pathological) conditions. For instance, specificity for tumor cells can be enhanced by selection of a spacer that is a good substrate for the tumor-associated enzymes, such as collagenase or cathepsin. Consequently, an increased amount of drugs is released in the vicinity of the target cell [149].

In addition, it is important to test whether carrier molecules preserve their cell specificity when applied in pathological circumstances. During fibrosis, the accessibility of target cells may be changed because of the matrix expansion and capillarization of the sinusoids (see section III.A), which may limit the usefulness of the carriers. However, we recently reported that the intrahepatic distribution of high–molecular weight proteins such as modified albumins is not markedly altered by the increased collagen deposition in the fibrotic rat livers [94]. The alteration in the total cell number is another important aspect that may affect the biodistribution of carriers in diseased livers. After induction of fibrosis, a twofold increase in KC and a 5 to 10 fold increase in stellate cells occurs [176, 177]. Another feature of diseases that should be taken into account in this respect is the up- or down-regulation of (target) receptors or the receptor redistribution to less accessible cell surface domains. For instance, the number of ASGP receptors present on hepatocytes is decreased in cirrhotic livers to 72% of normal values, including an abnormal cell surface distribution to the canalicular side of the membrane [178]. Patients with acute or chronic HBV infection exhibit a decreased ASGP receptor expression, and this receptor is practically absent in patients with HCC [179, 180]. In addition, the mannose receptors on KC, which are used as target receptors for neoglycoprotein carriers [181, 182], are down-regulated after leishmaniasis infection [183]. On the other hand, induction of receptor expression may also occur during disease processes. For example, the mannose 6-phosphate/ insulin-like growth factor II (M6P/IGF II) receptors and PDGF receptors are up regulated on the cell surface of hepatic stellate cells during fibrosis [96, 98, 184]. This disease-associated event will favor the disposition of carriers that bind to these receptors.

Another problem that may arise with respect to receptor binding is the presence of endogenous ligands at the target site. This may be a problem in particular in the diseased liver. The endogenous molecules may compete with the carrier for receptor binding dependent on the receptor affinity and concentration of the compounds. Also, a down-regulation of target receptors may occur after exposure to high concentrations of ligands [175]. If, in addition to this decrease in cell surface receptor density, the recruitment of new cell-surface receptors is low, the delivery of drug carriers by receptor mediated endocytosis will be hampered.

Another factor that should be taken into account is that binding of carriers to specific receptors may evoke a biological response that is connected to this receptor. In other words, receptors are not cellular markers but play an integral role in the regulation of many cellular functions, including growth, metabolism, differentiation, contraction, and migration. This aspect is reviewed by Feener and King [175]. The primary role of hormone and growth factor receptors, for instance, is to mediate a transmembrane signal transduction. For example, targeting to the brain is performed with an insulin fragment, which binds to the insulin receptor but does not have an effect on the glucose homeostasis, rather than with insulin itself. Insulin cannot be used as a carrier molecule, because the systemic delivery of insulin in nondiabetics patients can result in hypoglycemia because of a receptor-induced increased glucose uptake in the cells [185]. PDGF constitutes another example of targeting with a possible biological response. The PDGF receptors are induced on the cell membrane of HSC in liver fibrosis, and therefore they represent an attractive target for selective delivery to HSC. However, PDGF is one of the most potent mitogens for HSC identified so far [39, 80, 81], and therefore application of PDGF as a carrier molecule is not feasible. Therefore, we [186, 187] mimicked the binding of PDGF to the receptor and constructed an albumin carrier with cyclic peptides that only bind PDGF receptors with no intracellular signalling (see also section VI).

In plasma, the drug-carrier complex is exposed to normal clearance mechanisms by the liver, the kidney, and the reticuloendothelial system. The success of avoiding these nonspecific clearance mechanisms depends on the affinity of carriers for their target receptors and on their plasma concentrations. An excess of conjugate in plasma caused by saturation of its cellular uptake by means of the target receptors renders the conjugate present in plasma more susceptible to nonspecific removal.

Furthermore, before pharmacological actions of drug-carrier conjugates can be expected, a release of the drugs must occur from the intracellular vesicles into the cytosol. Therefore, drug-carrier complexes have to be endocytosed and subsequently be degraded by target cells to release the coupled drugs. At least three different kinds of endocytotic mechanisms can be involved in the internalization of the conjugate: fluid phase endocytosis, adsorptive endocytosis, and receptor-mediated endocytosis. These processes have been extensively described elsewhere [1, 188, 189]. The modified albumins commonly used for targeting of drugs to hepatic cells (see section VI) are internalized by receptor-mediated endocytosis. Consequently, targeted drugs have to survive the lysosomal conditions (proteolytic enzymes and low pH) and cross the lysosomal membrane before reaching the cytosol and exerting their pharmacological effects. Biological peptides internalized by means of the lysosomal route are in particular susceptible to degradation by proteolytic enzymes present in the endosomes and lysosomes. Eto and Takahashi [190], however, recently reported an inhibition of HBV replication by interferon that was targeted to the ASGP-receptor. So, endocytosed interferon was able to exert an antiviral effect even after lysosomal internalization. Furthermore, it is known that the A chain of the plant toxin ricin is able to translocate from the lysosome to cytosol [191, 192]. The knowledge of how ricin molecules exploit and avoid cellular degradation processes may be of great importance for targeted compounds that act on cytosolic targets. Endocytosed lipophilic compounds, such as most drugs, may have a better membrane passage than more hydrophilic compounds, such as, for instance targeted, genes and oligonucleotides. The latter compounds, whose therapeutic applications become increasingly overt, may also be particularly sensitive to the lysosomal pH and enzymes [17].

## VI. MODIFIED ALBUMINS

Albumin may be substituted with sugars, charged molecules, and (cyclic) peptides to obtain specificity for a certain cell type. The combination of receptor density on cells and the degree of substitution of receptor ligands on the albumin core protein greatly determines the cell-specific accumulation. To date, a set of carriers to all cell types of the liver is available (Table 4).

Targeting to hepatocytes is well established. Modification of albumin with at least 13 sugar moieties of lactose (lacHSA) leads to accumulation in hepatocytes after receptor-mediated endocytosis by the ASGP receptor through exposure of terminal galactose groups. The lactosylated albumin mimics terminal galactose groups in the antennary cluster of oligosaccharides as present in desialylated glycoproteins [4, 193, 194]. The ASGP receptor is a receptor that is only expressed on cell membranes of hepatocytes, which offers the possibility of highly specific targeting to this cell type.

**Table 4** Modified Albumins and Their Cell-specific Uptake in the Liver

| Modified albumin | PC | KC | EC | HSC |
|---|---|---|---|---|
| lacHSA | +++ | − | − | − |
| man$_{10}$HSA | − | ++ | + | − |
| sucHSA | − | − | +++ | − |
| acoHSA | − | − | +++ | − |
| M6P$_{28}$HSA | − | + | − | +++ |
| pCVI-HSA | − | − | − | +++ |
| pPB-HSA | + | − | − | ++ |

Mannosylated albumin (manHSA) is an albumin carrier with selectivity for KC as a result of the attachment of mannose residues in combination with the introduced negative charge of these groups [195, 196]. The attachment of mannose groups to the lysine residues of albumin leads to an increase of the net negative charge of the protein by preventing protonation of the lysine amino groups. Other albumin carriers developed for targeting to KC are albumin nanospheres and microspheres [197]. Similar to other particulate carriers, they have the advantage that drugs can be easily incorporated in the particle.

Mannosylated albumins can also bind to endothelial cells, because mannose receptors are present on the cell surface of sinusoidal endothelial cells [50]. In contrast to manHSA displaying selectivity for KC (previous paragraph), the manHSA that accumulated in endothelial cells were prepared without the introduction of negative charges [198]. Furthermore, targeting to the endothelial cells can be obtained by the attachment of negatively charged groups, succinic anhydride, or cis-aconitic anhydride, to albumin (sucHSA and acoHSA, respectively). These negatively charged albumins are specific substrates for the scavenger receptors (type A) on endothelial cells [199]. Scavenger receptors are also present on KC, but scavenger receptors display different substrate specificities. The polymeric negatively charged albumins are preferentially endocytosed in KC, whereas the negatively charged albumins in monomeric form display a higher affinity for the endothelial cells [200].

Recently, we developed neoglycoprotein carriers that accumulate in hepatic stellate cells. Modification of albumin with mannose 6-phosphate groups (M6P-HSA) resulted in a marked enhanced accumulation of this neoglycoprotein in stellate cells of fibrotic rat livers [201]. An increase in the mannose 6-phosphate substitution, from M6P$_{10}$-HSA to M6P$_{28}$-HSA, caused an increased accumulation in HSC from 15–70% of the total liver content of M6P$_x$-HSA. Recently performed *in vitro* studies clearly showed extensive internalization of M6P$_{28}$-HSA in activated rat HSC (unpublished data). Furthermore, it was shown that M6P$_{28}$-

HSA not only bound to rat HSC but also to HSC and endothelial cells of cirrhotic and normal *human* livers [201].

In addition, we took an alternative approach and developed two other albumin derivatives with specificity for HSC in fibrotic rat livers [186, 187, 202]. In these cases, albumin was derivatized with cyclic peptides containing the amino acid sequence that mimics the binding site of collagen type VI or PDGF to their receptors, yielding the carriers pCVI-HSA and pPB-HSA. Figure 4 schematically depicts the principle on which this type of drug targeting using albumin carriers is based. Because the receptor recognizing peptide moieties occupies specific receptors, a competition between the carrier and the endogenous ligands may occur. Because of this receptor-antagonizing effect of the carriers, a dual targeting is feasible with this type of carrier. This approach offers the possibility to construct a category of ligand peptides directed to various cytokine or growth factor receptors. In addition, this use of minimized proteins offers a wide spectrum of opportunities to deliver pharmacologically active compounds to diseased organs,

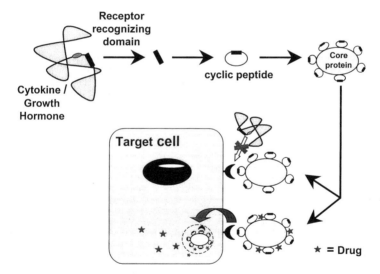

**Figure 4**  Schematic representation of the concept of drug delivery using receptor-recognizing cyclic peptides. The receptor recognizing domain of endogenous ligands is exposed on the core protein (albumin). The target cell of this modified protein in the case of liver fibrosis is the hepatic stellate cell. The cyclic peptide modified protein may bind to receptors present on the target cell membrane and compete with endogenous ligands (intrinsic activity of the carrier). Alternatively, the carrier may be used for the targeting and intracellular delivery of drugs (★).

because it takes advantage of the up-regulation of cytokine or growth factor receptors in diseased tissues.

## A. Chemical Preparation of the Modified Albumins

Different methods are used for the derivatization of albumin with sugar or peptide moieties. Because albumin contains 60 lysine molecules, a maximum of 60 moieties can be modified on derivatization of the free amine groups of lysine. In addition, coupling of moieties to carboxylic acid groups (present in glutamic acid and aspartic acid) and to sulfhydryl groups (cysteine) is also feasible. Attachment of the disaccharide lactose to albumin is achieved by reductive amination with cyanoborohydride according to Schwartz and Gray [203]. The aldehyde group of glucose reacts with $\varepsilon$-$NH_2$ groups of lysine, thereby exposing the galactose moiety. Other methods to couple sugars to proteins include the thioglycoside method according to Lee [204] and the coupling by means of thiophosgene activation of para-aminophenyl sugars as described by Kataoka and Tavassoli [205]. In contrast to the reductive amination and the thioglycoside method, sugar attachment according to the method of Kataoka and Tavassoli leads to an increased negative charge of the modified protein, because the introduction of sugar groups to the free $\varepsilon$-$NH_2$ groups prevents the protonation of these groups in albumin. We used the method of thiophosgene activation of para-aminophenyl sugars to attach mannose and mannose 6-phosphate to albumin and achieved cell-specific uptake of conjugates in KC and stellate cells.

Charged groups can also be coupled to albumin. Negatively charged albumins are obtained after the reaction of albumin with succinic acid and aconitic acid [206, 207], whereas positively charged albumins are obtained after the coupling of ethylene diamine to the carboxylic acid groups of albumin [208].

The coupling of peptide moieties to the free $\varepsilon$-$NH_2$ in albumin is also possible. We coupled cyclic peptides, prepared with S-S bonds between two final cystein residues, to the lysine groups of albumin by means of a succinimide-acetyl thioacetate (SATA) linker [187, 202]. This new method of preparing drug-targeting preparations, depicted in Figure 5, yielded carriers that selectively accumulated in HSC. Although the protein modification methods summarized here are well established and generally accepted, caution is warranted during synthesis of the carrier and the drug–carrier complexes. Because the *in vivo* tissue distribution of polymeric proteins can be completely different from the monomeric form [209, 210], it is necessary to obtain drug-targeting preparations that only contain the monomeric form. In addition, introduction of extra groups, in particular charged and hydrophobic moieties, to albumin may cause conformational changes of the albumin backbone. The attached groups might influence the three-dimensional structure of the albumin molecule because of electrostatic or hy-

pPB-HSA

drophobic interactions. Furthermore, the reaction conditions used for the coupling of groups can lead to denaturation or refolding of the protein. Alterations in the tertiary structure of albumin may negatively influence the applicability of the carrier through an enhanced KC-mediated clearance or increased immunogenicity. To examine this, the amount of α-helices and β-sheets in the modified albumins may be assessed with circular dichroism analysis [211, 212]. In conclusion, mild methods should be used to preserve the structure of the core protein, rendering the modified protein nonimmmunogenic, monomeric, and soluble.

## B. (Bio)Distribution Studies of the Carrier

### 1. *In Vivo* Studies

The *in vivo* behavior of a carrier is an essential issue in the concept of drug targeting to determine the pharmacological effects and side effects of therapeutic compounds. A number of experimental methods are available to examine this issue. Organ distribution studies can be performed to determine quantitatively the total amount of carriers present in the target organ compared with the other organs [94, 213, 214]. Usually, the carrier is radioactively labeled to allow detection *in vivo*. The initial distribution of the carrier can be assessed 10 minutes after injection, whereas at later time points redistribution and degradation rates can be studied. The whole body distribution of (radioactive labeled) molecules can also be determined with positron emission tomography (PET) [215, 216] or with gamma camera studies [217, 218]. In addition to these organ distribution studies, the accumulation of the carrier in various cell types in different organs can be assessed qualitatively with immunohistochemical techniques [94, 199]. For immunohistochemical analysis, tissue sections are stained with antibodies directed against the carrier to assess the localization of the modified albumin within the organs. To identify the cell type(s) involved in the uptake of carriers, the sections can be double stained with markers for the different cell types and the carrier. In the rat liver, the monoclonal antibodies HIS52, ED2, and the combination of desmin and GFAP are generally used to identify, respectively, endothe-

---

**Figure 5** Covalent coupling of cyclic peptide moieties to human serum albumin (HSA). The depicted cyclic peptide, C*SRNLIDC*, in which C* denotes the cyclizing cysteine residues, mimics the receptor binding site of PDGF-BB. First, a sulfhydryl group is introduced to the cyclic peptide by a reaction with succinimide-acetyl thioacetate (SATA). The primary amino groups of lysine in HSA are derivitized with maleimide-hexoyl-*N*-hydroxysuccinimide ester (MHS). Subsequently, the cyclic peptide is coupled to HSA. In this latter reaction, hydroxyl amine is used to remove the protecting acetate group from the sulfhydryl group of the cyclic peptide.

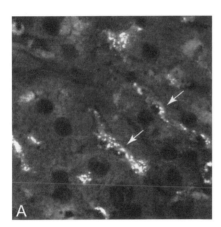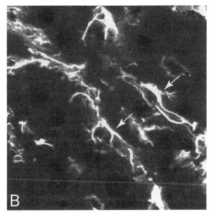

**Figure 6**  The co-localization of the carrier pCVI-HSA, identified with anti-HSA IgG (stained with FITC; [**A**]) and HSC, identified with antidesmin and GFAP IgG (stained with Tritc; [**B**]) in fibrotic rat livers as assessed with confocal scanning laser microscopy. Note the co-localization as indicated by arrows.

lial, Kupffer, and hepatic stellate cells [201]. Figure 6 shows an example of double positive cells in a fibrotic rat liver stained with an antibody against HSA to detect the stellate cell carrier pCVI-HSA in combination with the antibodies against GFAP and desmin to identify rat HSC. The combination of the radioactive and the immunohistochemical analysis is complementary and provides information about the total accumulation in the different organs and the cellular localization within these particular organs.

The hepatocellular distribution can also be assessed at various time points after injection of a radiolabeled carrier by the isolation of the different cell types [24]. Rat and human liver cells can be isolated after perfusion of the liver with collagenase and/or pronase [219–221]. Separation of the different cell types is performed by centrifugal elutriation, by density gradients (using Percoll, nycodenz, stractan, or sucrose), or by magnetic retention of the cells that are selectively recognized by antibodies attached to nonsoluble magnetic beads [222]. However, most of the obtained cell fractions are not 100% pure after isolation [220], and accurate determination of the uptake of ligands by a particular cell type may therefore be troublesome. In addition, isolation of liver cells from diseased rats, such as rats with liver fibrosis, is even more difficult with respect to purity and viability of cell fractions compared with normal rats. Another drawback in this method is that because pronase is necessary for the isolation of hepatic stellate cells, but at the same time affects the viability of hepatocytes [24], it is impossible to separate all cell types from one liver.

The behavior of conjugates *in vivo* is largely dictated by the properties of the carrier and differs from the free drug administered by the same route. Pharmacokinetic parameters such as volume of distribution, elimination half-life, and the clearance mechanisms of the carrier are important with respect to the pharmacological effects of the targeted drugs [6, 7, 223]. Alterations in matrix deposition, the activation state of certain cells, or the production of mediators induced by targeted drugs should be evaluated after treatment *in vivo*, because *in vitro* studies fail to reflect the complex interactions of mediators as they occur *in vivo*. In addition, *in vivo* studies are a prerequisite to evaluate the toxicity of the conjugated drug.

Because conjugates are destined to be applied in pathological conditions, the applicability of modified albumins should not only be tested in healthy animals but particularly in diseased animals. The microenvironment of the target cells may be altered in a disease state, whereas different morphological and cellular factors may interfere with the targeting concept (as discussed in section V). Drug targeting approaches that use neoglycoprotein carriers such as lacHSA are confronted with a decline in receptor density during liver diseases. On the other hand, the disposition of carriers aimed at receptors that are induced during disease processes, such as M6P-HSA and the cyclic peptide modified albumins (pCVI-HSA and pPB-HSA) [186, 201], may be favored in diseased livers.

## 2. *In Vitro* Studies

*In vitro* experiments are valuable for the assessment of the cellular handling of conjugates. Whether the carrier is internalized or remains at the cellular membrane is an important feature of the carrier. Binding of carrier molecules to cells is generally studied at 4°C, because internalization of compounds is small of less than 10°C. From 10–20°C internalization increases with the rise in temperature, whereas optimal internalization occurs at 37°C [224, 225]. The route of internalization can be assessed by the use of inhibitors that affect different levels in the endocytotic pathway [226–228]. NH$_4$Cl and chloroquine neutralize the acidic pH of the endosomes and lysosomes and completely stop ligand dissociation from the receptor and consequently the total uptake. Monensin also prevents the pH decline, but in addition it inhibits the release of receptor-ligand containing vesicles from the microtubules. Colchicine is a microtubule depolymerizing drug. Microtubules are important during endocytosis for both the intracellular organization of vesicles and their routing between sequential compartments. Consequently, colchicine inhibits endocytosis of the ligand, the movement of endocytotic vesicles, and receptor recycling. Leupeptin specifically inhibits the protease cathepsin and has been used to demonstrate specific proteolysis of compounds in the lysosomes. All these compounds do not affect the initial cell-surface binding of ligands to their receptors.

To study the *in vitro* behaviour of carrier molecules, cells may be derived from a cell line or from primary cultures of rat or human liver cells. In general, the latter cells better reflect the *in vivo* situation than the immortalized cells in a cell line culture. It should be noted, however, that during the isolation procedure the enzymes collagenase and pronase may destroy or damage the target receptors. *In vitro* preparations that approach the *in vivo* situation best are liver slices, in which the physiological context of the liver is maintained [229], and the model of the isolated perfused rat liver (IPRL) [230, 231]. These techniques allow us to study carrier-cell interactions within the organ, that is, in the presence of all other liver cells and their secreted mediators.

Carrier molecules have been extensively studied in animals like mice and rats, but little is known about the uptake and handling by human livers. Because clinical studies are usually performed at the latest stages of the drug development process, there is a need for alternative test methods using human tissue at an early stage of drug design. Human liver tissue can be obtained for scientific research on the basis of strict ethical guidelines including approval by a local ethical committee. Liver slices of tissue procured from organ donors after size reduction or split liver transplantation and diseased livers removed from patients with terminal liver cirrhosis have been extensively applied for pharmacological studies in our laboratory by Olinga et al. [229, 232]. Viability parameters and liver function tests clearly showed that slices can be incubated with ligands up to 72 h. Applying $M6P_{28}$-HSA to normal and cirrhotic liver slices from patients, we found an accumulation of the albumin carriers in nonparenchymal cells [201]. At present, an isolated perfused liver system is under development for the application to human livers [233]. This new method, referred to as the human liver lobe perfusion (HLLP), may be an appropriate tool to study liver uptake and handling of carriers in human tissue *ex vivo*. Using this system, Melgert et al. showed that an albumin-dexamethasone conjugate is taken up by the nonparenchymal cells of the human liver [233]. Moreover, these methods offer the possibility to compare experimental data on drug-targeting preparations obtained in animals to the human situation. Further optimization of these methods with human (pathological) tissue may lead to the implementation of these techniques in the research on drug-targeting approaches for (liver) diseases. This may help to predict success or failure of novel drug-targeting preparations in the patient.

## C. Drug-Albumin Conjugates

After a suitable carrier for the cell type of interest is developed, drugs can be coupled to this carrier either directly or through a spacer molecule. The bond between the carrier and the drug has to be stable in plasma but biodegradable when the drug has reached the target site. Spacer molecules used for drug targeting purposes are (endo- or exo-) peptidase-sensitive spacers, esterase-sensitive spacers, and acid-sensitive spacers [149, 150]. Acid-sensitive spacers are quite relevant, because in general drug targeting preparations are incorporated in endosomes with pH 5.0–6.0 after cellular internalization and subsequently end up in the lysosomes of target cells. In the latter organelles the drug-carrier complex can either be chemically destabilized at about pH 5.0 or be degraded by proteases. Application of acid-sensitive spacers will allow the release of drugs from endosomes, thus avoiding the lysosomal compartment. This may be particularly relevant for targeted peptides or oligonucleotides (sense or anti-sense). Another advantage of the use of spacers is that the coupled drug is released in its native form instead of linked to amino acids derived from the carrier, which can occur after proteolytic degradation [196]. Structural alteration in the drug may result in decreased pharmacological activity, although naproxen-lysine, intracellular released from a naproxen-albumin conjugate, was found to have a similar inhibitory effect on the synthesis of prostaglandin $E_2$ as native naproxen [196].

Adequate analytical methods are required to differentiate between the free drug, the drug-carrier conjugate, and possible metabolites to assess the amount of drug coupled to the conjugate and its concentrations in plasma and cells. In general, high-performance liquid chromatography (HPLC) techniques and enzyme-linked immunosorbant assay (ELISA) are used [234, 235]. A more recent and promising development is the use of micellar electrokinetic capillary chromatography (MECC) for sensitive monitoring of the drug, either covalently attached or present in the free form [236].

A number of conjugates for liver-targeting purposes have been reported in the literature. In some cases, the drug is coupled to native albumin instead of a modified albumin carrier. Examples are naproxen modified albumin (Nap-HSA), dexamethasone modified albumin, or methotrexate modified albumin (MTX-HSA). In these cases, coupling of drugs to albumin changes the chemical properties and hence the pharmacokinetic profile and whole-body distribution of both the drug and the carrier. In the next paragraphs, we will give a survey of the studies on the albumin-based drug-targeting concept either exploiting native or modified albumin as carriers in relation to the diseases of interest.

## 1. Liver Cirrhosis or Fibrosis

To date, no effective therapy is available for cirrhotic patients other than a liver transplantation. Targeting of antifibrotic drugs is necessary to optimize the therapy for this disease. Naproxen, a nonsteroidal anti-inflammatory drug (NSAID), has been studied in drug-targeting strategies in fibrotic rats. Naproxen inhibits cycloxygenase activity and consequently the synthesis of prostaglandins and thromboxanes from arachidonic acid. Covalent coupling of naproxen to albumin (Nap-HSA) resulted in a successful targeting of naproxen to the liver [196]. Three hours after intravenous injection of Nap-HSA, the hepatic contents of total naproxen was about 30 times higher for the conjugate compared with free naproxen [235]. Within the liver, Nap-HSA was taken up in endothelial cells and to a lesser extent in KC by way of scavenger receptors [196, 235]. This was also demonstrated in rats with liver fibrosis induced by bile duct ligation [237]. After receptor-mediated endocytosis and lysosomal degradation of the conjugate Nap-HSA, the metabolite Nap-lysine was released in the cells. It was assessed that Nap-lysine exhibited an equipotent inhibitory effect on the synthesis of prostaglandin $E_2$ as native naproxen [196]. In an acute *in vitro* model of hepatotoxicity, induced by endotoxin in rats pretreated with *Corynebacterium parvum*, Nap-HSA showed positive hepatoprotective effects. This protection was clearly superior to an equimolar amount of free naproxen [238]. Furthermore, Nap-HSA was able to protect cirrhotic rats, 4 weeks after ligation of the bile duct, from endotoxin-induced toxicity while preserving renal function. Thus, the renal side effects of naproxen were abolished by the coupling of naproxen to albumin, because urinary sodium and $PGE_2$ excretion were unaffected in rats treated with Nap-HSA but not in rats treated with native naproxen [239].

Melgert et al. studied the delivery of the corticosteroid dexamethasone to fibrotic livers [240]. Dexamethasone has more potent and broader anti-inflammatory effects compared with naproxen. It inhibits the release of inflammatory mediators like TNF-$\alpha$, IFN-$\gamma$, and IL-6 and acts as an NF$\kappa$B inhibitor [241, 242]. Dexamethasone coupled to albumin (Dexa-HSA) was specifically taken up by sinusoidal cells in fibrotic rat livers, whereas dexamethasone itself was mainly taken up by hepatocytes. *In vivo*, Dexa-HSA promoted survival in endotoxin-induced liver inflammation in rats [240]. *In vitro*, anti-inflammatory effects of the conjugate were measured in endotoxin-challenged liver slices. Dexa-HSA inhibited the release of nitric oxide and TNF-$\alpha$ in a dose-dependent manner (Melgert et al. unpublished data). To further enhance the delivery to KC at present dexamethasone is coupled to manHSA, and this conjugate is studied with respect to the pharmacokinetic profile and pharmacotherapeutic effects in fibrotic rats.

Endothelial cells and KC are the major target cells for anti-inflammatory drugs like naproxen and corticosteroids. To improve the antifibrotic therapy, it may also be an option to deliver hepatoprotective agents, for example, ursodeoxycholic acid [243], to hepatocytes using the carrier lacHSA. Palmerini et al. studied the targeting of formylcolchicine to $CCl_4$-treated rats applying lacHSA as carrier. Formylcolchicine-lacHSA showed a higher affinity for the liver than colchicine alone, and a higher antifibrotic activity was found as reflected by a reduction in alkaline phosphatase levels in plasma and the hydroxyproline content of the liver [244].

At present, attenuation of the hepatic stellate cell activities is an important goal of antifibrotic therapies, because this cell type is the major contributor to the extracellular matrix deposition in fibrotic livers. As mentioned before (section VI), we recently developed modified albumins (M6P$_{28}$-HSA, pCVI-HSA, pPB-HSA) that bind to HSC *in vitro* and accumulate in this cell type *in vivo* [186, 201]. M6P$_{28}$-HSA and pCVI-HSA were internalized in activated HSC after receptor-mediated endocytosis (unpublished data). Coupling of collagen synthesis inhibitors (prolyl hydroxylase inhibitors [93, 245]), inhibitors of cell activation (NFκB inhibitors, HOE077, or histone deacetylase inhibitors like trichostatin A [82, 246–249]), or inhibitors of portal hypertension [103] to these two HSC carriers will be the next step in the development of new antifibrotic therapies. The stellate cell carrier pPB-HSA is not taken up into the lysosomes of activated HSC (at least *in vitro*) and seems to stay at the outer cell surface (unpublished data). Therefore, this carrier may offer the possibility to deliver antifibrotic agents to the extracellular microenvironment of HSC. Metalloproteinase activators [93], IL-1 receptor antagonists, TGF-β neutralizing compounds, PDGF binding molecules, or other receptor antagonists exert their effects at the outer plasma membranes of HSC and may therefore be the drugs of choice to be targeted with the pPB-HSA carrier. Future *in vivo* studies will address the feasibility of the targeting of antifibrotic agents to HSC.

It should be realized, however, that therapeutic interference with only one cell type may not be enough to treat liver cirrhosis, because all hepatic cell types contribute to some extent to the development of the disease. Therefore, a combination of drug-targeting preparations to stellate cells, KC, endothelial cells, and/ or hepatocytes might improve the pharmacological therapies and compete with the liver transplantation technique. In addition to therapeutic applications, modified albumins may also be used for diagnostic purposes (an issue that will be addressed in section VI.C.4).

## 2. Infectious Liver Diseases

Drug-targeting preparations have been applied to patients with various infectious diseases with the aim to increase the therapeutic effects of drugs administered. Modified albumins have been used as carriers for the treatment of chronic hepatitis and visceral leishmaniasis. Because in chronic hepatitis the viruses reside in hepatocytes, drug-targeting preparations based on lacHSA were developed. Fiume et al. coupled araAMP (adenine arabinoside monophosphate, a phosphorylated nucleoside analogue active against HBV) and other nucleoside analogues such as acyclovir and 2'-3'-dideoxycytidine monophosphate [250, 251] to this carrier, which resulted in a rapid clearance of these conjugated drugs from the plasma into the hepatocytes. The conjugate araAMP-lacHSA was investigated in animals, and results of these studies have been extensively reviewed by Fiume et al. [4, 252]. Fiume et al. started a clinical trial with the conjugate araAMP-lacHSA (35 mg conjugate/kg/day corresponding with 1.5 mg araAMP/kg/day for 7 consecutive days) in patients with chronic hepatitis B infection [253]. In all patients, the plasma levels of HBV-DNA dropped significantly during the araAMP-lacHSA treatment but returned to pretreatment values when the administration was discontinued. This effect was also observed after short-term treatment with free araAMP. However, the conjugate did not produce side effects in patients during the treatment as observed after free araAMP treatment, and no antibody response to the conjugate was detected. These promising results obtained by treatment for 7 days prompted a second clinical study in which eight patients with chronic HBV were treated with araAMP-lacHSA for 28 days [173]. Pharmacokinetic analysis demonstrated a rapid hepatic uptake of araAMP-lacHSA after intravenous administration of the conjugate. During the 4 weeks of treatment, no neurotoxicity or other adverse effects were observed, whereas prolonged treatment with free araAMP is usually hampered because of these side effects. Also the immunogenic response to this modified albumin was low; only in one of the eight patients small amounts of antibodies were produced. From these studies, it was concluded that this conjugate allowed a more prolonged treatment of chronic hepatitis B than free araAMP, because of the lack of side effects after chronic application of the conjugate, which enhanced its chemotherapeutical index [173]. To date, no follow-up has been published.

To increase the antiparasitic effect of methotrexate [254, 255] and the more potent drugs doxorubicin [181] and muramyl dipeptide (MDP) [182], these drugs were coupled to manHSA. The beneficial effects of selective uptake of these drugs in KC was examined in mice infected with *Leishmania* parasites. These conjugates all inhibited the growth of these parasites in the macrophages predominantly in the liver but also in the spleen after repeated injections of the conjugate in leishmaniasis-infected mice.

The use of drug carriers to treat infectious liver diseases is at a stage at which clinical applications can be envisioned.

## 3. Liver Cancer

The therapeutic success of drug-delivery preparations to tumors in the liver will for a great part be determined by the specificity of the carrier system for the tumor cells and the possibility to deliver an effective amount of drug into the tumor [256]. The lack of tumor specificity of most conjugates is the major obstacle for the development of (neo)glycoprotein carriers for liver cancer targeting. To date, the highest tumor cell specificity has been obtained by the monoclonal and bispecific antibody carriers [257]. In addition to these tumor-directed strategies, targeting is aimed at the tumor vasculature, as described in section III.C, with bispecific antibodies. Modifying HSA with RGD peptides is now undertaken in our laboratory to exploit the resulting carrier for tumor vasculature targeting [140, 258].

The use of modified albumins for diagnostic purposes to assess the tumor burden is more successful. These studies are reviewed in the following section.

## 4.  Diagnostic Purposes

Early assessment of the onset and progression of liver diseases is an important factor determining the disease prognosis of most patients. Conventional liver function tests measuring hepatic protein synthetic functions, excretory functions, amino transferases, and bile duct enzymes are useful but may prove insensitive in monitoring the disease progression adequately. Noninvasive methods are needed to examine the effects of therapy or to determine the functional capacity of the liver. This is essential for HCC patients undergoing hepatectomy, for the prediction of the prognosis of patients with liver cirrhosis, and for the decision to perform a liver transplantation [259–262]. Many of the modified albumins that have been discussed earlier can also be used for liver imaging.

To evaluate hepatic function, a modified albumin that specifically binds to ASGP-receptors on hepatocytes, [99m]Tc-galactosyl neoglycoalbumin ([99m]Tc-NGA), has been developed [263–265]. The ASGP-receptor is affected by various hepatic diseases, and therefore [99m]Tc-NGA was considered to be useful as a diagnostic agent to determine the stage of the disease by noninvasive methods. The determination of receptor binding with [99m]Tc-NGA correlated well with conventional liver function tests (indocyanine green clearance or galactose elimination capacity) and morphological scores [266, 267]. Significant decreased receptor binding was measured in cirrhotic patients and reflected the severity of the disease. A gradual decline in receptor binding was observed when cirrhosis progressed to a more advanced state, which was reflected by a decreased total number of hepatocytes, a decreased receptor density, and an increased amount of fibrous tissue [266, 268, 269]. Therefore, this method provides a tool to evaluate liver cirrhosis even in a preclinical state [270].

The liver function was also studied in patients with viral hepatitis. During the acute phase of the infection, the ASGP-receptor concentration was markedly decreased, but during the course of the disease receptor concentrations increased to normal values. This correlated well with other laboratory tests for liver function and thus again demonstrates the potential of this imaging technique in evaluating the condition of patients with acute viral hepatitis [271].

The receptor density is also decreased in patients with primary or secondary liver cancer. The extent of this decrease expresses the nonfunctioning liver mass, because metastasis or hepatoma does not show uptake of [99m]Tc-NGA *in vivo*

[179]. Therefore, this method may also be valuable in the preoperative determination of the surgical procedure for hepatectomy [260].


## VII.  CONCLUSIONS

In this review, we have considered the technical, therapeutic, and diagnostic aspects of drug targeting to the liver. Albumin carriers are successfully applied to improve the therapy of chronic liver diseases. Some conjugates based on modified albumin as the carrier are now being studied in clinical settings. However, most clinical studies testing drug-targeting preparations have been conducted with liposomes and antibodies, but the obtained results justify further clinical evaluation [272, 273]. The clinical experience with these and other targeted therapies is currently expanding but is still limited compared with the total market of currently developed liver drugs.

As mentioned in section VI.C.2, good clinical results are obtained after prolonged administration of lacHSA-araAMP in patients with chronic HBV infection [173]. A disadvantage of albumin conjugates is that they have to be administered by an intravenous route. Because this administration route may be a burden to patients, less invasive methods may likely improve the compliance of patients during prolonged treatment. The ideal situation may be an oral administration of the drug-targeting preparations, but this is not feasible yet because most proteins are not absorbed in an unaltered form from the gastrointestinal tract. Drug-targeting preparations containing poly-L-lysine polymers as the core protein instead of albumin may be administered intramuscularly [274], an administration route commonly used to provide diabetics with insulin. Poly-L-lysine polymers also have the advantage that conjugates with this polymer yield a more constant quality as compared with HSA conjugates. Albumin can therefore be seen as a model protein; all modifications introduced to albumin can in general also be applied to other carrier systems, such as polymers or liposomes. Importantly, drug targeting also offers possibilities in more pathological or fundamental research areas. Targeting of pharmacologically active agents to an individual cell type offers the advantage of selective elimination of one cell type or blockade of a single process within this cell type. After specific inhibition of a process, the implications for the development of a particular disease can be studied. In this way, drug targeting allows us to gain more insight in the molecular basis of diseases *in vivo*. As a consequence, drug-targeting preparations may create new leads for novel therapeutic interventions. As discussed in this review, drug carriers may also create new opportunities for conventional therapeutic agents that are not sufficiently effective or that display serious extrahepatic side effects.

## ABBREVIATIONS

acoHSA, aconitylated HSA; ASGPR, asialoglycoprotein receptor; EC, endothelial cells; CMV, cytomegalovirus; Dexa-HSA, dexamethasone-HSA conjugate; GFAP, glial fibrillary acidic protein; HBV, hepatitis B virus; HCC, hepatocellular carcinoma; HCV, hepatitis C virus; HE, hematoxilin and eosin staining; HIV, human immunodeficiency virus; HLLP, human liver lobe perfusion; HSA, human serum albumin; HSC, hepatic stellate cells, IFN, interferon; IL-1, interleukin-1; KC, Kupffer cells; lacHSA, lactosylated HSA; manHSA, mannosylated HSA; M6P/IGFII, mannose 6-phosphate/insulin like growth factor II; M6P-HSA, mannose 6-phosphate modified HSA; MMP, matrix metalloproteinases; Nap-HSA, naproxen-HSA conjugate; NFκB, nuclear factor kappa B; PAS, periodic acid Schiff staining; PC, parenchymal cells or hepatocytes; pCVI-HSA, HSA modified with cyclic RGD peptides recognizing the collagen type VI receptor; PDGF, platelet derived growth factor; pPB-HSA, HSA modified with cyclic peptides recognizing PDGF receptors; RGD, Arg-Gly-Asp; sucHSA, succinylated HSA; $^{99m}$Tc-NGA, $^{99m}$Tc-galactosyl neoglycoalbumin; TIMP, tissue inhibitors of metalloproteinases; TNF-α, tumor necrosis factor-α; TGF-β, transforming growth factor-β.

## REFERENCES

1.  D. K. F. Meijer and G. Molema, Targeting of drugs to the liver, *Semin. Liver Dis.* *15*:202–256 (1995).
2.  D. K. F. Meijer, R. W. Jansen, and G. Molema, Drug targeting systems for antiviral agents: options and limitations, *Antiviral Res. 18*:215–258 (1992).
3.  T. J. C. Van Berkel, M. C. M. Van Dijk, M. K. Bijsterbosch, and P. C. N. Rensen, Drug targeting by neo-lipoproteins, *J. Contr. Rel. 41*:85–90 (1996).
4.  L. Fiume, C. Busi, G. Di Stefano, and A. Mattioli, Targeting of antiviral drugs to the liver using glycoprotein carriers, *Adv. Drug Deliv. Rev. 14*:51–65 (1994).
5.  R. Duncan, Drug targeting: Where are we now and where are we going? *J. Drug Target. 5*:1–4 (1997).
6.  Y. Takakura and M. Hashida, Macromolecular carrier systems for targeted drug delivery: Pharmacokinetic considerations on biodistribution, *Pharm. Res. 13*:820–831 (1996).
7.  M. Rowland and A. McLachlan. Pharmacokinetic considerations of regional administration and drug targeting: Influence of site of input in target tissue and flux of binding protein, *J. Pharmacokinet. Biopharm. 24*:369–387 (1996).
8.  V. J. Desmet, Organizational principles, *The Liver: Biology and Phathobiology*, 3rd ed. (I. M. Arias, J. L. Boyer, N. Fausto, W. B. Jakoby, D. Schachter, D. A. Shafritz, eds.), Raven Press, New York, 1994, pp. 3–14.
9.  M. Aumailley and B. Gayraud, Structure and biological activity of the extracellular matrix, *J. Mol. Med. 76*:253–265 (1998).

10. D. Schuppan, R. Somasundaram, W. Dieterich, T. Ehnis, and M. Bauer, The extracellular matrix in cellular proliferation and differentiation, *Ann. NY Acad. Sci. 733*: 87–102 (1994).

11. S. Milani, H. Herbst, D. Schuppan, C. Grappone, and O. E. Heinrichs, Cellular sources of extracellular matrix proteins in normal and fibrotic liver. Studies of gene expression by in situ hybridization, *J. Hepatol. 22*:71–76 (1995).

12. M. Müller and P. L. M. Jansen, Molecular aspects of hepatobiliary transport, *Am. J. Physiol. 272*:G1285–1303 (1997).

13. P. J. Meier, Molecular mechanisms of hepatic bile salt transport from sinusoidal blood into bile, *Am. J. Physiol. 269*:G801–812 (1995).

14. A. W. Wolkoff, Hepatocellular sinusoidal membrane organic anion transport and transporters, *Semin. Liver Dis. 16*:121–127 (1996).

15. K. Fluiter and T. J. C. Van Berkel, Scavenger receptor B1 (SR- B1) substrates inhibit the selective uptake of high-density-lipoprotein cholesteryl esters by rat parenchymal liver cells, *Biochem. J. 326*:515–519 (1997).

16. A. Rigotti, S. L. Acton, and M. Krieger, The class B scavenger receptors SR-BI and CD36 are receptors for anionic phospholipids, *J. Biol. Chem. 270*: 16221–16224 (1995).

17. G. Y. Wu and C. H. Wu, Receptor-mediated delivery of foreign genes to hepatocytes, *Adv. Drug Delivery Rev. 29*:243–248 (1998).

18. D. Schouten, M. Van der Kooij, J. Muller, M. N. Pieters, M. K. Bijsterbosch, and T. J. C. Van Berkel, Development of lipoprotein-like lipid particles for drug targeting: neo-high density lipoproteins. *Mol. Pharmacol. 44*:486–492 (1993).

19. C. A. Toth and P. Thomas, Liver endocytosis and Kupffer cells, *Hepatology 16*: 255–266 (1992).

20. E. A. L. Biessen, D. M. Beuting, H. F. Bakkeren, J. Kuiper, and T. J. C. Van Berkel, Evaluation of the particle size as a determinant of affinity for the hepatic galactose-particle and asialoglycoprotein receptor, *Hepatic Endocytosis of Lipids and Proteins* (E. Windler, H. Greten, eds.), W. Zuckschwerdt Verlag, München, 1992, pp. 157–165.

21. B. Smedsrod, J. Melkko, N. Araki, H. Sano, and S. Horiuchi. Advanced glycation end products are eliminated by scavenger-receptor-mediated endocytosis in hepatic sinusoidal Kupffer and endothelial cells, *Biochem. J. 322*:567–573 (1997).

22. P. D. Stahl, The mannose receptor and other macrophage lectins, *Curr. Opin. Immunol. 4*:49–52 (1992).

23. J. Gliemann, H. Henning Jensen, A. Nykjoer, and S. Kragh Moestrup, The $\alpha_2$-macroglobulin receptor/low density lipoprotein receptor-related protein ($\alpha_2$MR/LRP): A potential multifunctional receptor, *Hepatic Endocytosis of Lipids and Proteins* (E. Windler, H. Greten, eds.), W. Zuckschwerdt Verlag, München, 1992, pp. 273–281.

24. J. Kuiper, A. Brouwer, D. L. Knook, and T. J. C. Van Berkel, Kupffer and sinusoidal endothelial cells, *The Liver: Biology and Pathobiology*, 3rd ed. (I. M. Arias, J. L. Boyer, N. Fausto, W. B. Jakoby, D. A. Schachter, D. A. Shafritz, eds.), Raven Press, Ltd., New York, 1994, pp. 791–818.

25. G. M. Kindberg, H. Tolleshaug, T. Gjoen, and T. Berg, Lysosomal and endosomal

heterogeneity in the liver: a comparison of the intracellular pathways of endocytosis in rat liver cells. *Hepatology 13*:254–259 (1991).

26. A. W. Lohse, P. A. Knolle, K. Bilo, A. Uhrig, C. Waldmann, M. Ibe, E. Schmitt, G. Gerken, and K. H. Meyer zum Büschenfelde, Antigen-presenting function and B7 expression of murine sinusoidal endothelial cells and Kupffer cells. *Gastroenterology 110*:1175–1181 (1996).

27. P. Knolle, H. Lohr, U. Treichel, H. P. Dienes, A. Lohse, J. Schlaack, and G. Gerken, Parenchymal and nonparenchymal liver cells and their interaction in the local immune response, *Z. Gastroenterol. 33*:613–620 (1995).

28. K. Decker, Biologically active products of stimulated liver macrophages (Kupffer cells), *Eur. J. Biochem. 192*:245–261 (1990).

29. J. J. Maher and S. L. Friedman, Parenchymal and nonparenchymal cell interactions in the liver, *Semin. Liver Dis. 13*:13–20 (1993).

30. P. J. Winwood and M. J. P. Arthur, Kupffer cells: Their activation and role in animal models of liver injury and human liver disease, *Semin. Liver Dis. 13*:50–59 (1993).

31. B. Smedsrod, P. J. De Bleser, F. Braet, P. Lovisetti, K. Vanderkerken, E. Wisse, and A. Geerts, Cell biology of liver endothelial and Kupffer cells, *Gut 35*:1509–1516 (1994).

32. J. J. Widmann, R. S. Cotran, and H. D. Fahimi, Mononuclear phagocytes (Kupffer cells) and endothelial cells. Identification of two functional cell types in rat liver sinusoids by endogenous peroxidase activity, *J. Cell Biol. 52*:159–170 (1972).

33. G. Harms, C. D. Dijkstra, Y. C. Lee, and M. J. Hardonk, Glycosyl receptors in macrophage subpopulations of rat spleen and lymph node. A comparative study using neoglycoproteins and monoclonal antibodies ED1, ED2 and ED3, *Cell Tissue Res. 262*:35–40 (1990).

34. M. Tomita, K. Yamamoto, H. Kobashi, M. Ohmoto, and T. Tsujii, Immunohistochemical phenotyping of liver macrophages in normal and diseased human liver. *Hepatology 20*:317–325 (1994).

35. J. Y. Scoazes and G. Feldman, In situ immunophenotyping study of endothelial cells of the human hepatic sinusoid: Results and functional implications, *Hepatology 14*:789–797 (1991).

36. A. H. Stolpen, D. E. Golan, and J. S. Pober, Tumor necrosis factor and immune interferon act in concert to slow the lateral diffusion of proteins and lipids in human endothelial cell membranes, *J. Cell Biol. 107*:781–789 (1988).

37. R. Fraser, B. R. Dobbs, G. W. T. Rogers, and G. W. Rogers, Lipoproteins and the liver sieve: the role of the fenestrated sinusoidal endothelium in lipoprotein metabolism, atherosclerosis, and cirrhosis, *Hepatology 21*:863–874 (1995).

38. T. Mori, T. Okanoue, Y. Sawa, N. Hori, H. Kanaoka, Y. Itoh, F. Enjyo, M. Ohta, K. Kagawa, and K. Kashima, The change of sinusoidal endothelial cells in experimental liver cirrhosis—in vivo and in vitro study, *Cells of Hepatic Sinusoid* (D. L. Knook, E. Wisse, eds.), Kupffer Cell Foundation, Leiden, 1993, pp. 280–282.

39. R. Alcolado, M. J. P. Arthur, and J. P. Iredale, Pathogenesis of liver fibrosis, *Clin. Sci. 92*:103–112 (1997).

40. A. M. Gressner, Cytokines and cellular crosstalk involved in the activation of fat-storing cells, *J. Hepatol. 22*:28–36 (1995).

41. N. A. Essani, G. M. McGuire, A. M. Manning, and H. Jaeschke, Differential induction of mRNA for ICAM-1 and selections in hepatocytes, Kupffer cells and endothelial cells during endotoxemia, *Biochem. Biophys. Res. Commun. 211*:74–82 (1995).

42. H. Jaeschke, Cellular adhesion molecules: regulation and functional significance in the pathogenesis of liver diseases, *Amer. J. Physiol.-Gastrointest. L. 36*:G602–611 (1997).

43. R. Volpes, J. J. Van den Oord, and V. J. Desmet, Vascular adhesion molecules in acute and chronic liver inflammation, *Hepatology 15*:269–275 (1992).

44. J. J. Maher and R. F. McGuire, Extracellular matrix gene expression increases preferentially in rat lipocytes and sinusoidal endothelial cells during hepatic fibrosis in vivo, *J. Clin. Invest. 86*:1641–1688 (1990).

45. M. G. Irving, F. J. Roll, S. Huang, and D. M. Bissell, Characterization and culture of sinusoidal endothelium from normal rat liver: Lipoprotein uptake and collagen phenotype, *Gastroenterology 691*:239–246 (1995).

46. W. R. Jarnagin, D. C. Rockey, V. E. Koteliansky, S. S. Wang, and D. M. Bissell, Expression of variant fibronectins in wound healing: Cellular source and biological activity of the EIIIA segment in rat hepatic fibrogenesis, *J. Cell Biol. 127*:2037–2048 (1994).

47. B. Smedsrod, H. Pertoft, S. Gustafson, and T. C. Laurent, Scavenger functions of the liver endothelial cell, *Biochem. J. 266*:313–327 (1990).

48. T. J. C. Van Berkel, A. Van Velzen, J. K. Kruijt, H. Suzuki, and T. Kodama, Uptake and catabolism of modified LDL in scavenger-receptor class A type I/II knock-out mice, *Biochem. J. 331*:29–35 (1998).

49. M. van Oosten, E. van de Bilt, T. J. C. Van Berkel, and J. Kuiper. New scavenger receptor-like receptors for the binding of lipopolysaccharide to liver endothelial and Kupffer cells, *Infect. Immun. 66*:5107–5112 (1998).

50. S. Magnusson and T. Berg, Extremely rapid endocytosis mediated by the mannose receptor of sinusoidal endothelial rat liver cells, *Biochem. J. 257*:651–656 (1989).

51. N. Forsberg and S. Gustafson, Characterization and purification of the hyaluronan-receptor on liver endothelial cells, *Biochim. Biophys. Acta 1078*:12–18 (1991).

52. J. Melkko, T. Hellevik, L. Risteli, J. Risteli, and B. Smedsrod, Clearance of NH2-terminal propeptides of types I and III procollagen is a physiological function of the scavenger receptor in liver endothelial cells, *J. Exp. Med. 179*:405–412 (1994).

53. T. Hellevik, I. Martinez, R. Olsen, B. H. Toh, P. Webster, and B. Smedsrod, Transport of residual endocytosed products into terminal lysosomes occurs slowly in rat liver endothelial cells, *Hepatology 28*:1378–1389 (1998).

54. I. Kosugi, H. Muro, H. Shirasawa, and I. Ito, Endocytosis of soluble IgG immune complex and its transport to lysosomes in hepatic sinusoidal endothelial cells, *J. Hepatol. 16*:106–114 (1992).

55. H. Muro, H. Shirasawa, I. Kosugi, and S. Nakamura, Defect of Fc receptors and phenotypical changes in sinusoidal endothelial cells in human liver cirrhosis, *Am. J. Pathol. 143*:105–120 (1993).

56. R. L. Roberts and A. Sandra, Receptor-mediated endocytosis of insulin by cultured endothelial cells. *Tissue Cell 24*:603–611 (1992).

57. S. Irie and M. Tavassoli, Liver endothelium desialylates ceruloplasmin, *Biochem. Biophys. Res. Commun. 140*:94–100 (1986).

58. S. Irie and M. Tavassoli, Desialylation of transferrin by liver endothelium is selective for its triantennary chain, *Biochem. J. 263*:491–496 (1989).

59. M. Pinzani, Novel insights into the biology and physiology of the Ito cell, *Pharmacol. Ther. 66*:387–412 (1995).

60. M. Pinzani, F. Marra, and V. Carloni, Signal transduction in hepatic stellate cells, *Liver 18*:2–13 (1998).

61. M. L. Hautekeete and A. Geerts, The hepatic stellate (Ito) cell: Its role in human liver disease, *Virchows Arch. 430*:195–207 (1997).

62. A. M. Gressner, Mediators of hepatic fibrogenesis, *Hepatogastroenterology 43*:92–103 (1996).

63. G. Ramadori, The stellate cell (Ito-cell, fat-storing cell, lipocyte, perisinusoidal cell) of the liver. New insights into pathophysiology of an intriguing cell, *Virchows Arch. [B] 61*:147–158 (1991).

64. S. L. Friedman, G. H. Millward-Sadler, and M. J. P. Arthur, Liver fibrosis and cirrhosis, *Wright's Liver and Biliary Disease. Pathophysiology, Diagnosis and Management*, 3rd ed. (G. H. Millward-Sadler, R. Wright, M. J. P. Arthur, eds.), W. B. Saunders Company Ltd., London Philadelphia Toronto Sydney Tokyo, 1992, pp. 822–881.

65. M. J. P. Arthur, D. A. Mann, and J. P. Iredale, Tissue inhibitors of metalloproteinases, hepatic stellate cells and liver fibrosis, *J. Gastroenterol. Hepatol. 13*:S33–38 (1998).

66. A. D. Burt, Cellular and molecular aspects of hepatic fibrosis, *J. Pathol. 170*:105–114 (1993).

67. A. Mallat, Hepatic stellate cells and intrahepatic modulation of portal pressure, *Digestion 59*:416–419 (1998).

68. N. Kawada, The hepatic perisinusoidal stellate cell, *Histol. Histopathol. 12*:1069–1080 (1997).

69. G. A. Ramm, R. S. Britton, R. O'Neill, W. S. Blaner, and B. R. Bacon, Vitamin A-poor lipocytes: a novel desmin-negative lipocyte subpopulation, which can be activated to myofibroblasts, *Am. J. Physiol. 269*: G532–541 (1995).

70. G. Ballardini, P. Groff, L. Badiali de Giorgi, D. Schuppan, and F. B. Bianchi, Ito cell heterogeneity: desmin-negative Ito cells in normal rat liver, *Hepatology 19*: 440–446 (1994).

71. K. Neubauer, T. Knittel, S. Aurisch, P. Fellmer, and G. Ramadori, Glial fibrillary acidic protein—A cell type specific marker for Ito cells in vivo and in vitro, *J. Hepatol. 24*:719–730 (1996).

72. G. Buniatian, B. Hamprecht, and R. Gebhardt, Glial fibrillary acidic protein as a marker of perisinusoidal stellate cells that can distinguish between the normal and myofibroblast-like phenotypes, *Biol. Cell 87*:65–73 (1996).

73. A. Geerts, P. J. De Bleser, M. L. Hautekeete, T. Niki, and E. Wisse, Fat-storing (Ito) cell biology, *The Liver: Biology and Pathobiology*, 3rd ed (I. M. Arias, J. L. Boyer, N. Fausto, W. B. Jacoby, D. A. Schachter, D. A. Shafritz, eds.), Raven Press, Ltd., New York, 1994, pp. 819–838.

74. H. Enzan, H. Himeno, T. Saibara, S. Onishi, Y. Yamamoto, and H. Hara, Immuno-

histochemical identification of Ito cells and their myofibroblastic transformation in adult human liver, *Virchows Arch. 424*:249–256 (1994).

75. T. Knittel, L. Muller, B. Saile, and G. Ramadori, Effect of tumour necrosis factor-alpha on proliferation, activation and protein synthesis of rat hepatic stellate cells, *J. Hepatol. 27*:1067–1080 (1997).

76. A. M. Gressner, S. Lotfi, G. Gressner, E. Haltner, and J. Kropf, Synergism between hepatocytes and Kupffer cells in the activation of fat storing cells (perisinusoidal lipocytes), *J. Hepatol. 19*:117–132 (1993).

77. H. Moshage, Cytokines and the hepatic acute phase response, *J. Pathol. 181*:257–266 (1997).

78. F. W. Ruscetti, M. C. Birchenallroberts, J. M. McPherson, and R. H. Wiltrout, Transforming growth factor beta(1), *Cytokines* (A. R. Miresluis, R. Thorpe, eds.), Academic Press Inc. San Diego, 1998, pp. 415–432.

79. C. Hellerbrand, B. Stefanovic, F. Giordano, E. R. Burchard, and D. A. Brenner, The role of TGFβ1 in initiating hepatic stellate cell activation in vivo. *J. Hepatol. 30*:77–87 (1999).

80. L. Wong, G. Yamasaki, R. J. Johnson, and S. L. Friedman, Induction of β-platelet-derived growth factor receptor in rat hepatic lipocytes during cellular activation in vivo and in culture, *J. Clin. Invest. 94*:1563–1569 (1994).

81. S. L. Friedman and M. J. P. Arthur, Activation of cultured rat hepatic lipocytes by Kupffer cell conditioned medium. Direct enhancement of matrix synthesis and stimulation of cell proliferation via induction of platelet-derived growth factor receptors, *J. Clin. Invest. 84*:1780–1785 (1989).

82. K. S. Lee, M. Buck, K. Houglum, and M. Chojkier, Activation of hepatic stellate cells by TGF alpha and collagen type I is mediated by oxidative stress through c-myb expression, *J. Clin. Invest. 96*:2461–2468 (1995).

83. A. Lalazar, L. Wong, G. Yamasaki, and S. L. Friedman, Early genes induced in hepatic stellate cells during wound healing. *Gene 195*:235–243 (1997).

84. M. Ohata, M. Lin, M. Satre, and H. Tsukamoto, Diminished retinoic acid signaling in hepatic stellate cells in cholestatic liver fibrosis, *Am. J. Physiol. 272*:G589–596 (1997).

85. L. RubbiaBrandt, G. Mentha, A. Desmouliere, A. M. A. Costa, E. Giostra, G. Molas, H. Enzan, and G. Gabbiani, Hepatic stellate cells reversibly express alpha-smooth muscle actin during acute hepatic ischemia, *Transplant Proc. 29*:2390–2395 (1997).

86. M. J. P. Arthur. Fibrosis and altered matrix degradation, *Digestion 59*:376–380 (1998).

87. M. Guido, M. Rugge, L. Chemello, G. Leandro, G. Fattovich, G. Giustina, M. Cassaro, and A. Alberti, Liver stellate cells in chronic viral hepatitis: the effect of interferon therapy, *J. Hepatol. 24*:301–307 (1996).

88. C. Schmitt, C. Royer, A. M. Steffan, N. Labouret, C. Caussin, M. C. Navas, D. Jaeck, A. Kim, and F. Stoll-Keller, Interaction between human hepatic stellate cells and hepatitis C virus. *Hepatology 28*:438A [abstract] (1998).

89. S. J. Johnson, A. W. Burr, K. Toole, C. L. Dack, J. Mathew, and A. D. Burt, Macrophage and hepatic stellate cell responses during experimental hepatocarcinogenesis, *J. Gastroenterol. Hepatol. 13*:145–151 (1998).

90. Y. N. Park, C. P. Yang, O. Cubukcu, S. N. Thung, and N. D. Theise, Hepatic stellate cell activation in dysplastic nodules: Evidence for an alternate hypothesis concerning human hepatocarcinogenesis, *Liver 17*:271–274 (1997).

91. J. Wu and Å Danielsson, Inhibition of hepatic fibrogenesis: a review of pharmacologic candidates, *Scand. J. Gastroenterol. 29*:385–391 (1994).

92. A. M. Gressner, Perisinusoidal lipocytes and fibrogenesis, *Gut 35*:1331–1333 (1994).

93. D. Schuppan, D. Strobel, and E. G. Hahn, Hepatic fibrosis—Therapeutic strategies, *Digestion 59*:385–390 (1998).

94. L. Beljaars, K. Poelstra, G. Molema, and D. K. F. Meijer, Targeting of sugar- and charge-modified albumins to fibrotic rat livers: The accessibility of hepatic cells after chronic bile duct ligation, *J. Hepatol. 29*:579–588 (1998).

95. H. Senoo, S. Smeland, E. Stang, N. Roos, T. Berg, K. R. Norum, and R. Blomhoff, Stellate cells take up retinol-binding protein, *Cells of the Hepatic Sinusoid* (D. L. Knook, E. Wisse, eds.), Kupffer Cell Foundation, Leiden, 1993, pp. 423–425.

96. P. J. De Bleser, P. Jannes, S. C. Van Buul-Offers, C. M. Hoogerbrugge, C. F. H. Van Schravendijk, T. Niki, V. Rogiers, J. L. Van den Brande, E. Wisse, and A. Geerts, Insulinlike growth factor-II/mannose 6-phosphate receptor is expressed on CCl4-exposed rat fat-storing cells and facilitates activation of latent transforming growth factor-beta in cocultures with sinusoidal endothelial cells, *Hepatology 21*: 1429–1437 (1995).

97. P. J. De Bleser, C. D. Scott, T. Niki, G. Xu, E. Wisse, and A. Geerts, Insulin-like growth factor II/mannose 6-phosphate-receptor expression in liver and serum during acute CCl4 intoxication in the rat, *Hepatology 23*:1530–1537 (1996).

98. J. K. Weiner, A. P. Chen, and B. H. Davis, E-box-binding repressor is down-regulated in hepatic stellate cells during up-regulation of mannose 6-phosphate insulin-like growth factor-II receptor expression in early hepatic fibrogenesis, *J. Biol. Chem. 273*:15913–15919 (1998).

99. A. M. Tiggelman, C. Linthorst, W. Boers, H. S. Brand, and R. A. Chamuleau, Transforming growth factor-beta-induced collagen synthesis by human liver myofibroblasts is inhibited by alpha 2-macroglobulin. *J. Hepatol. 26*:1220–1228 (1997).

100. C. A. Kawser, J. P. Iredale, P. J. Winwood, and M. J. P. Arthur, Rat hepatic stellate cell expression of alpha 2-macroglobulin is a feature of cellular activation: implications for matrix remodelling in hepatic fibrosis, *Clin. Sci. 95*:179–186 (1998).

101. G. A. Ramm, R. S. Britton, R. O. O'Neill, and B. R. Bacon, Identification and characterization of a receptor for tissue ferritin on activated rat lipocytes, *J. Clin. Invest. 94*:9–15 (1994).

102. V. Carloni, R. G. Romanelli, M. Pinzani, G. Laffi, and P. Gentilini, Expression and function of integrin receptors for collagen and laminin in cultured human hepatic stellate cells, *Gastroenterology 110*:1127–1136 (1996).

103. F. Ballet, Hepatic circulation: Potential for therapeutic intervention, *Pharmacol. Ther. 47*: 281–328 (1990).

104. M. Rojkind and P. Greenwel, Homeostasis of the liver ecosystem: alterations in cirrhosis. *Cells of the Hepatic Sinusoid* 4th ed. (D. L. Knook, E. Wisse, eds.), Kupffer Cell Foundation, Leiden, 1993, pp. 205–208.

105. J. M. Crawford, Cellular and molecular biology of the inflamed liver. *Curr. Opin. Gastroenterol. 13*:175–185 (1997).

106. G. Majno, Chronic inflammation. Links with angiogenesis and wound healing, *Am. J. Pathol. 153*:1035–1039 (1998).

107. H. Sprenger, A. Kaufmann, H. Garn, B. Lahme, D. Gemsa, and A. M. Gressner, Induction of neutrophil-attracting chemokines in transforming rat hepatic stellate cells, *Gastroenterology 113*:277–285 (1997).

108. S. Suzuki, L. H. Toledo Pereyra, F. Rodriguez, and F. Lopez, Role of Kupffer cells in neutrophil activation and infiltration following total hepatic ischemia and reperfusion, *Circ. Shock 42*:204–209 (1994).

109. H. Jaeschke, C. W. Smith, M. G. Clemens, P. E. Ganey, and R. A. Roth, Mechanisms of inflammatory liver injury: Adhesion molecules and cytotoxicity of neutrophils, *Toxicol. Appl. Pharmacol. 139*:213–226 (1996).

110. A. Casini, E. Ceni, R. Salzano, P. Biondi, M. Parola, A. Galli, M. Foschi, A. Caligiuri, M. Pinzani, and C. Surrenti, Neutrophil-derived superoxide anion induces lipid peroxidation and stimulates collagen synthesis in human hepatic stellate cells: role of nitric oxide, *Hepatology 25*:361–367 (1997).

111. B. N. Cronstein, S. C. Kimmel, R. I. Levin, F. Martiniuk, and G. Weissmann, A mechanism for the antiinflammatory effects of corticosteroids: the glucocorticoid receptor regulates leukocyte adhesion to endothelial cells and expression of endothelial-leukocyte adhesion molecule 1 and intercellular adhesion molecule 1, *Proc. Natl. Acad. Sci. USA 89*:9991–9995 (1992).

112. J. P. Iredale, Matrix turnover in fibrogenesis, *Hepatogastroenterology 43*:56–71 (1996).

113. G. Ramadori, T. Knittel, and B. Saile, Fibrosis and altered matrix synthesis, *Digestion 59*:372–375 (1998).

114. D. Schuppan, T. Cramer, M. Bauer, T. Strefeld, E. G. Hahn, and H. Herbst, Hepatocytes as a source of collagen type XVIII endostatin, *Lancet 352*:879–880 (1998).

115. I. Ogata, S. Mochida, T. Tomiya, and K. Fujiwara, Minor contribution of hepatocytes to collagen production in normal and early fibrotic Rat Livers, *Hepatology 14*:361–367 (1991).

116. M. Rojkind, Fibrogenesis in cirrhosis. Potential for therapeutic intervention, *Pharmacol. Ther. 53*:81–104 (1992).

117. G. P. Anderson, Resolution of chronic inflammation by therapeutic induction of apoptosis, *Trends Pharmacol. Sci. 17*:438–442 (1996).

118. J. P. Iredale, R. C. Benyon, J. Pickering, M. McCullen, M. Northrop, S. Pawley, C. Hovell, and M. J. P. Arthur, Mechanisms of spontaneous resolution of rat liver fibrosis—Hepatic stellate cell apoptosis and reduced hepatic expression of metalloproteinase inhibitors, *J. Clin. Invest. 102*:538–549 (1998).

119. W. R. Gong, A. Pecci, S. Roth, B. Lahme, M. Beato, and A. M. Gressner, Transformation-dependent susceptibility of rat hepatic stellate cells to apoptosis induced by soluble Fas ligand, *Hepatology 28*:492–502 (1998).

120. B. Saile, T. Knittel, N. Matthes, P. Schott, and G. Ramadori, Cd95/cd95l-mediated apoptosis of the hepatic stellate cell: a mechanism terminating uncontrolled hepatic stellate cell proliferation during hepatic tissue repair, *Am. J. Pathol. 151*:1265–1272 (1997).

121. R. S. Koff, Solving the mysteries of viral hepatitis, *Sci. Am. Sci. Med. 1*:24–33 (1994).
122. R. G. Gish and E. B. Keeffe, Recent developments in the treatment of chronic hepatitis B virus infection, *Exp. Opin. Invest. Drugs 4*:95–115 (1995).
123. V. Carreño, J. Bartolomé, and I. Castillo, Long-term effect of interferon therapy in chronic hepatitis B, *J. Hepatol. 20*:431–435 (1994).
124. P. C. N. Rensen, R. L. A. de Vrueh, and T. J. C. Van Berkel, Targeting hepatitis B therapy to the liver: Clinical pharmacokinetic considerations, *Clin. Pharmacokinet. 31*:131–165 (1996).
125. J. C. Arnold, J. G. Ogrady, G. Otto, B. Kommerell, G. J. M. Alexander, and R. Williams, CMV reinfection/reactivation after liver transplantation, *Transplant Proc. 23*:2632–2633 (1991).
126. T. Martelius, L. Krogerus, K. Hockerstedt, C. Bruggeman, and I. Lautenschlager, Cytomegalovirus infection is associated with increased inflammation and severe bile duct damage in rat liver allografts, *Hepatology 27*:996–1002 (1998).
127. F. T. Hufert, J. Schmitz, M. Schreiber, H. Schmitz, P. Rácz, and D. D. Von Laer, Human Kupffer cells infected with HIV-1 in vivo, *J. Acquir. Immune Defic. Syndr. 6*:772–777 (1993).
128. P. J. Swart, E. Beljaars, A. Pasma, C. Smit, H. Schuitemaker, and D. K. F. Meijer, Comparative pharmacokinetic, immunologic and hematologic studies on the anti-HIV-1/2 compounds aconitylated and succinylated HSA, *J. Drug Target 4*:109–116 (1996).
129. M. Kuipers, M. Witvrouw, J. Este, E. De Clercq, D. Meijer, and P. Swart, Anti-HIV-1 activity of combinations and covalent conjugates of negatively charged albumins and AZT, *Clin. Pharmacokinet. 13*:131–140 (1998).
130. G. T. Strickland and S. L. Hoffman, Strategies for the control of malaria, *Sci. Am. Sci. Med. July/August*:24–33 (1994).
131. A. Mukhopadhyay, G. Chaudhuri, S. K. Arora, S. Sehgal, and S. K. Basu, Receptor-mediated drug delivery to macrophages in chemotherapy of leishmaniasis, *Science 244*:705–770 (1989).
132. G. Chaudhuri, A. Mukhopadhyay, and S. K. Basu, Selective delivery of drugs to macrophages through a highly specific receptor. An efficient chemotherapeutic approach against Leishmaniasis, *Biochem. Pharmacol. 38*:2995–3002 (1989).
133. D. J. Wyler, Molecular and cellular basis of hepatic fibrogenesis in experimental *Schistosomiasis mansoni* infection, *Mem. Inst. Oswaldo Cruz 87*:117–125 (1992).
134. M. A. Adson, Primary hepatocellular cancers—Western experience, *Survey of the Liver and Biliary Tract* (L. H. Blumgart, ed.), Churchill Livingstone Publisher, New York, 1988, pp. 1153–1165.
135. D. C. Snover, Thoughts on the development of hepatocellular carcinoma in cirrhotic and noncirrhotic livers, *Am. J. Clin. Pathol. 105*:3–5 (1996).
136. P. V. M. Shekhar, C. J. Aslakson, and F. R. Miller, Molecular events in metastatic progression, *Semin. Cancer Biol. 4*:193–204 (1993).
137. A. J. Hayes, L. Y. Li, and M. E. Lippmann, Science, medicine, and the future. Antivascular therapy: A new approach to cancer treatment, *BMJ 318*:853–856 (1999).
138. X. Huang, G. Molema, S. King, L. Watkins, T. S. Edgington, and P. E. Thorpe,

Tumor infarction in mice by antibody-directed targeting of tissue factor to tumor vasculature, *Science* 275:547–550 (1997).

139. B. J. Kroesen, W. Helfrich, G. Molema, and L. de Leij, Bispecific antibodies for treatment of cancer in experimental animal models and man, *Adv. Drug Delivery Rev.* 31:105–129 (1998).

140. G. Molema, D. K. F. Meijer, and L. F. M. H. de Leij, Tumor vasculature targeted therapies. Getting the players organized, *Biochem. Pharmacol.* 55:1939–1945 (1998).

141. G. Molema and A. W. Griffioen, Rocking the foundations of solid tumor growth by attacking the tumor's blood supply, *Immunol. Today* 19:392–394 (1998).

142. D. J. A. Crommelin, G. Scherphof, and G. Storm, Active targeting with particulate carrier systems in the blood compartment, *Adv. Drug Deliv. Rev.* 17:49–60 (1995).

143. D. K. F. Meijer, G. Molema, F. Moolenaar, D. De Zeeuw, and P. J. Swart, (Glyco)-protein drug carriers with an intrinsic therapeutic activity: The concept of dual targeting, *J. Contr. Rel.* 39:163–172 (1996).

144. M. E. Kuipers, J. G. Huisman, P. J. Swart, M. de Béthune, R. Pauwels, E. De Clercq, H. Schuitemaker, and D. K. F. Meijer, Mechanism of anti-HIV activity of negatively charged albumins: Biomolecular interaction with the HIV-1 envelope protein gp120, *J. Acq. Immun. Defic. Synd. Hum. R.* 11:419–429 (1996).

145. P. J. Swart, M. E. Kuipers, C. Smit, R. Pauwels, M. de Béthune, E. De Clercq, H. Huisman, and D. K. F. Meijer, Antiviral effects of milk proteins: Acylation results in polyanionic compounds with potent activity against human immunodeficiency virus type 1 and 2 in vitro, *AIDS Res. Hum. Retroviruses* 12:769–775 (1996).

146. K. Poelstra, W. W. Bakker, P. A. Klok, J. A. Kamps, M. J. Hardonk, and D. K. F. Meijer, Dephosphorylation of endotoxin by alkaline phosphatase in vivo, *Am. J. Pathol.* 151:1163–1169 (1997).

147. P. Caliceti, O. Schiavon, T. Hirano, S. Ohashi, and F. M. Veronese, Modification of physico-chemical and biopharmaceutical properties of superoxide dismutase by conjugation to the co-polymer of divinyl ether and maleic anhydride, *J. Contr. Rel.* 39:27–34 (1996).

148. R. Duncan, Drug-polymer conjugates: Potential for improved chemotherapy, *Anti-Cancer Drugs* 3:175–210 (1992).

149. H. Soyez, E. Schacht, and S. Van Der Kerken, The crucial role of spacer groups in macromolecular prodrug design, *Adv. Drug Deliv. Rev.* 21:81–106 (1996).

150. E. J. F. Franssen, F. Moolenaar, D. De Zeeuw, and D. K. F. Meijer, Low molecular weight proteins as carriers for renal drug targeting: Naproxen coupled to lysozyme via the spacer L-lactic acid, *Pharm. Res.* 10:963–969 (1993).

151. E. J. F. Franssen, J. Koiter, C. A. M. Kuipers, A. P. Bruins, F. Moolenaar, D. De Zeeuw, W. H. Kruizinga, R. M. Kellogg, and D. K. F. Meijer, Low molecular weight proteins as carriers for renal drug targeting. Preparation of drug-protein conjugates and drug-spacer derivatives and their catabolism in renal cortex homogenates and lysosomal lysates, *J. Med. Chem.* 35:1246–1259 (1992).

152. G. L. Scherphof and J. A. A. M. Kamps, Receptor versus non-receptor mediated clearance of liposomes, *Adv. Drug Delivery Rev.* 32:81–97 (1998).

153. G. Gregoriadis and A. T. Florence, Liposomes in drug delivery. Clinical, diagnostic and ophthalmic potential, *Drugs* 45:15–28 (1993).

154. J. Wu and M. A. Zern, Modification of liposomes for liver targeting, *J. Hepatol.* *24*:575–763 (1996).

155. M. K. Bijsterbosch and T. J. C. Van Berkel, Lactosylated high density lipoprotein: a potential carrier for the site-specific delivery of drugs to parenchymal liver cells, *Mol. Pharmacol.* *41*:404–411 (1992).

156. P. C. N. Rensen, M. C. M. Van Dijk, E. C. Havenaar, M. K. Bijsterbosch, J. K. Kruijt, and T. J. C. Van Berkel, Selective liver targeting of antivirals by recombinant chylomicrons- a new therapeutic approach to hepatitis B, *Nature Med.* *1*:221–225 (1995).

157. S. S. Davis. Biomedical applications of nanotechnology—Implications for drug targeting and gene therapy, *Trends Biotechnol.* *15*:217–224 (1997).

158. C. Kosmas, H. Linardou, and A. A. Epenetos, Review: Advances in monoclonal antibody tumour targeting, *J. Drug Target* *1*:81–91 (1993).

159. R. J. Kreitman and I. Pastan, Immunotoxins for targeted cancer therapy, *Adv. Drug Deliv. Rev.* *31*:53–88 (1998).

160. W. Kramer, G. Wess, A. Enhsen, E. Falk, A. Hoffmann, G. Neckermann, G. Schubert, and M. Urmann, Modified bile acids as carriers for peptides and drugs, *J. Contr. Rel.* *46*:17–30 (1997).

161. T. Reynolds, Polymers help guide cancer drugs to tumor targets—and keep them there, *J. Natl. Cancer Inst.* *87*:1582–1584 (1995).

162. E. A. L. Biessen, D. M. Beuting, H. Vietsch, M. K. Bijsterbosch, and T. J. C. Van Berkel, Specific targeting of the antiviral drug 5-Iodo 2′-deoxyuridine to the parenchymal liver cell using lactosylated poly-L-lysine, *J. Hepatol.* *21*:806–815 (1994).

163. E. Bonfils, C. Depierreux, P. Midoux, N. T. Thuong, M. Monsigny, and A. C. Roche, Drug targeting: synthesis and endocytosis of oligonucleotide-neoglycoprotein conjugates, *Nucl. Acids Res.* *20*:4621–4629 (1992).

164. S. M. Moghimi and S. S. Davis, Innovations in avoiding particle clearance from blood by Kupffer cells: Cause for reflection, *Crit. Rev. Ther. Drug Carrier Syst.* *11*:31–59 (1994).

165. E. Forssen and M. Willis, Ligand-targeted liposomes, *Adv. Drug Del. Rev.* *29*:249–271 (1998).

166. D. D. Spragg, D. R. Alford, R. Greferath, C. E. Larsen, K. D. Lee, G. C. Gurtner, M. J. Cybulsky, P. F. Tosi, C. Nicolau, and M. A. Gimbrone, Immunotargeting of liposomes to activated vascular endothelial cells: A strategy for site-selective delivery in the cardiovascular system, *Proc. Natl. Acad. Sci. USA* *94*:8795–8800 (1997).

167. J. A. A. M. Kamps, H. W. M. Morselt, P. J. Swart, D. K. F. Meijer, and G. L. Scherphof, Massive targeting of liposomes, surface-modified with anionized albumins, to hepatic endothelial cells, *Proc. Natl. Acad. Sci. USA* *94*:11681–11685, (1997).

168. T. M. Allen, L. Murray, S. MacKeigan, and M. Shah, Chronic liposome administration in mice: Effects on reticuloendothelial function and tissue distribution, *J. Pharmacol. Exp. Ther.* *229*:267–275 (1984).

169. G. Storm, C. Oussoren, P. A. M. Peeters, and Y. Barenholz, Tolerability of liposomes in vivo, *Liposome Technology* 2nd (G. Gregoriadis, ed.), CRC Press, Boca Raton, 1993, pp. 345–383.

170. L. W. Seymour, Soluble polymers for lectin-mediated drug targeting, *Adv. Drug Deliv. Rev. 14*:89–111 (1994).

171. L. Fiume, C. Busi, P. Preti, and G. Spinosa, Conjugates of ara-AMP with lactosaminated albumin: A study on their immunogenicity in mouse and rat, *Cancer Drug Delivery 4*:145–150 (1987).

172. L. Fiume, A. Mattioli, C. Busi, G. Spinosa, and T. Wieland, Conjugates of adenine 9-β-D-arabinofuranoside monophosphate (ara-AMP) with lactosaminated homologous albumin are not immunogenic in the mouse, *Experientia 38*:1087–1089 (1982).

173. M. T. Cerenzia, L. Fiume, W. D. Venon, B. Lavezzo, M. R. Brunetto, A. Ponzetto, G. Di Stefano, C. Busi, A. Mattioli, G. B. Gervasi, et al., Adenine arabinoside monophosphate coupled to lactosaminated human albumin administered for 4 weeks in patients with chronic type B hepatitis decreased viremia without producing significant side effects, *Hepatology 23*:657–661 (1996).

174. Y. Kato and Y. Sugiyama, Targeted delivery of peptides, proteins, and genes by receptor-mediated endocytosis, *Crit. Rev. Ther. Drug Carrier Syst. 14*:287–331 (1997).

175. E. P. Feener and G. L. King, The biochemical and physiological characteristics of receptors, *Adv. Drug Deliv. Rev. 29*:197–213 (1998).

176. I. Grinko, A. Geerts, and E. Wisse, Experimental biliary fibrosis correlates with increased numbers of fat-storing and Kupffer cells, and portal endotoxemia, *J. Hepatol. 23*:449–458 (1995).

177. J. E. Hines, S. J. Johnson, and A. D. Burt, In vivo responses of macrophages and perisinusoidal cells to cholestatic liver injury, *Am. J. Pathol. 142*:511–518 (1993).

178. J. B. Burgess, J. U. Baenziger, and W. R. Brown, Abnormal surface distribution of the human asialoglycoprotein receptor in cirrhosis, *Hepatology 15*:702–706 (1992).

179. I. Virgolini, C. Müller, W. Klepetko, P. Angelberger, H. Bergmann, J. O'Grady, and H. Sinzinger, Decreased hepatic function in patients with hepatoma or liver metastasis monitored by a hepatocyte specific galactosylated radioligand, *Br. J. Cancer 61*:937–941 (1990).

180. T. Sawamura, H. Nakada, H. Hazama, Y. Shiozaki, Y. Sameshima, and Y. Tashiro, Hyperasialoglycoproteinemia in patients with chronic liver diseases and/or liver cell carcinoma. Asialoglycoprotein receptor in cirrhosis and liver cell carcinoma, *Gastroenterology 87*:1217–1221 (1984).

181. R. Sett, K. Sarkar, and P. K. Das, Macrophage-directed delivery of doxorubicin conjugated to neoglycoprotein using leishmaniasis as the model disease, *J. Infect. Dis. 168*:994–999 (1993).

182. K. Sarkar and P. K. Das, Protective effect of neoglycoprotein-conjugated muramyl dipeptide against *Leishmania donovani* infection: the role of cytokines. *J. Immunol. 158*:5357–5365 (1997).

183. N. Basu, R. Sett, and P. K. Das, Down-regulation of mannose receptors on macrophages after infection with *Leishmania donovani. Biochem. J. 277*:451–455 (1991).

184. M. G. Bachem, D. Meyer, W. Schäfer, U. Riess, R. Melchior, K. M. Sell, and A. M. Gressner, The response of rat liver perisinusoidal lipocytes to polypeptide growth regulator changes with their transdifferentiation into myofibroblast-like cells in culture, *J. Hepatol. 18*:40–52 (1993).

185.  M. Fukuta, H. Okada, H. Iinuma, S. Yanai, and H. Toguchi, Insulin fragments as a carrier for peptide delivery across the blood-brain barrier, *Pharm. Res. 11*:1681–1688 (1994).

186.  L. Beljaars, K. Poelstra, B. Weert, G. Molema, D. Schuppan, and D. K. F. Meijer, The development of novel albumin carriers to hepatic stellate cells by application of cyclopeptide moieties recognizing collagen type VI and platelet derived growth factor receptors. *Hepatology 28*:313A [abstract] (1998).

187.  L. Beljaars, G. Molema, A. Geerts, P. J. De Bleser, B. Weert, D. K. F. Meijer, and K. Poelstra, A cyclic peptide recognizing the platelet derived growth factor receptor as a homing device for targeting to hepatic stellate cells in fibrotic rat livers, submitted (2001).

188.  S. Mukherjee, R. N. Ghosh, and F. R. Maxfield. Endocytosis, *Physiol. Rev. 77*: 759–803 (1997).

189.  R. J. Fallon and A. L. Schwartz, Receptor-mediated delivery of drugs to hepatocytes, *Adv. Drug Deliv. Rev. 4*:49–65 (1989).

190.  T. Eto and H. Takahashi, Enhanced inhibition of hepatitis B virus production by asialoglycoprotein receptor-directed interferon, *Nature Med. 5*:577–581 (1999).

191.  B. Hazes and R. J. Read, Accumulating evidence suggests that several AB-toxins subvert the enoplasmic reticulum-associated protein degradation pathway to enter target cells, *Biochemistry 36*:11051–11054 (1997).

192.  A. Rapak, P. O. Falnes, and S. Olsnes, Retrograde transport of mutant ricin to the endoplasmic reticulum with subsequent translocation to the cytosol, *Proc. Natl. Acad. Sci. USA 94*:3783–3788 (1997).

193.  P. Van der Sluijs, H. P. Bootsma, B. Postema, F. Moolenaar, and D. K. F. Meijer, Drug targeting to the liver with lactosylated albumins. Does the glycoprotein target the drug, or is the drug targeting the glycoprotein? *Hepatology 6*:723–728 (1986).

194.  M. Nishikawa, H. Hirabayashi, Y. Takakura, and M. Hashida, Design for cell-specific targeting of proteins utilizing sugar-recognition mechanisms: effect of molecular weight of proteins on targeting efficiency, *Pharm. Res. 12*:209–214 (1995).

195.  R. W. Jansen, G. Molema, T. L. Ching, R. Oosting, G. Harms, F. Moolenaar, M. J. Hardonk, and D. K. F. Meijer, Hepatic endocytosis of various types of mannose-terminated albumins. What is important, sugar recognition, net charge or the combination of these features, *J. Biol. Chem. 266*:3343–3348 (1991).

196.  E. J. F. Franssen, R. W. Jansen, M. Vaalburg, and D. K. F. Meijer, Hepatic and intrahepatic targeting of an anti-inflammatory agent with human serum albumin and neoglycoproteins as carrier molecules, *Biochem. Pharmacol. 45*:1215–1226 (1993).

197.  P. Walday, H. Tolleshaug, T. Gjoen, G. M. Kindberg, T. Berg, T. Skotland, and E. Holtz, Biodistribution of air-filled albumin microspheres in rats and pigs. *Biochem. J. 299*:437–443 (1994).

198.  Y. C. Lee, C. P. Stowell, and M. J. Krantz, 2-imino -2-methoxyethyl l-thioglycosides: new reagents for attaching sugars to proteins, *Biochemistry 15*:3956–3963 (1976).

199.  R. W. Jansen, P. Olinga, G. Harms, and D. K. F. Meijer, Pharmacokinetic analysis and cellular distribution of the anti-HIV compound succinylated human serum albumin (Suc-HSA) in vivo and in the isolated perfused rat liver, *Pharm. Res. 10*:1611–1614 (1993).

200. R. W. Jansen, G. Molema, G. Harms, J. K. Kruijt, T. J. C. Van Berkel, M. J. Hardonk, and D. K. F. Meijer, Formaldehyde treated albumin contains monomeric and polymeric forms that are differently cleared by endothelial and Kupffer cells of the liver: evidence for scavenger receptor heterogeneity, *Biochem. Biophys. Res. Commun. 180*:23–32 (1991).

201. L. Beljaars, G. Molema, B. Weert, H. Bonnema, P. Olinga, G. M. M. Groothuis, D. K. F. Meijer, and K. Poelstra, Albumin modified with mannose 6-phosphate: a potential carrier for selective delivery of anti-fibrotic drugs to rat and human hepatic stellate cells, *Hepatology 29*:1486–1493 (1999).

202. L. Beljaars, G. Molema, D. Schuppan, A. Geerts, P. J. De Bleser, B. Weert, D. K. F. Meijer, and K. Poelstra, Successful targeting to rat hepatic stellate cells using albumin modified with cyclic peptides that recognize the collagen type VI receptor. *J. Biol. Chem. 275*:12743–12751 (2000).

203. B. A. Schwartz and G. R. Gray, Proteins containing reductively aminated disaccharides: synthesis and characterization, *Arch. Biochem. Biophys. 181*:542–549 (1977).

204. M. J. Krantz, N. A. Holtzman, C. P. Stowell, and Y. C. Lee, Attachment of thiogly-cosides to proteins: Enhancement of liver membrane binding, *Biochemistry 15*: 3963–3968 (1976).

205. M. Kataoka and M. Tavassoli, Synthetic neoglycoproteins: a class of reagents for detection of sugar-recognizing substances, *J. Histochem. Cytochem. 32*:1091–1094 (1984).

206. R. W. Jansen, G. Molema, R. Pauwels, D. Schols, E. De Clercq, and D. K. F. Meijer, Potent in vitro anti-human immunodeficiency virus-1 activity of modified human serum albumins, *Mol. Pharmacol. 39*:818–823 (1991).

207. R. W. Jansen, D. Schols, R. Pauwels, E. De Clercq, and D. K. F. Meijer, Novel, negatively charged, human serum albumins display potent and selective in vitro anti-human immunodeficiency virus type 1 activity, *Mol. Pharmacol. 44*:1003–1007 (1993).

208. J. N. Purtell, A. J. Pesce, D. H. Clyne, W. C. Miller, and V. E. Pollak, Isoelectric point of albumin: Effect on renal handling of albumin, *Kidney Int. 16*:366–376 (1979).

209. T. L. Wright, F. J. Roll, A. L. Jones, and R. A. Weisiger, Uptake and metabolism of polymerized albumin by rat liver, *Gastroenterology 94*:443–452 (1988).

210. L. Fiume, C. Busi, G. Di Stefano, and A. Mattioli, Coupling of antiviral nucleoside analogs to lactosaminated human albumin by using the imidazolides of their phosphoric esters, *Anal. Biochem. 212*:407–411 (1993).

211. S. W. Provencher and J. Glockner, Estimation of globular protein secondary structure from circular dichroism, *Biochemistry 20*:33–37 (1981).

212. T. E. Creighton, Protein folding, *Biochem. J. 270*:1–16 (1990).

213. L. Fiume, C. Busi, S. Corzani, G. Di Stefano, G. B. Gervasi, and A. Mattioli, Organ distribution of a conjugate of adenine arabinoside monophosphate with lactosaminated albumin in the rat, *J. Hepatol. 20*:681–682 (1994).

214. M. E. Kuipers, P. J. Swart, M. Schutten, C. Smit, J. H. Proost, A. D. M. E. Osterhaus, and D. K. F. Meijer, Pharmacokinetics and anti-HIV-1 efficacy of negatively charged human serum albumins in mice, *Antiviral Res. 33*:99–108 (1997).

215. L. L. Boles Ponto and J. A. Ponto, Uses and limitations of positron emission tomography in clinical pharmacokinetics/dynamics (Part I), *Clin. Pharmacokinet. 22*: 211–222 (1992).

216. L. L. Boles Ponto and J. A. Ponto, Uses and limitations of positron emission tomography in clinical pharmacokinetics/dynamics (Part II), *Clin. Pharmacokinet. 22*: 274–283 (1992).

217. M. V. Pimm, A. C. Perkins, J. Strohalm, K. Ulbrich, and R. Duncan, Gamma scintigraphy of a $^{123}$I-labelled N-(2-hydroxypropyl)methacrylamide copolymer-doxorubicin conjugate containing galactosamine following intravenous administration to nude mice bearing hepatic human colon carcinoma, *J. Drug Target 3*:385–390 (1996).

218. M. Haas, A. C. A. Kluppel, E. S. Wartna, F. Moolenaar, D. K. F. Meijer, P. E. De Jong, and D. De Zeeuw, Drug targeting to the kidney: Renal delivery and degradation of a naproxen-lysozyme conjugate in vivo, *Kidney Int. 52*:1693–1699 (1997).

219. F. Braet, R. De Zanger, T. Sasaoki, M. Baekeland, P. Janssens, B. Smedsrod, and E. Wisse, Assessment of a method of isolation, purification, and cultivation of rat liver sinusoidal endothelial cells, *Lab. Invest. 70*:944–952 (1994).

220. A. Geerts, T. Niki, K. Hellemans, D. De Craemer, K. Van den Berg, J. -M. Lazou, G. Stange, M. Van de Winkel, and P. J. De Bleser, Purification of rat hepatic stellate cells by side scatter-activated cell sorting, *Hepatology 27*:590–598 (1998).

221. S. L. Friedman and F. J. Roll, Isolation and culture of hepatic lipocytes, Kupffer cells, and sinusoidal endothelial cells by density gradient centrifugation with Stractan, *Anal. Biochem. 161*:207–218 (1987).

222. Q. G. Dong, S. Bernasconi, S. Lostaglio, R. W. De Calmanovici, I. Martin Padura, F. Breviario, C. Garlanda, S. Ramponi, A. Mantovani, and A. Vecchi, A general strategy for isolation of endothelial cells from murine tissues. Characterization of two endothelial cell lines from the murine lung and subcutaneous sponge implants, *Arterioscler. Thromb. Vasc. Biol. 17*:1599–1604 (1997).

223. D. D. Breimer, An integrated pharmacokinetic and pharmacodynamic approach to controlled drug delivery, *J. Drug Target 3*:411–415 (1996).

224. H. Tomoda, Y. Kishimoto, and Y. C. Lee, Temperature effect on endocytosis and exocytosis by rabbit alveolar macrophages, *J. Biol. Chem. 264*:15445–15450 (1989).

225. P. H. Weigel and J. A. Oka, Temperature dependence of endocytosis mediated by the asialoglycoprotein receptor in isolated rat hepatocytes. Evidence for two potentially rate-limiting steps. *J. Biol. Chem. 256*:2615–2617 (1981).

226. J. S. Goltz, A. W. Wolkoff, P. M. Novikoff, R. J. Stockert, and P. Satir, A role for microtubules in sorting endocytic vesicles in rat hepatocytes, *Proc. Natl. Acad. Sci. USA 89*:7026–7030 (1992).

227. B. L. Clarke and P. H. Weigel, Differential effects of leupeptin, monensin and colchicine on ligand degradation mediated by the two asialoglycoprotein receptor pathways in isolated rat hepatocytes, *Biochem. J. 262*:277–284 (1989).

228. T. Wileman, R. L. Boshans, P. Schlesinger, and P. Stahl, Monensin inhibits recycling of macrophage mannose-glycoprotein receptors and ligand delivery to lysosomes, *Biochem. J. 220*:665–675 (1984).

229. P. Olinga, D. K. F. Meijer, M. J. H. Slooff, and G. M. M. Groothuis, Liver slices in in vitro pharmacotoxicology with special reference to the use of human liver tissue, *Toxicol. In Vitro 12*:77–100 (1998).

230. D. K. F. Meijer, K. Keulemans, and G. J. Mulder, Isolated perfused rat liver technique, *Methods Enzymol. 77*:81–94 (1981).

231.  D. K. F. Meijer and P. J. Swart, Isolated perfused liver as a tool to study the disposi-
      tion of peptides, liver first-pass effects, and cell-specific drug delivery, *J. Contr.
      Rel. 46*:139–156 (1997).
232.  P. Olinga and G. M. M. Groothuis, Human liver slices and isolated hepatocytes in
      drug disposition and transplantation research, *NCA Newsletter 5*:5–7 (1998).
233.  B. N. Melgert, P. Olinga, K. Poelstra, M. J. H. Slooff, D. K. F. Meijer, and G. M.
      M. Groothuis, Human liver lobe perfusion (HLLP) as a tool for studying cellular
      distribution and handling of liver targeting preparations. *Hepatology 28*:188A [ab-
      stract] (1998).
234.  L. Fiume, A. Mattioli, C. Busi, S. Corzani, and G. Di Stefano. High-pressure liquid
      chromatographic method for determining adenine arabinoside monophosphate cou-
      pled to lactosaminated albumin in plasma, *Pharm. Acta Helv. 66*:230–232 (1991).
235.  B. N. Melgert, E. Wartna, C. Lebbe, C. Albrecht, G. Molema, K. Poelstra, J. Re-
      ichen, and D. K. F. Meijer. Targeting of naproxen covalently linked to HSA to
      sinusoidal cell types of the liver, *J. Drug Target 5*:329–342 (1998).
236.  C. Albrecht, R. Reichen, J. Visser, D. K. F. Meijer, and W. Thormann, Differentia-
      tion between naproxen, naproxen-protein-conjugates and naproxen-lysine in
      plasma via micellar electrokinetic capillary chromatography—A new approach in
      the bioanalysis of drug targeting preparations, *Clin. Chem. 43*:2083–2090 (1997).
237.  C. Albrecht, B. N. Melgert, J. Reichen, K. Poelstra, and D. K. F. Meijer, Effect
      of chronic bile duct obstruction and LPS upon targeting of naproxen to the liver
      using naproxen-albumin conjugate, *J. Drug Target 6*:105–117 (1998).
238.  C. Lebbe, J. Reichen, E. Wartna, H. Saegesser, K. Poelstra, and D. K. F. Meijer,
      Targeting naproxen to non-parenchymal liver cells protects against endotoxin in-
      duced liver damage, *J. Drug Target 4*:303–310 (1997).
239.  C. Albrecht, D. K. F. Meijer, C. Lebbe, H. Sagesser, B. N. Melgert, K. Poelstra,
      and J. Reichen, Targeting naproxen coupled to human serum albumin to nonparen-
      chymal cells reduces endotoxin-induced mortality in rats with biliary cirrhosis,
      *Hepatology 26*:1553–1559 (1997).
240.  B. N. Melgert, K. Poelstra, V. K. Jack, G. Molema, and D. K. F. Meijer, Dexameth-
      asone coupled to albumin is taken up by non parenchymal liver cells and protects
      against endotoxin in rats with liver cirrhosis. *Hepatology 28*:440A [abstract] (1998).
241.  M. E. DeVera, B. S. Taylor, Q. Wang, R. A. Shapiro, T. R. Billiar, and D. A.
      Geller, Dexamethasone suppresses iNOS gene expression by upregulating I-kappa
      B alpha and inhibiting NF-kappa B, *Am. J. Physiol. Gastrointest. L. 36*:G1290–
      1296 (1997).
242.  R. Hoffmann, H. P. Henninger, A. Schulze Specking, and K. Decker, Regulation
      of interleukin-6 receptor expression in rat Kupffer cells: modulation by cytokines,
      dexamethasone and prostaglandin E2, *J. Hepatol. 21*:543–550 (1994).
243.  I. Makino and H. Tanaka, From a choleretic to an immunomodulator: Historical
      review of ursodeoxycholic acid as a medicament, *J. Gastroenterol. Hepatol. 13*:
      659–664 (1998).
244.  C. A. Palmerini, C. Saccardi, A. Floridi, and G. Arienti, Formylcolchicine bound
      to lactosaminated serum albumin is a more active antifibrotic agent than free colchi-
      cine, *Clin. Chim. Acta 254*:149–157 (1996).
245.  M. Bickel, K.-H. Baringhaus, M. Gerl, V. Günzler, J. Kanta, L. Schmidts, M. Stapf,

G. Tschank, K. Weidmann, U. Werner, et al., Selective inhibition of hepatic collagen accumulation in experimental liver fibrosis in rats by a new prolyl 4-hydroxylase inhibitor, *Hepatology 28*:404–411 (1998).

246. J. I. Lee and G. J. Burckart, Nuclear factor kappa B: Important transcription factor and therapeutic target, *J. Clin. Pharmacol. 38*:981–993 (1998).

247. Y. Matsumura, I. Sakaida, K. Uchida, T. Kimura, T. Ishihara, and K. Okita, Prolyl 4-hydroxylase inhibitor (HOE 077) inhibits pig serum-induced rat liver fibrosis by preventing stellate cell activation, *J. Hepatol. 27*:185–192 (1997).

248. Y. J. Wang, S. S. Wang, M. Bickel, V. Guenzler, M. Gerl, and D. M. Bissell, Two novel antifibrotics, HOE 077 and safironil, modulate stellate cell activation in rat liver injury: Differential effects in males and females, *Am. J. Pathol. 152*:279–287 (1998).

249. T. Niki, K. Rombouts, P. J. De Bleser, K. de Smet, V. Rogiers, D. Schuppan, M. Yoshida, G. Gabbiani, and A. Geerts, A histone deacetylase inhibitor, trichostatin A, suppresses myofibroblastic differentation of rat hepatic stellate cells in primary culture. *Hepatology 29*:858–867 (1999).

250. A. Ponzetto, L. Fiume, B. Forzani, S. Y. Song, C. Busi, A. Mattioli, C. Spinelli, M. Marinelli, A. Smedile, E. Chiaberge, et al., Adenine arabinoside monophosphate and acyclovir monophosphate coupled to lactosaminated albumin reduce woodchuck hepatitis virus viremia at doses lower than do the unconjugated drugs, *Hepatology 14*:16–24 (1991).

251. F. E. Zahm, N. d'Urso, F. Bonino, and A. Ponzetto, Treatment of woodchuck hepatitis virus infection in vivo with 2′,-3′-dideoxycytidine (ddC) and 2′,-3′-dideoxycytidine monophosphate coupled to lactosaminated human serum albumin (L-HSA ddCMP), *Liver 16*:88–93 (1996).

252. L. Fiume, G. Di Stefano, C. Busi, A. Mattioli, F. Bonino, M. Torrani-Cerenzia, G. Verme, M. Rapicetta, M. Bertini, and G. B. Gervasi, Liver targeting of antiviral nucleoside analogues through the asialoglycoprotein receptor. *J. Viral Hepatol. 4*: 363–370 (1997).

253. L. Fiume, F. Bonino, A. Mattioli, E. Chiaberge, M. R. T. Cerenzia, C. Busi, M. R. Brunetto, and G. Verme, Inhibition of hepatitis B virus replication by vidarabine monophosphate conjugated with lactosaminated serum albumin, *Lancet 332*:13–15 (1988).

254. P. Chakraborty, A. Bhaduri, and P. K. Das, Neoglycoproteins as carriers for receptor-mediated drug targeting in the treatment of experimental visceral leishmaniasis, *J. Protozool. 37*:358–364 (1990).

255. R. Sett, H. S. Sarkar, and P. K. Das, Pharmacokinetics and biodistribution of methotrexate conjugated to mannosyl human serum albumin, *J. Antimicrob. Chemother. 31*:151–159 (1993).

256. R. K. Jain, Barriers to drug delivery in solid tumors, *Sci. Am. July*:58–65 (1994).

257. M. W. Fanger, P. M. Morganelli, and P. M. Guyre, Bispecific antibodies, *Crit. Rev. Immunol. 12*:101–124 (1992).

258. W. Arap, R. Pasqualini, and E. Ruoslahti, Cancer treatment by targeted drug delivery to tumor vasculature in a mouse model, *Science 279*:377–380 (1998).

259. R. C. Stadalnik, M. Kudo, W. C. Eckelman, and D. R. Vera, In vivo functional imaging using receptor-binding radiopharmaceuticals. Technetium 99m-galactosyl-neoglycoalbumin as a model, *Invest. Radiol. 28*:64–70 (1993).

260. A. H. Kwon, S. K. Ha Kawa, S. Uetsuji, T. Inoue, Y. Matsui, and Y. Kamiyama, Preoperative determination of the surgical procedure for hepatectomy using technetium-99m-galactosyl human serum albumin (99mTc-GSA) liver scintigraphy, *Hepatology 25*:426–429 (1997).
261. M. Kudo, D. R. Vera, and R. C. Stadalnik, Hepatic receptor imaging using radiolabeled asialoglycoprotein analogs, *Neoglycoconjugates: Preparation and Applications* (Y. C. Lee, R. T. Lee, eds.), Academic Press, Inc., San Diego, 1994, pp. 373–402.
262. D. R. Vera, R. C. Stadalnik, C. E. Metz, and N. R. Pimstone, Diagnostic performance of a receptor-binding radiopharmacokinetic model, *J. Nucl. Med. 37*:160–164 (1996).
263. D. R. Vera, R. C. Stadalnik, and K. A. Krohn, Technetium-99m galactosyl-neoglycoalbumin: Preparation and preclinical studies, *J. Nucl. Med. 26*:1157–1167 (1985).
264. D. R. Vera, K. A. Krohn, R. C. Stadalnik, and P. O. Scheibe, Tc-99m galactosyl-neoglycoalbumin: In vitro characterization of receptor-mediated binding, *J. Nucl. Med. 25*:779–787 (1984).
265. D. R. Vera, K. A. Krohn, R. C. Stadalnik, and P. O. Scheibe, Tc-99m-galactosyl-neoglycoalbumin: In vivo characterization of receptor-mediated binding to hepatocytes, *Radiology 151*:191–196 (1984).
266. N. R. Pimstone, R. C. Stadalinik, D. R. Vera, D. P. Hutak, and W. L. Trudeau, Evaluation of hepatocellular function by way of receptor-mediated uptake of a Technetium-99m-labeled asialoglycoprotein analog, *Hepatology 20*:917–923 (1994).
267. R. Mastai, S. Laganière, I. R. Wanless, L. Giroux, B. Rocheleau, and P.-M. Huet, Hepatic sinusoidal fibrosis induced by cholesterol and stilbestrol in the rabbit: 2. Hemodynamic and drug disposition studies, *Hepatology 24*:865–870 (1996).
268. I. Virgolini, C. Müller, P. Angelberger, J. Höbart, H. Bergmann, and H. Sinzinger, Functional liver imaging with $^{99}$Tc$^m$-galactosyl-neoglycoalbumin (NGA) in alcoholic liver cirrhosis and liver fibrosis, *Nucl. Med. Commun. 12*:507–517 (1991).
269. Y. Kubota, S. Kitagawa, K. Inoue, S. K. Ha-Kawa, M. Kojima, and Y. Tanaka, Hepatic functional scintigraphic imaging with $^{99m}$Technetium galactosyl serum albumin, *Hepatogastroenterology 40*:32–36 (1993).
270. E. S. Woodle, D. R. Vera, R. C. Stadalnik, and R. E. Ward, Tc-NGA imaging in liver transplantation: preclinical studies, *Surgery 102*:55–62 (1987).
271. I. Virgolini, C. Müller, J. Höbart, W. Scheithauer, P. Angelberger, H. Bergmann, J. O'Grady, and H. Sinzinger, Liver function in acute viral hepatitis as determined by a hepatocyte-specific ligand: $^{99m}$Tc-galactosyl-neoglycoalbumin, *Hepatology 15*:593–598 (1992).
272. H. G. Eichler, Clinical experience of targeted therapy, *Biotherapy 3*:77–85 (1991).
273. D. D. Breimer, Future challenges for drug delivery research, *Adv. Drug Deliv. Rev. 33*:265–268 (1998).
274. L. Fiume, G. Di Stefano, C. Busi, and A. Mattioli, A conjugate of lactosaminated poly-l-lysine with adenine arabinoside monophosphate, administered to mice by intramuscular route, accomplishes a selective delivery of the drug to the liver. *Biochem. Pharmacol. 47*:643–650 (1994).

# 8
# Gene Delivery with Artificial Viral Envelopes

**Frank L. Sorgi**
*OPTIME Therapeutics, Petaluma, California*

**Lucie Gagné**
*University of California, San Francisco, California*

**Hans Schreier**
*H. Schreier Consulting, Langley, Washington*

## I. INTRODUCTION

The first generation of gene carriers consisted of modified viral vectors, including retroviruses, adenovirus, and more recently adeno-associated virus (AAV), and herpes simplex and lentiviruses [1]. These viruses have developed cellular mechanisms through evolution to strike and enter target cells with an accuracy and efficiency unsurpassed by any synthetic delivery system, which, theoretically, would render them as ideal gene delivery systems. However, concomitantly, the mammalian host has developed powerful defense mechanisms, specifically inflammatory reactions and immune responses, to combat viral assaults. These defense mechanisms may be dose- and/or therapy-limiting when recombinant adenovirus or AAV (or others) are used as carriers for cDNA. Furthermore, at least early generations of viral vectors, although designed to be replication incompetent, have been found to replicate by complementation or recombination, and expression of viral genes, even at very low levels, may trigger T-lymphocyte responses and destruction of the transfected target cells [2].

As a means to overcome these problems, alternative nonviral DNA carrier systems have been developed [3], and of those cationic liposomes have been the most widely used and characterized [4]. A number of such liposomal systems have advanced to the clinic in recent years [3], although their *in vivo* usefulness has been limited by poor transfection efficacy, particularly in cystic fibrosis trials [5]. This can, at least in part, be attributed to a strong and irreversible electrostatic interaction of net cationic complexes with plasma components and the vascular wall lining (other factors, e.g., intrinsic plasmid transfection efficacy, may play a role as well). Although some applications that use direct injection into the target tissue (for instance, accessible tumor sites such as melanoma, head and neck cancer) may be suitable targets for cationic delivery systems, their general use, particularly indications that require systemic application, will likely be extremely limited.

Alternative nonviral systems that are not based on cationic lipid complexation, but rather exhibit a net negative surface, should, therefore, be more suitable if they can be designed such that they are both target-selective and efficient in the delivery of gene products.

## II. ARTIFICIAL VIRAL ENVELOPES

For several years, we have been interested in exploiting physiological and pathophysiological pathways as potential avenues for drug targeting and delivery [6]. Chander and Schreier [7] proposed the use of synthetic lipid vesicles that mimic the fusogenic envelope of retroviruses to exploit the target selectivity and the efficiency of delivery of such viruses. These liposomes, which were called artificial viral envelopes (AVE), were strongly negatively charged ($-40$ to $-50$ mV), contained a high fraction of cholesterol, and a lipid composition similar to the major components of the lipid membrane of enveloped viruses [8] (i.e., phosphatidylcholine [PC], phosphatidylserine [PS], phosphatidylethanolamine [PE], and sphingomyelin).

Because of their net negative surface charge, targeting moieties such as viral binding proteins [7, 9] can be inserted into, or attached to, the outer surface of the liposome. Constructs were built with both recombinant HIV gp160 [7] and with GPI-anchored gp120 [9] and shown to bind to CD4-positive target cells in a specific and receptor concentration-dependent fashion. Accordingly, when these envelopes were loaded with ricin A, they were shown to kill CD4+ cells in a selective and dose-dependent fashion [6]. Receptor binding was inhibited in the presence of anti-gp120 antibody, anti-CD4 antibody, or soluble CD4-IgG and abolished when the GPI anchor was enzymatically cleaved with phospholipase C [9].

Figure 1 shows an AVE into which the fusion protein of the respiratory syncytial virus (RSV) has been inserted.

## III. DNA CONDENSATION AND LOADING

The major obstacle for the use of this type of liposomes in gene delivery was their intrinsically poor encapsulation efficiency of ultra-large payload compounds

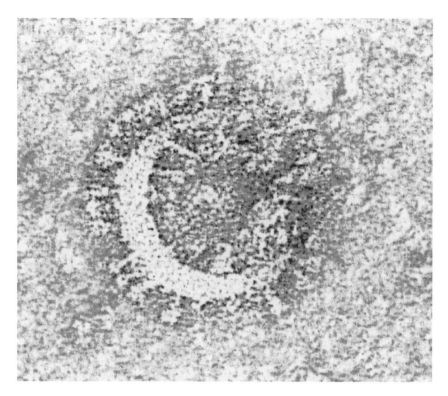

**Figure 1** Electron micrograph of an artificial viral envelope. In a unilamellar lipid bilayer of a diameter of approximately 200 nm are surface glycoproteins of the respiratory syncytial virus (RSV) inserted as described in detail by Chander & Schreier [7] (electron micrographic image by G. Erdos, EM Core Laboratory, University of Florida, Gainesville, FL).

such as plasmids that range in size from 3–11 kb. Analogous to the membrane design work by Chander and Schreier [7], Sorgi et al. [10] proposed to mimic the condensation of DNA by protamine sulfate. The physiological role of protamine is to condense DNA into the head of sperm. This has, indeed, provided a solution to the problem of poor encapsulation efficiency.

Plasmid DNA is strongly anionic because of the phosphate groups contained within each nucleotide base. Protamine sulfate is a 32-mer protein containing 21 arginine residues and is, therefore, strongly basic [10]. DNA, when titrated with protamine, forms a highly condensed structure on neutralization of 90% of the DNA charge. The resulting zeta-potential curve of such a titration is shown in Figure 2.

The neutralization and condensation event occurs at a DNA/protamine ratio (w/w) of approximately 1.4. At that point, the complex begins to exhibit a net

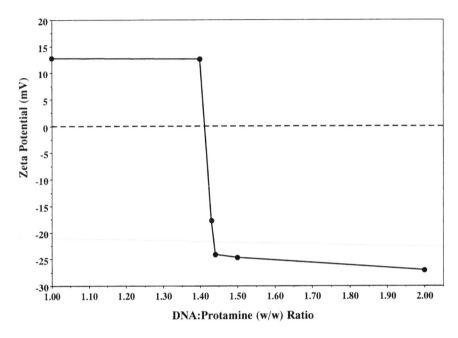

**Figure 2** Zeta potential of DNA-protamine sulfate complexes. Plasmid DNA was condensed with protamine sulfate in increasing ratios (w/w). The surface charge (zeta-potential) was determined using a Nicomp 380 ZLS Zetasizer (Particle Sizing Systems, Santa Barbara, CA).

cationic surface charge, thereby facilitating the interaction with the anionic liposome formulations. Highly condensed net-cationic DNA facilitates the electrostatic coating of a negatively charged lipid bilayer about the condensed DNA.

In addition to protamine sulfate, a number of other cationic polymers have been used to condense DNA [11].

## IV. PLASMID DELIVERY WITH ARTIFICIAL VIRAL ENVELOPES

Because of their "physiological" composition, artificial viral envelope–DNA complexes are essentially noncytotoxic, exhibit a quasilinear expression dose response, and deliver their payload in a serum-independent fashion to cells in culture.

The surface characteristics of net anionic DNA-liposome complexes are distinctly different from classic cationic liposome-DNA complexes and, therefore, their interaction with biological fluids and membranes is different. In Figure 3, the transfection efficacy of both types of complexes in the presence of serum

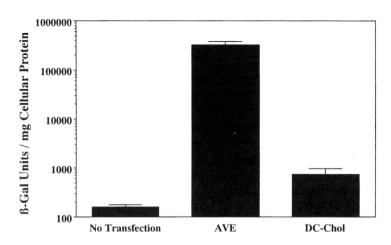

**Figure 3** β-GAL expression 48 h after transfection of HepG2 cells. AVE consisted of dioleylphosphatidylethanolamine, dioleylphosphatidylcholine, and cholesterol-glutarate in equimolar concentrations. Lipids were hydrated in 10 mM Tris (pH 8.0) and sonicated for approximately 5 min. The average size as measured by laser particle sizing (Nicomp Model 370) was 100–200 nm. Cationic DC-Chol liposomes were prepared as described by Sorgi et al. [10]. DNA was condensed with protamine sulfate at a ratio of 1.3:1 (w/w) and subsequently complexed with the AVE formulation to give a lipid/DNA ratio of 3:1 (w/w). HepG2 cells grown in 24-well plates were transfected at 40% confluency with 100 μL of the DNA-AVE formulation in Eagle's MEM, supplemented with 10% fetal bovine serum; 48 h after incubation, cells were analyzed for β-galactosidase transgene expression (Galacto-LightPlus™ Reagent kit, Tropix).

is shown. Plasmid complexed with DC-Chol [12], which is a powerful and widely used transfection agent in serum-free medium, transfects over two orders of magnitude lower than the net anionic DNA-liposome complex. Transfection in the presence of serum results in charge neutralization of the cationic liposome DC-Chol, whereas the anionic DNA-liposome complex is resistant to serum inactivation, resulting in high levels of transfection.

Another effect related to the charge characteristics of AVE is the apparent absence of cytotoxicity as indicated by complete absence of cell death. This has been shown in a concentration-dependent (Figure 4) and a time-dependent fashion (Figure 5). The cellular protein concentration curve remains flat over a range of 5–40 μg DNA, corresponding to 15–120 μg of total lipid (Figure 4). No rounded-up or floating cells were found by microscopic observation at any concentration or time point.

Luciferase expression increased linearly with increasing concentration of DNA and lipid, respectively, to a maximum, after which there was a continuous decrease, resulting in a bell-shaped dose-efficacy relationship. Although cationic

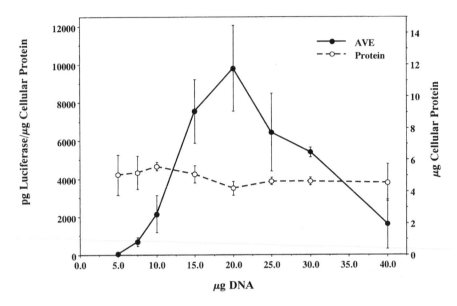

**Figure 4** Luciferase expression 48 h after transfection of HepG2 cells with an AVE/pLuc complex in 10% serum as a function of DNA concentration. Experimental conditions were as described in Figure 3. Luciferase transgene expression was analyzed with a Promega luciferase assay system, and samples were read in a Tropix TR717 microplate luminometer. Cellular protein was measured using a Bio-Rad protein assay kit.

liposome DNA formulations express generally at lower DNA concentrations than used here [3], there is always a distinct dose-related increase in cell death that eventually becomes the dose-limiting factor.

Unexpectedly, and in sharp contrast to cationic liposomes, the time course of gene expression of net anionic DNA-liposome complexes is characterized by a distinct lag phase. Cationic liposomes generate a fairly rapid expression response that usually peaks within 24–48 h after transfection and slowly decays thereafter [3]. In contrast, with AVE-DNA complexes, an extended lag phase of up to 36 h was noted, followed by an increase and plateau of gene expression up to 64 h (Figure 5). It is currently unknown what causes the lag phase. One or several factors could play a role, including: (1) the degree of condensation exerted by the protamine; (2) the kinetics of lipid and protamine uncoating, independently or as a combined event; (3) the release of the coated or uncoated DNA from the endosomal compartment; and finally (4) the kinetics of DNA entry into the nucleus. Similar to the finding shown in Figure 4, cellular protein concentrations remain constant within a narrow range over a time frame of at least 64 h after transfection, again indicating the absence of cytotoxic events.

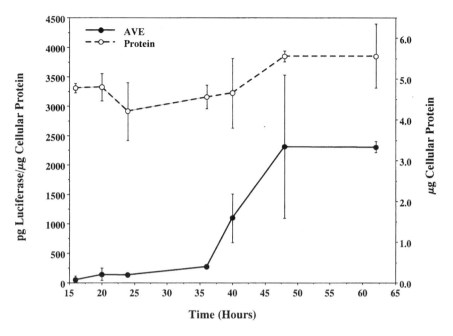

**Figure 5** Luciferase expression 48 h after transfection of HepG2 cells with an AVE/ pLuc complex in 10% serum as a function of time. Experimental conditions were as in Figure 4.

## V. CONCLUSIONS

In conclusion, we have designed a synthetic vesicular DNA carrier that physically and functionally mimics an enveloped virus particle. To achieve an acceptable degree of encapsulation within the vesicle, we use a process that is essentially inverse to the preparation of cationic lipid-DNA complexes. A suitable DNA condensing agent is introduced that, at a certain critical concentration, conveys a weak net cationic charge to the condensed DNA that then interacts spontaneously with a liposome containing one or more anionic components. These DNA formulations behave distinctly different from classic cationic liposome DNA complexes *in vitro* in as much as they have been shown to be nontoxic, to display a "traditional" linear dose response, and to be serum-insensitive.

Interestingly, Kaneda and co-workers (see Chapter 9) have systematically compared *in vivo* transgene expression (after direct injection into muscle and a liver lobe) of DNA liposome formulations akin to the ones used here with DC-Chol liposomes [13]. When DC-Chol (i.e., cationic) liposomes were admixed to the net anionic DNA-liposome complexes, there was a gradual decline in the apparent transfection efficacy leading to an eventual collapse of expression when cationic liposome alone was used. This observation corresponds well with our observations *in vitro* and supports the prediction that net anionic DNA-liposome complexes may become clinically useful as *in vivo* gene delivery systems, provided other rate-limiting factors such as efficient targeting, endosomal release, and higher intrinsic expression efficacy can be accomplished eventually.

## REFERENCES

1.  K. Roemer and T. Friedmann. Concepts and strategies for human gene therapy, *Eur. J. Biochem. 208*:211–225 (1992).
2.  R. G. Crystal. The gene as the drug, *Nature Med. 1*:15–17 (1995).
3.  F. L. Sorgi and H. Schreier. Non-viral vectors for gene delivery, *Biopharmaceutical Drug Design and Development* (S. Wu-Pong, and Y. Rojanasakul, eds.), The Humana Press, Totowa, N.J., 1999, pp. 107–142.
4.  X. Gao and L. Huang. Cationic liposome-mediated gene transfer, *Gene Therapy 2*: 710–722 (1995).
5.  H. Schreier, and S. M. Sawyer. Liposomal DNA vectors for cystic fibrosis gene therapy. Current applications, limitations, and future directions, *Adv. Drug Del. Rev. 19*:73–87 (1996).
6.  H. Schreier, M. Ausborn, S. Günther, V. Weissig, and R. Chander. (Patho)physiologic pathways to drug targeting: artificial viral envelopes, *J. Mol. Recognit. 8*:59–62 (1995).
7.  R. Chander and H. Schreier. Artificial viral envelopes containing recombinant HIV gp160. *Life Sci. 50*:481–489 (1992).
8.  R. C. Aloia, F. C. Jensen, C. C. Courtain, P. W. Mobley, and L. M. Gordon. Lipid composition and fluidity of the human immunodeficiency virus, *Proc. Natl. Acad. Sci. USA 85*:900–904 (1988).
9.  H. Schreier, P. Moran, and I. W. Caras. Specific targeting of liposomes to cells using a GPI-anchored ligand; influence of liposome composition on intracellular trafficking, *J. Biol. Chem. 269*:9090–9098 (1994).
10. F. L. Sorgi, S. Bhattacharya, and L. Huang. Protamine sulfate enhances lipid-mediated gene transfer, *Gene Therapy 4*:961–968 (1997).
11. X. Gao and L. Huang. Potentiation of cationic liposome-mediated gene delivery by polycations, *Biochemistry 35*:1027–1036 (1996).
12. X. Gao and L. Huang. A novel cationic liposome reagent for efficient transfection of mammalian cells, *Biochem. Biophys. Res. Comm. 179*:280–285 (1991).
13. Y. Saeki, N. Matsumoto, Y. Nakano, M. Mori, K. Awai, and Y. Kaneda. Development and characterization of cationic liposomes conjugated with HVJ (Sendai virus): reciprocal effect of cationic lipid for in vitro and in vivo gene transfer, *Hum. Gene Ther. 8*:2133–2141 (1997).

# 9

# Evolution of Viral Liposomes

## Improvements and Applications

**Yasufumi Kaneda, Yoshinaga Saeki, and Ryuichi Morishita**
*Osaka University, Suita, Osaka, Japan*

## I. SUMMARY

Toward human gene therapy, we have developed an efficient liposome vector based on cell fusion. The liposome was decorated with HVJ (Sendai virus) envelope fusion proteins to introduce DNA directly into the cytoplasm and contained DNA and DNA-binding nuclear protein inside the particle to enhance expression of the gene. The HVJ liposome was highly efficient for the introduction of oligonucleotides into cells *in vivo* and the transfer of genes less than 100 kb without damaging cells. Most animal organs were found to be suitable targets for the fusogenic viral liposomes, and numerous gene therapy strategies that used this system were successful in animals.

For more efficient gene delivery, we investigated lipid components of the liposomes and reached the conclusion that the use of anionic liposomes with a virus-mimicking lipid composition (HVJ-AVE liposome) resulted in the highest gene delivery efficiency in animal organs such as liver, muscle, kidney, and heart. We then produced HVJ-cationic liposomes by adding cationic lipids to HVJ liposomes. HVJ-cationic liposomes facilitated efficient entrapment of DNA and yielded 100 to 800 times higher gene expression *in vitro* than the conventional HVJ-anionic liposome. However, the use of cationic lipid reduced transgene expression dramatically in organs such as muscle and liver. We then found that HVJ-cationic liposomes were more useful for gene expression in restricted compartments and for gene therapy of disseminated cancers. Therefore, appropriate HVJ liposomes should be used for different targets and objectives.

## II. INTRODUCTION

For gene therapy of various human disorders, viral and nonviral (synthetic) methods of gene transfer have been developed. Generally, viral methods are more efficient than nonviral methods for gene transfection into cells. However, viral vectors are limited in terms of the size and content of the heterologous DNA to be transferred. In addition, these vectors are safety hazards because of the co-introduction of essential genetic elements from the parent viruses, leaky expression of viral genes, immunogenicity, and alterations of host genomic structure.

A number of nonviral gene transfer methods have been developed, and liposomes are the best studied [1]. There are two classes of liposome-mediated methods, the electrostatic model and the internal model. The electrostatic model is a complex of cationic liposomes and negatively charged macromolecules (usually DNA). Efficient transfection has been achieved by use of cationic liposomes–DNA complexes *in vitro* [2, 3], whereas *in vivo* gene transfer is much more difficult because liposomes are unstable *in vivo* and inhibited by circulating proteins. The internal type is a true liposome in which macromolecules are encapsulated within a lamellar lipid structure. This type of liposome is thought to be more stable *in vivo*; however, it is less effective at gene transfer than electrostatic methods that use cationic lipids [4, 5]. Liposome-mediated gene transfer is less toxic and less immunogenic; liposomes can deliver a variety of macromolecules such as DNA, RNA, proteins, and drugs; and liposomes are available for modification by other molecules. However, most nonviral methods, including liposomes, are less efficient for gene transfer, especially *in vivo*, compared with some of the viral vectors.

Therefore, to develop an *in vivo* gene transfer vector with high efficiency and low toxicity, the limitations of one vector system should be compensated by the strengths of another type of vector system. From this standpoint, we have developed a novel hybrid gene transfer vector by combining viral and nonviral vectors.

## III. THE DEVELOPMENT OF HVJ LIPOSOMES

To introduce DNA directly into the cytoplasm without degradation, liposomes were constructed with a fusogenic envelope derived from HVJ (hemagglutinating

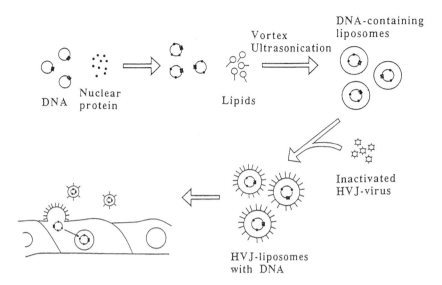

**Figure 1**  Procedure of gene transfer by Sendai virus (HVJ)-liposomes. DNA and nuclear proteins are incorporated into liposomes by vortex-ultrasonication-annealing method, and the liposomes are fused with UV-inactivated Sendai virus. The resulting fusigenic viral liposomes can fuse with cell membrane to introduce DNA and nuclear protein complex directly into the cytoplasm. Nuclear protein such as HMG-1 can enhance the expression of foreign DNA in the nucleus.

virus of Japan; Sendai virus) (Figure 1) [6]. DNA-loaded liposomes consisting of phosphatidylserine, phosphatidylcholine, and cholesterol were prepared by vortexing or reverse-phase evaporation. The trapping efficiency of DNA into liposomes was 10–20%. Namely, 40 to 60 molecules of plasmid DNA and more than one half million copies of a 20mer oligonucleotide can be enclosed in a single liposome. The liposomes were fused with UV-inactivated HVJ to form the fusogenic viral-liposome, HVJ liposome (400–500 nm in diameter). HVJ liposomes can fuse with the cell surface, because HVJ fusion occurs at neutral and acidic pH. We have studied the direct introduction of macromolecules into the cytoplasm by HVJ, which is a paramyxovirus that is 300 nm in diameter and contains two distinct glycoproteins on the envelope that are involved in cell fusion [7]. HN protein is required for binding to receptor consisting of sialoglycoproteins or sialolipids and removal of sugars by its neuraminidase activity, and F can interact with the lipid bilayer of the cell membrane to induce cell fusion. F protein is produced as a fusion-inactive form F0, and is activated by proteolytic cleavage to F1 and F2. The hydrophobic region of F1 can interact with cholesterol to induce cell fusion. HVJ is able to fuse with liposomes [8] most likely because of the direct interaction of F1 polypeptide with lipid, although liposomes have no receptors for the virus.

## IV. THE ADVANTAGES OF HVJ-LIPOSOMES

### A. Efficient Delivery of DNA, Oligonucleotides (ODN), and Protein to Cells

The liposome can enclose DNA less than 100 kb, although the trapping efficiency was 10–20% of added DNA amounts. We succeeded in transducing cultured mouse cells by introduction of a 45-kb plasmid containing the thymidine kinase gene [9]. Recently, the full-length cDNA of human dystrophyn was introduced and expressed in the skeletal muscle and diaphragm of the mdx mouse [10].

Fusion is complete in 10 to 30 min at 37°C. Gene transfer by cationic liposome requires a much longer time, generally 5 to 20 h. This short incubation time for gene transfer by HVJ liposome is beneficial for *in vivo* use. Another advantage of fusion-mediated delivery is protection of molecules from degradation in endosomes and lysosomes. When FITC-ODN was introduced into vascular smooth muscle cells (VSMC) by use of HVJ liposomes, fluorescence was detected in the nuclei 5 min after the transfer, and the fluorescence was stable in the nucleus for at least 72 h [11]. In contrast, fluorescence was observed in the cellular components (most likely in the endosomes) and not in the nucleus when FITC-ODN was transferred directly without HVJ liposomes, and the fluorescence was not detected 24 h after the transfer. When a basic fibroblast growth factor (FGF) antisense ODN was introduced into VSMC by HVJ liposome, cationic liposome, or direct transfer without vector, the concentrations of antisense-FGF required to reduce cellular DNA synthesis to 75% were 0.1, 10, and 20 μmols respectively [12]. Thus, the delivery of antisense ODN by HVJ liposomes was about 50 times more effective than lipofection and 100 times more effective than direct transfer. The vector was also useful for introduction of RNA, proteins, and drugs.

### B. Penetration of the Vector into Tissues

Gene expression was observed in medial layer cells of intact rabbit carotid artery by filling the lumen with HVJ liposome for 10 min under 150 mmHg [13]. We estimate that the vector can penetrate endothelial cells to reach vascular smooth muscle cells. This penetration ability is conferred by the liposome itself, not by HVJ, because the medial layer of the artery was stained by incubation with Evans Blue–containing liposomes without HVJ. The liposome contains phosphatidylserine, phosphatidylcholine, and cholesterol and has a negative charge conferred by phosphatidylserine. When the negative charge of the liposomes was reduced, the penetration activity decreased. Therefore, we estimate that the negative charge of the liposomes may be a key for tissue penetration of liposomes.

## C. No Apparent Toxicity and Low Antigenicity

Whereas cationic liposomes generally induce cytotoxicity because of persistent interactions with cells, anionic liposomes are not as cytotoxic, even in the presence of the HVJ envelope. We have not observed significant cell damage *in vivo* or detected dysfunction of the target organ when $10^{10}$ to $10^{11}$ particles of HVJ liposome were injected into the portal vein of a rat [14]. Before human trials, we are testing the safety of HVJ liposomes in various monkey tissues. When FITC-labeled oligonucleotides were transferred to monkey eye tissue by use of HVJ liposome, oligonucleotides were introduced specifically to the trabecular meshwork of the monkey eye and were retained there for 10 days. No pathological changes were observed in the eyes of 24 monkeys, and no abnormal behavior was observed [15].

Regarding antigenicity of HVJ liposomes, antibodies against HVJ were detected 1 week after injection of HVJ liposomes into the portal vein of rats. The antibody titer was less than 1/1000 that of the polyclonal antibody produced by rabbits immunized with adjuvant-conjugated HVJ. Moreover, when HVJ liposomes containing a marker gene were injected into the portal vein of rats on day 7 after injection of empty HVJ liposomes, gene expression was detected at the same level as that without preinjection of empty HVJ liposome. Cytotoxic T cells against HVJ could not be induced in rats after repeated injections of HVJ liposomes [14]. Thus, gene expression *in vivo* can be sustained by repeated injection of HVJ liposomes.

## V. IMPROVEMENT OF THE CURRENT VECTOR SYSTEM

### A. AVE Liposome: A New Anionic HVJ Liposome

To increase efficiency of gene delivery, we investigated the lipid components of liposomes. First, we evaluated several neutral lipids in various combinations by luciferase expression in cultured cells 24 h after transfer. The neutral phospholipid composition of conventional HVJ liposomes was only phosphatidylcholine. We further investigated lipid components of HVJ-liposome by changing the components of phospholipids, by changing the ratio of phospholipids to total cholesterol, and by changing the ratio of cationic lipid to cholesterol. We conclude that for the most efficient gene expression in cultured cells, phosphatidylcholine (ePC), dioleoylphosphatidylethanolamine (DOPE), and sphingomyelin (eSph) should be present at a molar ratio of 1:1:1. Then, negatively charged lipids were evaluated for more efficient gene delivery, and we concluded that phosphatidylserine should be present at a 10% molar ratio of the total lipids. We further found that the ratio of phospholipids to cholesterol should be 1:1. This new anionic liposome was called HVJ-AVE liposome [16].

AVE stands for artificial viral envelope [17]. In fact, the lipid components of the AVE liposomes are similar to the HIV envelope and mimic the red blood cell membrane [17]. HVJ-AVE liposomes gave gene expression in liver and muscle that was 5 to 10 times greater than observed with conventional HVJ liposomes [16]. HVJ-AVE liposomes were effective for gene delivery to isolated rat heart by way of the coronary artery. LacZ gene expression was observed in the entire heart, whereas no expression was observed with empty HVJ-AVE liposomes.

## B. Development of HVJ-cationic Liposomes

To examine whether the use of cationic liposomes can augment the transfection efficiency of the HVJ-liposomes method, we prepared two different dried lipid mixtures [16]; a conventional anionic mixture containing PS and a new mixture containing positively charged DC-cholesterol [18]. Approximately 10–20% of plasmid DNA was recovered from the conventional anionic PS liposomes, whereas 50–60% was recovered from the cationic DC liposomes. The presence of a cationic lipid, instead of an anionic lipid, in the lipid mixture facilitated enclosing negatively charged plasmid DNA. HVJ, the PS liposomes, and the HVJ-PS liposomes were negatively charged, whereas DC liposomes and the HVJ-DC liposomes were shown to have a net positive charge and net neutral charge, respectively. These two types of HVJ-liposome complexes were then used to transfect BHK-21 cells *in vitro*. Approximately 20-fold greater luciferase expression was obtained with HVJ-DC-Chol-liposomes than HVJ-PS liposomes. When we used the sized unilamellar DC liposomes, five times greater (100 times greater than HVJ-PS liposomes) luciferase activity was obtained. Without HVJ, however, the sized unilamellar DC liposomes yielded only trace luciferase activity when added to BHK-21 cells cultured in medium containing 10% fetal bovine serum. Thus, we found that liposomes containing DC-Chol greatly enhanced transfection.

We then examined the effect of phospholipid composition on the transfection activity. Liposomes containing various combinations of phospholipids were tested for transfection activity on BHK-21 and HeLa-S3 cells. As described in HVJ-AVE liposomes, the cationic liposomes containing all of ePC, DOPE, and eSph in equal molar amounts showed the highest transfection efficiency both with BHK-21 and HeLa-S3 cells. The same results were obtained with Ltk-, HEK 293, and NB-1 cells. We also examined other phospholipids, but none was observed to be more effective. We then examined the effect of the cholesterol/phospholipid ratio on the transfection efficiency. The phospholipid composition (ePC:DOPE:eSph = 1:1:1) and DC-Chol content (10% of total lipid) were kept constant, and the relative proportion of cholesterol derivatives (Chol and DC-Chol) to phospholipid was varied. The liposomes composed of an equimolar amount of phospholipid and cholesterol derivatives showed the highest transfection activity. Finally, the effect of cationic lipid content on the transfection activity was examined. The content of cationic DC-Chol in liposomes varied from 0–50%, and luciferase expression of the transfected cells was examined. The liposomes containing 10% DC-Chol showed the greatest transfection activity in all cell types examined. As the positive charge on liposomes was increased, they captured larger amounts of HVJ and showed greater cytotoxicity.

With the optimized lipid composition (opDC; ePC:DOPE:eSph:Chol: DC-Chol = 5:5:5:12:3), the HVJ-cationic liposomes showed 100 to 800 times greater transfection efficiency *in vitro* compared with the conventional HVJ-PS liposomes. The presence of serum (10% FCS) in the transfection mixture did not decrease luciferase activity significantly. Even 70% FCS reduced the activity by less than 40%. LacZ gene expression showed that transfection efficiency of BHK-21 cells by optimized HVJ-cationic liposomes (opDC) and by conventional HVJ-cationic liposomes (DC) was 90–100% and 50–60%, respectively. With conventional HVJ-anionic liposomes (PS), LacZ expression was found in only 1–3% of the cells. The optimized HVJ-cationic liposomes were also much more effective for the transfer of FITC-labeled ODNs to cultured cells [16].

Thus, the HVJ-cationic liposomes were shown to be useful for *in vitro* delivery of macromolecules. We then investigated the transfection activity of the new gene transfer vehicles for muscle and liver of adult mice *in vivo* [16]. The results are shown in Table 1. Total luciferase expression in organs transfected with the HVJ-cationic liposomes (DC and opDC) was 10 to 150 times lower than that of the conventional HVJ-PS liposomes, which were less efficient for *in vitro* transfection. These findings suggest that it is difficult to optimize *in vivo* transfection strategies with *in vitro* experiments. We then examined four new types of liposomes, AVE, AVE + DC10 (contains 10% bPS and 10% DC-Chol), AVE + DC20 (contains 10% bPS and 20% DC-Chol), and AVE-PS (contains neither bPS nor DC-Chol). AVE yielded the greatest luciferase expression both in muscle and liver (six to seven times greater than PS, 60 to 300 times greater than opDC). AVE-PS and AVE + DC10 liposomes, which have a net neutral charge, showed intermediate luciferase activities. AVE + DC20 liposomes, which have an excessive amount of cationic lipid were not more difficult than opDC liposomes. However, we recently found that HVJ-cationic liposomes were more effective for *in vivo* gene transfer in some special cases. High expression of LacZ gene was obtained in restricted parts of the chick embryos after injection of HVJ-cationic liposomes, whereas HVJ-anionic liposomes were ineffective. When HVJ-liposomes containing LacZ gene were administered to rat lung by a jet nebulizer, more efficient gene expression in the tracheal and bronchial epithelium was observed compared with that by HVJ-anionic liposomes.

**Table 1** Effect of Lipid Compositions of Liposomes on the Transfection Activity of the HVJ Liposomes *in Vivo*

| Liposomes[d] | Lipid compositions (% molar ratio) | | | | | | Muscle[a] luciferase activity[b] | | Liver[c] luciferase activity[b] | |
|---|---|---|---|---|---|---|---|---|---|---|
| | ePC | DOPE | eSph | bPS | Chol | DC-chol | mean (RLU) | %CV | mean (RLU) | %CV |
| PS | 48.8 | 0 | 0 | 9.9 | 41.3 | 0 | 108,333 | 32.0 | 624,442 | 25.0 |
| DC | 45.2 | 0 | 0 | 0 | 45.9 | 8.9 | 3,421 | 16.1 | 4,083 | 16.2 |
| opDC | 16.7 | 16.7 | 16.7 | 0 | 40 | 10 | 11,550 | 26.1 | 13,150 | 18.5 |
| AVE | 13.3 | 13.3 | 13.3 | 10 | 50 | 0 | 702,554 | 26.7 | 3,841,542 | 21.9 |
| AVE + DC10 | 13.3 | 13.3 | 13.3 | 10 | 40 | 10 | 147,050 | 25.7 | 1,107,950 | 42.1 |
| AVE + DC20 | 13.3 | 13.3 | 13.3 | 10 | 30 | 20 | 8,800 | 35.9 | 18,542 | 41.4 |
| AVE-PS | 16.7 | 16.7 | 16.7 | 0 | 50 | 0 | 80,083 | 46.6 | 462,166 | 35.5 |

[a] 100 μL of each HVJ-liposome suspension was injected directly into quadriceps ($n = 6$).

[b] Luciferase activities were examined 24 h after transfection.

[c] 200 μL of each HVJ-liposome suspension was injected into a lobe of the liver under the perisplanchnic membrane ($n = 3$).

[d] All liposomes were prepared by vortexing-filtration method with 15-μmol lipid and 200-μg pAct-Luc-NII, and then incubated with 30,000 HAU of inactivated HVJ. HVJ-liposome complexes were purified and total volumes of each suspension was adjusted to 1 mL.

## VI.  APPLICATION OF FUSOGENIC VIRUS LIPOSOMES TO GENE THERAPY OF INTRACTABLE DISEASES

The use of HVJ liposome gene delivery for clinical trials has been evaluated in several animal models, and we believe that this gene delivery system is promising for the treatment of cancers, chronic diseases, and organ transplantation [19].

### A.  Restenosis

Arterial restenosis occurs in 30–40% of patients after angioplasty. The dilated vessel subsequently narrows again because of neointimal hyperplasia. We introduced HVJ liposome containing ec-NOS expression vector into injured rat carotid artery [20]. At 2 weeks after the transfer, histological analysis revealed a 70% reduction in neointimal area. In contrast, no inhibition of neointima formation was observed in the control vector transfection group.

NO plays multiple roles in the arterial wall, which are dilatation of the artery, inhibition of platelet aggregation, prevention of adhesion of leukocyte with endothelium, and suppression of growth of VSMC. Multiple factors are involved in the induction of restenosis, so that the strategy to augment NO production may be the most practical approach for its treatment.

### B.  Ischemic Heart Disease

A number of cytokines and adhesion molecules appear to be involved in myocardial damage after ischemia and reperfusion, and most of those molecules are upregulated by NFκB. We attempted to inhibit NFκB by the introduction of double-stranded ODN containing the NFκB-binding sequence GGGATTTCCC. This double-stranded ODN was named NFκB decoy, which can trap released NFκB. By the introduction of NFκB decoy into rat coronary artery 30 min after ischemia, the infarcted area was significantly reduced [21]. In those hearts, the expression of IL-6 and VCAM was inhibited, and migration of neutrophils was blocked by NFκB decoy, but not by scrambled decoy. Thus, NFκB decoy treatment coupled with HVJ liposomes is very promising for the inhibition of myocardial infarction after reperfusion.

### C.  Malignant Glioma

HVJ liposomes have been effective for the treatment of cancers. Glioblastoma is a severe type of brain tumor associated with rapid and extensive metastasis, and a gene therapy for disseminated glioblastoma is desirable. All mice injected with glioblastoma died, and the survival time was dependent on the number of cancer cells injected. After dissemination of mouse glioblastoma cells into mouse brain, HSV-TK gene was administered with HVJ liposomes followed by intra-peritoneal injection of gancyclovir. With HVJ-cationic liposomes, 80% of the tumor-bearing mice were alive [22], whereas HVJ-anionic liposomes were less effective, and all the mice died after retrovirus-mediated gene transfer. Thus, HVJ-cationic liposomes appear to be a more effective tool for cancer gene therapy than HVJ-anionic liposomes.

## D. Organ Transplantation

Organ transplantation is one of the most suitable targets for gene therapy. We have reported that vein grafts after transfer of antisense oligonucleotides against cell cycle regulators were successful in rabbits [23]. We also reported that cardiac function was protected by transfer of the HSP70 gene to the ischemic heart during transplantation [24], and chronic rejection induced by arteriosclerosis after cardiac transplantation was successfully prevented by antisense-cdk2 kinase oligo-nucleotides by use of HVJ liposome [25].

## ACKNOWLEDGMENTS

The authors thank Dr. Hans Schreier for his interest in our work and for critically reading the manuscript.

## REFERENCES

1. F. D. Ledley, Nonviral gene therapy: the promise of genes as pharmaceutical products. *Hum. Gene Ther. 6*:1129 (1995).
2. P. L. Felgner, T. R. Gadek, M. Holm, R. Roman, H. S. Chan, M. Wenz, J. P. Northrop, G. M. Ringold, and H. Danielsen, Lipofection: a highly efficient, lipid-mediated DNA-transfection procedure. *Proc. Natl. Acad. Sci. USA 84*:7413 (1987).
3. J-P. Behr, B. Demeneix, J-P. Loeffler, and J. Perez-Mutul, Efficient gene transfer into mammalian primary endocrine cells with lipopolyamine-coated DNA. *Proc. Natl. Acad. Sci. USA 86*:6982 (1989).
4. W. M. Berting, M. Gareis, V. Paspaleeva, A. Zimmer, J. Kreuter, E. Numberg, and P. Harrer, Use of liposomes, viral capsids, and nanoparticles as DNA carriers. *Biotechnol. Appl. Biochem. 13*:390 (1991).
5. J. Y. Legendre, and F. C. Szoka Jr. Delivery of plasmid DNA into mammalian cell lines using pH-sensitive liposomes: comparison with cationic liposomes. *Pharm. Res. 9*:1235 (1992).
6. Y. Kaneda, Virus (Sendai virus envelopes) mediated gene transfer, *Cell Biology: A Laboratory Handbook* (J. E. Celis, ed.) vol. 3. Academic Press, San Diego, 1994, pp. 50–57.

7.  Y. Okada, Sendai-virus induced cell fusion. *Methods in Enzymology* (N. Duzugnes, ed.), Academic Press, San Diego, Vol. 221, 1993, pp. 18–41.

8.  M. Nakanishi, T. Uchida, H. Sugawa, M. Ishiura, and Y. Okada, Efficient introduction of contents of liposomes into cells using HVJ (Sendai virus), *Exp. Cell Res. 159*:399 (1985).

9.  Y. Kaneda, T. Uchida, J. Kim, M. Ishiura, and Y. Okada, The improved efficient method for introducing macromolecules into cells using HVJ (Sendai virus) liposomes with gangliosides, *Exp. Cell Res. 173*:56 (1987).

10.  I. Yanagihara, K. Inui, G. Dickson, G. Turner, T. Piper, Y. Kaneda, and S. Okada, Expression of full-length human dystrophin cDNA in mdx mouse muscle by HVJ-liposome injection, *Gene Ther. 3*:549 (1996).

11.  Y. Kaneda, R. Morishita, and N. Tomita, Increased expression of DNA cointroduced with nuclear protein in adult rat liver, *J. Molec. Med. 73*:289 (1995).

12.  R. Morishita, G. H. Gibbons, E. Ellison, M. Nakajima, L. Zhang, Y. Kaneda, T. Ogihara, and V. J. Dzau, Single intraluminal delivery of antisense cdc2 kinase and proliferating-cell nuclear antigen oligonucleotides results in chronic inhibition of neointimal hyperplasia, *Proc. Natl. Acad. Sci. USA 90*:8474 (1993).

13.  Y. Yonemitsu, Y. Kaneda, R. Morishita, K. Nakagawa, Y. Nakashima, and K. Sueishi, Characterization of in vivo gene transfer into the arterial wall mediated by the Sendai Virus (Hemagglutinating Virus of Japan) liposomes: An effective tool for the in vivo study of arterial diseases, *Laboratory Investigation 75*:313 (1996).

14.  T. Hirano, J. Fujimoto, T. Ueki, H. Yamamoto, T. Iwasaki, R. Morishita, Y. Sawa, Y. Kaneda, H. Takahashi, and E. Okada, Persistent gene expression in rat liver in vivo by repetitive transfections using HVJ-liposome, *Gene Ther. 5*:459 (1998).

15.  M. Hangai, H. Tanihara, Y. Honda, and Y. Kaneda, Introduction of DNA into the rat and primate trabecular meshwork by fusogenic liposomes. *Invest. Ophthalmol. Vis. Sci. 39*:509 (1998).

16.  Y. Saeki, N. Matsumoto, Y. Nakano, M. Mori, K. Awai, and Y. Kaneda, Development and characterization of cationic liposomes conjugated with HVJ (Sendai virus); Reciprocal effect of cationic lipid for in vitro and in vivo gene transfer, *Human Gene Ther. 8*:1965 (1997).

17.  R. Chander and H. Schreier, Artificial viral envelopes containing recombinant human immunodeficiency virus (HIV) gp160, *Life Sci. 50*:481 (1992).

18.  X. Gao and L. Huang, A novel cationic liposome reagent for efficient transfection of mammalian cells, *Biochem. Biophys. Res. Commun. 179*:280 (1991).

19.  V. J. Dzau, M. Mann, R. Morishita, and Y. Kaneda, Fusigenic viral liposome for gene therapy in cardiovascular diseases, *Proc. Natl. Acad. Sci. USA 93*:11421 (1996).

20.  H. E. von der Leyen, G. H. Gibbons, R. Morishita, N. P. Lewis, L. Zhang, M. Nakajima, Y. Kaneda, J. P. Cooke, and V. J. Dzau, Gene therapy inhibiting neointimal vascular lesion; In vivo transfer of endothelial cell nitric oxide synthase gene, *Proc. Natl. Acad. Sci. USA 92*:1137 (1995).

21.  R. Morishita, T. Sugimoto, M. Aoki, I. Kida, N. Tomita, A. Moriguchi, K. Maeda, Y. Sawa, Y. Kaneda, J. Higaki, and T. Ogihara, In vivo transfection of cis element decoy against NFκB binding site prevented myocardial infarction as gene therapy, *Nat. Med. 3*:894 (1997).

22. E. Mabuchi, K. Shimizu, Y. Miyao, Y. Kaneda, H. Kishima, M. Tamura, K. Ikenaka, and T. Hayakawa, Gene delivery by HVJ-liposome in the experimental gene therapy of murine glioma, *Gene Therapy* 4:768 (1997).

23. M. Mann, G. H. Gibbons, R. S. Kernoff, F. P. Diet, P. S. Tsao, J. P. Cooke, Y. Kaneda, and V. J. Dzau, Genetic engineering of vein grafts resistant to atherosclerosis, *Proc. Natl. Acad. Sci. USA* 92:4502 (1995).

24. K. Suzuki, Y. Sawa, Y. Kaneda, H. Ichikawa, R. Shirakura, and H. Matsuda, In vivo gene transfection with heat shock protein 70 enhances myocardial tolerance to ischemia-reperfusion injury in rat, *J. Clin. Invest.* 99:1645 (1997).

25. J. Suzuki, M. Isobe, R. Morishita, K. Aoki, S. Horie, Y. Okubo, Y. Kaneda, Y. Sawa, H. Matsuda, T. Ogihara, and M. Sekiguchi, Prevention of graft coronary atherosclerosis by antisense cdk2 kinase oligonucleotides, *Nat. Med.* 3:900 (1997).

# 10
# Targeting of Viral Vectors for Cancer Gene Therapy

**Dirk M. Nettelbeck and Rolf Müller**
*Philipps University, Marburg, Germany*

## I. INTRODUCTION

One of the major problems of conventional cancer therapy is the lack of tumor specificity, which limits the applicable therapeutic dose and renders most advanced malignancies refractory to treatment. In view of this pressing need for treatment modalities, the application of genetic therapy to cancer has attracted great attention [1]. Obviously, the success of gene therapy depends on the stable delivery of the therapeutic gene to the target tissue, its cellular uptake, nuclear translocation, and efficient gene expression. Viruses have developed specific and efficient strategies during evolution to achieve these goals and should, therefore, in principle represent suitable tools for gene therapy. Indeed, recombinant viruses have been shown in numerous studies to be efficient gene therapeutic vectors. Most studies performed so far have made use of adenoviral vectors, because these are easy to handle, give rise to high titers, and are derived from nonpathogenic or low pathogenic viruses [2–4]. Adenoviruses have been used either as replication-incompetent vectors (because of the deletion of essential viral genes) or as replication-competent oncolytic agents. For both strategies, the retargeting of viral particles to the site of disease is of paramount importance to be able to achieve a sufficiently high dose at the site of disease in the absence of serious side effects, both prerequisites for a successful clinical application. The different approaches used to achieve this goal will be the focus of the first part of this chapter. We will then discuss the possibility to achieve high levels of transcription in defined cell populations through specific cellular promoters or regulatory elements, including both natural and designer promoters with improved specificity and activity. Finally, we will discuss the approaches that aim at exploiting tumor-specific genetic alterations to restrict viral vector replication to tumor cells.

## II. RETARGETING OF ADENOVIRAL VECTORS

Natural-occurring adenoviruses show little target cell specificity. *In vivo*, a major problem is the strong preference for liver cells and the rapid elimination by phagocytic cells. Therefore, to make adenoviral vectors suitable for systemic delivery an alteration of the natural tropism is obligatory. For this purpose, the viral surface (capsid) must be modified in a way that the molecules engaged in natural host cell recognition are neutralized and substituted by target-specific mechanisms. Such approaches are facilitated by the fact that the adenoviral particle structure and replicative cycle have been characterized in detail.

The building block of the adenoviral capsid is the icosahedron. Each of its 20 triangular surfaces consists of 12 hexon proteins, and each of the 12 vertices is built by a protein complex consisting of a penton base and fiber protein. The protruding knob domain of the fiber protein is responsible for binding of the adenoviral particles to the host cell, in the case of serotypes 2 and 5, to the coxsackie adenovirus receptor (CAR) [5–7]. Internalization by receptor-mediated endocytosis is mediated by binding of the penton base RGD-motif to the cellular integrins $\alpha_v\beta_3$ or $\alpha_v\beta_5$ [8]. Primary tumors contain low levels of CAR [9–12], which emphasizes the need for retargeting. It is now clear from a number of studies that efficient postbinding cell entry is not impaired in retargeted adenoviruses, because targeting does not affect downstream steps required for productive infection, which represents a problem in the targeting of retroviruses. Consequently, greatly increased transduction efficiencies can be achieved by retargeted adenoviruses. The different approaches used for the retargeting of adenoviruses are discussed in the following.

### A. Immunological Approaches for the Modification of Adenoviral Tropism

Ablation of endogenous tropism of adenoviruses has been achieved by Fab fragments of neutralizing monoclonal antibodies or neutralizing recombinant single-chain antibody fragments (scFv) in the absence of cross-linking of viral particles or activation of Fc-dependent immunological effector functions. The introduction of a novel tropism was then brought about by linkage of a target cell binding antibody, antibody fragment, or ligand to the neutralizing domain. Various ligands or antibody fragments have been chemically linked to neutralizing anti-knob Fab:

1. Folate for targeting diverse tumors (transduction inhibited in folate receptor-negative control cells, transduction like wild-type virus in folate receptor-positive cells) [13]
2. Fibroblast growth factor-2 (FGF-2) for CAR-/integrin-independent targeting of several tumor cell lines (>100-fold enhanced transduction), Kaposi's sarcoma (>40-fold enhanced transduction), endothelial cells (>30-fold enhanced transduction), and smooth muscle cells (>90-fold enhanced transduction) [9, 14–17]
3. Antiepidermal growth factor (EGF) antibody for targeting of glioblastoma (up to >60-fold enhanced, CAR-independent transduction) [10]
4. Anti-EpCAM antibody for CAR-independent targeting of diverse carcinomas (10-fold increase in activity) [18]
5. Anti-CD40 antibody for targeting of dendritic cells (>100-fold enhanced transduction) [19]

In addition, fusion proteins of neutralizing anti-knob scFv and EGF have been used for CAR- and integrin-independent transduction of diverse EGF receptor overexpressing tumor cell lines (up to >10-fold enhanced transduction compared with wild-type adenovirus, >200-fold compared with wild-type adenovirus plus neutralizing antiknob scFv) [20]. Furthermore, recombinant bispecific single-chain diabodies (scDb) [21] derived from neutralizing antiknob scFv and antiendoglin (CD105) scFv have been successfully used for the CAR- and integrin-independent targeting of endothelial cells (D.M.N., D. Miller, S.J. Watkins, R.M. and R. Kontermann, unpublished observations).

*In vivo*, neutralizing antibodies have been reported to reduce the liver uptake of $^{99m}$Tc-labeled Ad5-knob to some extent [22]. FGF2-complexed adenovirus showed increased transduction of ovarian carcinoma cells after interperitoneal injection and in conjunction with the HSV-tk/GCV effector system the same therapeutic effect as untargeted adenovirus at a 10-fold higher titer and reduced liver transduction and liver toxicity after IV injection [23, 24]. These observations indicate that the immunological approach for the transductional retargeting of adenoviruses might also be a promising strategy for the generation of viral vectors that are suitable for systemic delivery.

In a related approach a bimodular 35-mer oligopeptide consisting of the knob-binding domain of human MHCI molecule fused to gastrin-releasing peptide (GRP) has been used for the GRP-dependent and enhanced transduction of various tumor cell lines [25]. GRP receptors are overexpressed on several carcinomas and melanoma cells and may therefore represent a useful target. Finally, the coupling of peptides derived by phage display to the adenovirus surface by means of bifunctional polytheleneglycol molecules resulted in an increased and CAR-independent transduction of primary human airway epithelium cells [26].

## B.  Genetic Approaches for Modification
of Adenoviral Tropism

An alternative strategy for the modification of adenoviral tropism is the genetic
modification of the viral capsid. Three different structural components of the
capsid have been the target of such modifications:

1.  The hexon for knob-independent transduction of smooth muscle cells
    by incorporation of an integrin-recognizing RGD-peptide [27].
2.  The penton for targeting to specific integrins by modification of se-
    quences flanking the RGD motif present in the penton base [28].
3.  The fiber: Because the fiber seems to be primarily responsible for bind-
    ing of wild-type adenoviruses to cells, most studies focus on modifica-
    tion of the fiber. This work is summarized below.

The crucial role of the fiber in target cell selection has been demonstrated
by the altered tropism of chimeric adenoviruses endowed with fibers derived from
other adenoviral strains (pseudotyping) [29]. Thus, exchange of the Ad5 fiber
with the Ad3 fiber resulted in strongly increased transduction of human embry-
onic lung fibroblast, squamous cell carcinoma, and monocyte cell lines, which
are relatively refractory to transduction by Ad5 [28]. The same modification was
shown in another study to endow the virus with a specificity for EBV-transformed
B-lymphocytes [30], which may be useful for therapy of lymphoma and purging
of donor marrow. Furthermore, substitution of the Ad2 fiber with the Ad17 fiber
led to an up to 20-fold increase in infection of primary rat neurons and human
endothelial cells [31]. Finally, exchange of the Ad5 fiber with the Ad35 fiber
resulted in an increased transduction of CD34+ cells, which was CAR and integrin independent [32].

In an alternative scenario, 7-mer or 20-mer oligolysine sequences (which
interact with heparan sulfate) have been fused to the C terminus of the fiber
protein. This modification led to a dramatically increased and CAR-independent
transduction of endothelial cells, smooth muscle cells, and macrophages *in vitro*
and *in vivo* (up to > 100-fold) [33, 34], of myeloma cells and AML blast cells
*in vitro* [35, 36], and of cultured glioma cells *in vitro* [37, 38]. A similar approach
has been used to fuse an RGD peptide (RDG-4C) to the C terminus of the fiber,
which led to an increased transduction of endothelial cells *in vitro* and *in vivo*
[33, 39]. The same RGD-containing peptide has also been incorporated into the
HI-loop of the fiber protein, resulting in an up to 1000-fold increased and CAR-
independent transduction of primary endothelial cells and ovarian and head and
neck cancer cells *in vitro* [40, 41].

Compared with the immunological approach the genetic modification has
advantages. Thus, the production of two agents (virus and bispecific molecule)

is not necessary, and the new ligand is covalently linked to the vector. On the other hand, there are potential disadvantages associated with this approach. Thus, only few ligands have so far been successfully incorporated into the adenovirus capsid. Especially in the case of antibodies, genetic modifications may prove difficult. Furthermore, incorporation of a peptide ligand does not abrogate the natural tropism. Additional modifications of the knob domain responsible for CAR binding [7, 42] will be necessary but might not be sufficient in view of the integrin-binding motif in the penton base. Once the function of molecules and domains involved in target recognition have been solved, it should, however, be possible to construct a truly retargeted adenovirus. Finally, a producer cell line with a "pseudoreceptor" is required when the endogenous tropism is ablated [43, 44], which is a further complication in generating viruses with genetically modified fibers. Thus, the major advantage of the immunological approach seems to be wider applicability (at least at present) and the relative ease of construction (in particular, if multiple ligands are to be tested).

The strategies just described are in principle also applicable to other viral vectors, and a number of studies describing such approaches have been published, mainly for adeno-associated viruses (AAV) [45–47] and retroviruses [48–50]. It can be anticipated that this field will rapidly expand as the mechanisms of virus-host cell interactions are unraveled in detail.

## III. TRANSCRIPTIONAL TARGETING

Achieving high levels of transcription in defined cell populations through specific cellular promoters or regulatory elements would be a major step toward the development of a stringently targeted vector. Some of the recently described experimental gene therapy protocols, indeed, make use of natural tissue-specific or otherwise selective promoters, but in many instances these promoters suffer from a lack of activity, specificity, or both. Several groups have therefore focussed on the design of improved promoters for cancer gene therapy. The applicability of most promoter systems is presumably not dependent on a particular type of vector and in a number of cases have also been used in the context of viral vectors. It should, however, be noted that the viral environment can have an impact on the regulation of the inserted transgene. In such cases, the inclusion of transcriptional insulator sequences [51–53] may be useful, as shown, for instance, in case of an adenovirally transduced *erbB-2* promoter-driven transgene [54]. In the following, we will give a brief description of various natural promoters used in cancer gene therapy and an account of recent successful developments in the area of "designer" promoters.

## A.  Tissue-specific Promoters

A great number of tissue-specific promoters has been isolated and characterized, and some of these have been used in experimental cancer gene therapy (see Table 1). Obviously, tissue-specific promoters are also active in the normal tissue from which the tumor originated. Unless the loss of these normal cells is acceptable (as in the case of melanocytes), additional specificity mechanisms need to be considered to make the resulting vector useful for a systemic delivery. This can be achieved, for example, by introducing an additional level of specificity, such as a specificity for proliferating cells or a selectivity for other conditions that are characteristic of tumor cells (see later).

## B.  Tumor Endothelium-directed Promoters

Targeting the tumor vasculature by gene therapy has numerous potential advantages compared with direct tumor cell targeting. The tumor blood vessels are more readily accessible to vectors, endothelial cells (ECs) are not known to undergo mutations or to gain resistance to treatment, and the endothelium represents a target that is largely independent of tumor type. Therefore, vectors are being designed for the selective targeting of the tumor vasculature. An important part of this strategy are promoters that are up-regulated in tumor ECs. These are exemplified by those of the genes encoding vascular endothelial growth factor receptors KDR/FLK-1 [55], E-selectin [55, 56], and the transforming growth factor-β binding protein endoglin/CD105 [57] (see also Table 1).

## C.  Tumor-selective Promoters

A set of genes has been identified that shows little or no activity in the cells of an adult organism under nonpathological conditions, but that is turned on or up-regulated in certain types of tumors (see Table 1). This applies, for instance, to the promoter of the gene-encoding oncofetal α-fetoprotein (AFP) [58–62], which under nonpathological conditions is active specifically in the fetal liver but becomes reactivated in hepatoma cells. Another example is the promoter of the carcinoembryonic antigen (CEA) [63–68] gene, which is reactivated in different types of adenocarcinoma.

Other promoters that may be useful for the design of targeted vectors are those induced by disease-specific conditions, such as hypoxia (see Table 1). Thus, a promoter responding to hypoxia through activation by the hypoxia-inducible factor-1 (HIF-1) has been successfully used for experimental cancer gene therapy [69]. Disease-specific conditions in tumor cells can also directly result from genetic alterations or from altered signaling pathways. The latter are presumably

**Table 1** Natural Promoters Used in Cancer Gene Therapy

| | Promoter | Target | References |
|---|---|---|---|
| Tissue-specific | Tyrosinase, tyrosinase related protein-1 (TRP-1) | Melanocytes | [123–128] |
| | Prostate-specific antigen (PSA) | Prostate | [129–133] |
| | Albumin | Liver | [58, 134, 135] |
| | Muscle creatinine kinase (MCK) | Muscle | [136–138] |
| | Myelin basic protein (MBP) | Oligodendrocytes, glioma cells | [139, 140] |
| | GFAP | Glioma cells | [139, 141–143] |
| | NSE | Neurons | [143, 144] |
| Tumor endothelium-directed | KDR (human Flk-1) | | [55] |
| | E-selectin | | [55, 56] |
| | Endoglin | | [57] |
| Tumor-selective | α-Fetoprotein (AFP) | Liver tumor | [58–62] |
| | Carcinoembryonal antigen (CEA) | Adenocarcinomas (breast, lung, colorectal carcinoma) | [63–68, 145] |
| | erbB2 | Breast and pancreatic cancer | [54, 70–72] |
| | Mucin-1 (muc-1; DF3) | Breast cancer | [72–74] |
| | α- and β-lactalbumin | Breast cancer | [146] |
| | Osteocalcin | Osteosarcoma, prostate cancer | [147–149] |
| | Secretory leukoprotease inhibitor (SLPI) | Ovarial/cervical carcinoma | [150, 151] |
| | Hypoxia response element (HRE) | Solid tumors | [69] |
| | Glucose-regulated protein 78 (Grp 78; BIP) | Solid tumors | [152, 153] |
| | L-plastin | Cancer cells | [154] |
| | Hexokinase II | Cancer cells | [155] |
| Treatment-responsive | Early growth response-1 gene (egr-1) | Radiation-induced | [79–85, 156] |
| | Tissue plasminogen activator (t-PA) | | |
| | Multiple drug resistance gene 1 (mdr-1) | Chemotherapy-induced | [86] |
| | Heat shock protein 70 (hsp 70) | Heat-induced | [157] |
| Cell cycle-regulated | E2F-1 | Malignant cells with disrupted pRb pathway | [88] |
| | cyclin A, cdc25C | Proliferating cells | [89, 90] |

the reason for the frequent up-regulation of the *erbB-2* promoter in a range of adenocarcinomas, a feature that has also been exploited in experimental therapeutic models [54, 70–72] (see also Table 1). Another promoter often found up-regulated in human adenocarcinomas and successfully used in experimental cancer gene therapy is the mucin-1 gene (*muc-1*) promoter [72–74].

Tumor-specific genomic rearrangements have also been exploited for the design of tumor-specific artificial promoters. A rhabdomyosarcoma-specific translocation leads to the expression of a chimeric transcription factor consisting of the DNA binding and the transactivation domains derived from PAX3 and FKHR, respectively. This novel transcriptional activator is tumor-cell specific, because PAX3 expression is normally restricted to prenatal development. Thus, an artificial promoter, consisting of multiple PAX3-binding sites in front of a minimal adenoviral E1B promoter, could be used to drive the expression of the desired transgene in a rhabdomyosarcoma-specific fashion [75]. Translocations generating transcription factors with novel properties are frequently found in human cancers, suggesting that analogous strategies might be useful for other malignancies.

Finally, the expression of viral genes is a hallmark of certain tumors, such as Burkitt's lymphoma. To direct gene expression selectively to Burkitt's lymphoma cells, regulatory elements responsive to the Epstein-Barr virus–(EBV–) encoded transcriptional activators, EBNA-1 or EBNA-2, have been used to construct promoters with an extremely high degree of specificity [76–78]. Obviously, this kind of approach might be applicable to a variety of human tumors that have been associated with specific oncogenic viruses.

## D.  Treatment-responsive Promoters

Another interesting scenario is the use of promoters that are inducible by certain conventional cancer therapy modalities (see Table 1). Such promoters are interesting for the temporally and spatially restricted expression of genes whose products are able to sensitize the tumor cells to the actual treatment. These proteins—like TNF-$\alpha$—are often highly toxic, which precludes their systemic application and prevents the accumulation of therapeutically desirable levels at the site(s) of the tumor. In this context, promoters that are induced by therapeutic doses of ionizing radiation, such as the *egr-1* promoter [79–85], or by chemotherapeutic agents, such as the promoter of the P-glycoprotein/multidrug resistance-1 gene (*mdr-1*) [86], have attracted some attention (see Table 1). Other promoters that might fulfil a similar purpose are those activated through the transcription factor nuclear factor $\kappa$B (NF$\kappa$B) in response to ionizing radiation and certain chemotherapeutic agents.

## E. Cell Cycle–Regulated Promoters

Unrestrained cell proliferation is a hallmark of cancer. This is exemplified by the dysfunction of the retinoblastoma protein (Rb) pathway, which controls the $G_1$ restriction point, in nearly all tumor cells [87]. As a consequence, the transcription factor E2F loses its $G_0/G_1$-specific repressor function and is constitutively active. In an attempt to exploit this finding for cancer gene therapy, the E2F-regulated promoter of the E2F-1 gene has been incorporated into an adenoviral vector and shown to confer tumor cell–specific transgene expression in an animal model of human glioma [88]. Because hyperproliferation is characteristic of most tumor cells and tumor ECs, the promoters of cell cycle genes regulated by other mechanisms, such as *cyclin A* or *cdc25C*, also represent potentially interesting tools for cancer gene therapy [89, 90] (see Table 1).

## F. Enhancing the Activity of Tissue-specific or Other Selective Promoters

Frequently, tissue-specific or other selective promoters are inefficient activators of transcription, which severely limits their applicability. Several strategies for improving promoter strength while maintaining specificity have been described (see later). The first and simplest approach is to eliminate from a natural promoter all the regions that do not contribute to its transcriptional strength or specificity and to multimerize the positive regulatory promoter elements or enhancer domains. This strategy has been used in several instances with variable success. Alternatively, promoters with activating point mutations have been used, but such promoters have been found only in a few specific cases, like the AFP [91], or *mdr-1* [92] genes. A third strategy involves the construction of chimeric promoters by combining the transcription regulatory elements from different promoters that are specific for the same tissue. In one example [93], 5–20 DNA elements involved in muscle-specific transcriptional activation (E-box, MEF-2 site, TEF-1 site, and SRE) were assembled in random order in such a way that they were exposed on the same side of the double-helix. These synthetic cassettes were linked to a minimal chicken α-actin promoter. One of these combinations was found to be considerably more active than the CMV immediate early promoter/enhancer both in cell culture and *in vivo*.

However, the use of recombinant transcriptional activators (RTAs) appears to be the most generally applicable system at present. The construction of RTAs is based on the modular structure of transcription factors, which allows for the combination of DNA binding and transactivation domains derived from different proteins. For example, RTAs have been used to establish a positive feedback loop initiated by transcription from a weak cell type–specific promoter [94]. Such a promoter drives the simultaneous expression of the desired effector/reporter gene product and a strong artificial transcriptional activator (VP16-LexA or Gal4-VP16), which subsequently stimulates transcription through appropriate LexA or Gal4 binding sites in the promoter. This approach has been successfully used with several promoters, including the EC-specific vWF promoter, the gastrointestinal cell–specific sucrase isomaltase promoter, and the AFP promoter. In each case a dramatic enhancement of transcriptional activity without affecting cell type specificity was achieved.

## G.  Dual-specificity Promoters

It would obviously be advantageous if the specificities of different promoters could be combined within one transcriptional control unit. Two different strategies that seem to be generally applicable for the construction of dual specificity promoters have been described. The functionality of both systems has been demonstrated for the specific transcriptional targeting of proliferating melanoma cells. In the first approach, the transgene is driven by an artificial heterodimeric transcription factor, the DNA-binding subunit of which is expressed from a tissue-specific promoter, whereas the transactivating subunit is transcribed from a cell cycle–regulated promoter [95]. As a result, gene expression occurs preferentially in the proliferating cells of a specific type of tissue. The selectivity of this strategy has been demonstrated for the expression of a transgene in proliferating melanoma cells using *cyclin A* and the tyrosinase promoter elements. In the second approach [158], a chimeric transcription factor (Gal4/NF-Y), consisting of the transactivation domain of NF-Y and the DNA-binding domain of Gal4, is expressed from a tissue-specific promoter. Gal4/NF-Y binds to a second promoter consisting of a *cyclin A* minimal promoter harboring a cell cycle–regulated repressor site (CDE-CHR) [89] downstream of multiple Gal4 binding sites. As a result, the stimulatory activity of Gal4/NF-Y is restrained in resting cells by the recruitment of the CDF-1 repressor [96] to the CDE-CHR.

## IV.  TARGETING OF VIRAL REPLICATION TO TUMORS

The major problem of gene therapy is the low transduction efficiency *in vivo*. For the application to cancer this problem is exacerbated by the inefficient vascularization and high interstitial pressure in malignant tumors, which limit the accessibility for any kind of vector. One possibility to address this problem is the use of viruses that are able to replicate *in vivo* and thereby spread throughout the tumor tissue. This approach is not only useful for the improved delivery of therapeutic proteins but seems particularly appealing if combined with the intrinsic ability of viruses to kill cells (oncolysis). In both cases, tumor selectivity is a prerequisite. The attempts made in this direction are discussed in the following.

## A. Targeting of Tumor-specific Dysfunctions in Cells Cycle Control or Apoptosis by Adenovirus Mutants

Two of the most commonly affected genetic loci in human tumors are the tumor suppressor genes p53 and pRb or other components in the respective pathways. Both protein activities must be eliminated from the host cell in case of a productive adenoviral infection. This is normally brought about by proteins encoded by the adenoviral E1A and E1B genes. Therefore, adeno-viral mutants lacking one of these proteins should be unable to replicate efficiently in normally cells, but should be able to do so in p53- or pRb-deficient tumor cells.

ONYX-015 is a mutant adenovirus lacking the E1B 55K gene product, which is required for inactivation of p53. This virus has been described to replicate specifically in p53-deficient cells [97]. The latter observation is, however, controversial at present [98–101], which could be because wild-type p53 might be inactive in certain tumor cells because of other genetic alterations (such as mdm2 amplification, loss of p14$^{ARF}$ function, or HPV E6 expression) [102]. In addition, different assays have been used in different studies, and the role of p53 in preventing viral replication and cell lysis may be different [98, 102]. Nevertheless, tumor regression after intratumoral injection and tumor growth inhibition after IV injection have been shown in human tumor xenograft models [97, 103–105]. In addition, recent clinical trials have shown that the virus is well-tolerated treatment and safe (no dose limiting toxicity up $10^{-11}$ pfu) and that oncolytic activity can be achieved in humans [102, 106].

AdΔ24 is an adenoviral mutant targeting pRb-dysfunction [107]. It contains a 24-bp deletion in the pRb-binding domain of E1A and therefore depends on a deregulated pRb pathway for replication. This mutant has been shown to replicate in proliferating or $G_1$-arrested glioma and osteosarcoma cell lines (in which the pRb/p16$^{INK4a}$ pathway is disrupted), but not in normal $G_1$-arrested fibroblasts and to inhibit tumor growth after intratumoral injection [107].

Several recent studies have combined replication competent adenoviruses with therapeutic genes [108–112], such as HSV-tk, cyctosine deaminase (CD), or a tk-CD fusion gene, for prodrug conversion (ganciclovir or/and 5-fluorouracil). This approach led to a marked enhancement of the cytopathic effect on tumor cells and sensitized the tumor cells to radiation *in vitro* and *in vivo* (human tumor xenotransplants in mice). A 100% cure was observed with a ''trimodal'' regimen of viral oncolysis plus double suicide therapy plus radiation therapy [111, 112]. In addition, both suicide systems were capable of suppressing viral replication *in vitro*, which is an important safety issue. A similar strategy has been used for the generation of an oncolytic mutant HSV virus expressing cyclophosphamide and ganciclovir activating genes [113].

Finally, replication-competent viral vector systems have been combined with standard cancer therapy in animal models, resulting in synergistic effects. This has, for example, been shown for ONYX-015 plus cisplatin or ionizing radiation [108, 112] and for an AdE1B 55k-deleted virus expressing HSV-tk (plus ganciclovir) in combination with the topoisomerase inhibitor topotecan [114].

## B. Transcriptional Targeting of Viral Replication

An alternative approach to prevent viral replication in cells outside the tumor is the transcriptional targeting of viral replication that is the expression of an essential viral gene under the control of a promoter that is preferentially or specifically active in tumor cells. This strategy has been successfully used for the construction of replication-competent prostate carcinoma-specific adenoviruses. This was achieved by placing the *E1A* gene under the control of either the PSA or the kallikrein-2 promoter or by driving the expression of both *E1A* and *E1B* by the prostate-specific promoters of the PSA, probasin, and kallikrein-2–encoding genes, respectively [115–117]. Replication-competent hepatoma-specific adenoviruses have been generated by placing the *E1A* gene under the control of the AFP promoter [118, 119].

In an analogous fashion, replication-competent hepatoma-directed herpesviruses have been generated by placing the viral ICP4 gene under the control of the human albumin gene promoter/enhancer [120]. This virus showed hepatoma-specific replication *in vitro* and *in vivo* and tumor growth inhibition [121]. In addition the viral γ34.5 gene was placed under the control of the *B-myb* promoter. This modified herpesvirus specifically replicated in proliferating fibroblasts showed strongly reduced neurotoxicity and inhibited tumor growth *in vivo* [122].

Numerous transcriptionally targeted replication-competent adenoviruses and herpes viruses have been shown to be highly specific (up to 10,000-fold relative to nontarget cells) in cell culture experiments. In addition, animal models have demonstrated their safety and efficacy. It thus appears that the transcriptional targeting of viral replication holds some promise as a strategy for cancer gene therapy.

## REFERENCES

1.  K. W. Culver and R. M. Blaese, Gene therapy for cancer, *Trends Genet. 10*:174 (1994).
2.  G. Ross, R. Erickson, D. Knorr, A. G. Motulsky, R. Parkman, J. Samulski, S. E.

Straus, and B. R. Smith, Gene therapy in the United States: a five-year status report, *Hum. Gene Ther.* 7:1781 (1996).

3.  W. F. Anderson, Human gene therapy, *Nature 392*:25 (1998).

4.  G. J. Nabel, Development of optimized vectors for gene therapy, *Proc. Natl. Acad. Sci. USA 96*:324 (1999).

5.  L. J. Henry, D. Xia, M. E. Wilke, J. Deisenhofer, and R. D. Gerard, Characterization of the knob domain of the adenovirus type 5 fiber protein expressed in Escherichia coli, *J. Virol. 68*:5239 (1994).

6.  J. M. Bergelson, J. A. Cunningham, G. Droguett, J. E. Kurt, A. Krithivas, J. S. Hong, M. S. Horwitz, R. L. Crowell, and R. W. Finberg, Isolation of a common receptor for Coxsackie B viruses and adenoviruses 2 and 5, *Science 275*:1320 (1997).

7.  I. Kirby, E. Davison, A. J. Beavil, C. P. Soh, T. J. Wickham, P. W. Roelvink, I. Kovesdi, B. J. Sutton, and G. Santis, Mutations in the DG loop of adenovirus type 5 fiber knob protein abolish high-affinity binding to its cellular receptor CAR, *J. Virol. 73*:9508 (1999).

8.  T. J. Wickham, P. Mathias, D. A. Cheresh, and G. R. Nemerow, Integrins alpha v beta 3 and alpha v beta 5 promote adenovirus internalization but not virus attachment, *Cell 73*:309 (1993).

9.  C. K. Goldman, B. E. Rogers, J. T. Douglas, B. A. Sosnowski, W. Ying, G. P. Siegal, A. Baird, J. A. Campain, and D. T. Curiel, Targeted gene delivery to Kaposi's sarcoma cells via the fibroblast growth factor receptor, *Cancer Res. 57*: 1447 (1997).

10. C. R. Miller, D. J. Buchsbaum, P. N. Reynolds, J. T. Douglas, G. Y. Gillespie, M. S. Mayo, D. Raben, and D. T. Curiel, Differential susceptibility of primary and established human glioma cells to adenovirus infection: targeting via the epidermal growth factor receptor achieves fiber receptor-independent gene transfer, *Cancer Res. 58*:5738 (1998).

11. S. Hemmi, R. Geertsen, A. Mezzacasa, I. Peter, and R. Dummer, The presence of human coxsackievirus and adenovirus receptor is associated with efficient adenovirus-mediated transgene expression in human melanoma cell cultures, *Hum. Gene Ther. 9*:2363 (1998).

12. Y. Li, R. C. Pong, J. M. Bergelson, M. C. Hall, A. I. Sagalowsky, C. P. Tseng, Z. Wang, and J. T. Hsieh, Loss of adenoviral receptor expression in human bladder cancer cells: a potential impact on the efficacy of gene therapy, *Cancer Res. 59*: 325 (1999).

13. J. T. Douglas, B. E. Rogers, M. E. Rosenfeld, S. I. Michael, M. Feng, and D. T. Curiel, Targeted gene delivery by tropism-modified adenoviral vectors, *Nat. Biotech. 14*:1574 (1996).

14. B. E. Rogers, J. T. Douglas, C. Ahlem, D. J. Buchsbaum, J. Frincke, and D. T. Curiel, Use of a novel cross-linking method to modify adenovirus tropism, *Gene. Ther. 4*:1387 (1997).

15. B. E. Rogers, J. T. Douglas, B. A. Sosnowski, W. Ying, G. Pierce, D. J. Buchsbaum, D. DellaManna, A. Baird, and D. T. Curiel, Enhanced in vivo gene delivery to human ovarian cancer xenografts utilizing a tropism-modified adenovirus vector, *Tumor Targeting 3*:25 (1998).

16.  P. N. Reynolds, R. C. Miller, C. K. Goldman, J. Doukas, B. A. Sosnowski, B. E. Rogers, J. Gómez-Navarro, G. F. Pierce, D. T. Curiel, and J. T. Douglas, Targeting adenoviral infection with basic fibroblast growth factor enhances gene delivery to vascular endothelial and smooth muscle cells, *Tumor Targeting 3*:156 (1998).

17.  J. Doukas, D. K. Hoganson, M. Ong, W. Ying, D. L. Lacey, A. Baird, G. F. Pierce, and B. A. Sosnowski, Retargeted delivery of adenoviral vectors through fibroblast growth factor receptors involves unique cellular pathways, *FASEB J. 13*:1459 (1999).

18.  H. Haisma, H. Pinedo, T. D. de Gruije, S. A. Luykx-de Bakker, B. Sosnowski, W. Ying, V. Beusechem, B. Tillman, W. Gerritsen, and D. Curiel, Tumor-specific gene transfer via an adenoviral vector targeted to the pan-carcinoma antigen EpCAM, *Gene. Ther. 6*:1469 (1999).

19.  B. W. Tillman, A. van Rijswijk, I. van der Meulen-Muileman, R. J. Scheper, H. M. Pinedo, T. J. Curiel, W. R. Gerritsen, and D. T. Curiel, Maturation of dendritic cells accompanies high-efficiency gene transfer by a CD40-targeted adenoviral vector, *J. Immunol. 162*:6378 (1999).

20.  S. J. Watkins, V. V. Mesyanzhinov, L. P. Kurochkina, and R. E. Hawkins, The 'adenobody' approach to viral targeting: specific and enhanced adenoviral gene delivery, *Gene. Ther. 4*:1004 (1997).

21.  S. Brüsselbach, T. Korn, T. Völkel, R. Müller, and R. Kontermann, Enzyme recruitment and tumor cell killing in vitro by a secreted bispecific single-chain diabody, *Tumor Targeting 4*:115 (1999).

22.  K. R. Zinn, J. T. Douglas, C. A. Smyth, H. G. Liu, Q. Wu, V. N. Krasnykh, J. D. Mountz, D. T. Curiel, and J. M. Mountz, Imaging and tissue biodistribution of 99mTc-labeled adenovirus knob (serotype 5), *Gene Ther. 5*:798 (1998).

23.  C. Rancourt, B. E. Rogers, B. A. Sosnowski, M. Wang, A. Piche, G. F. Pierce, R. D. Alvarez, G. P. Siegal, J. T. Douglas, and D. T. Curiel, Basic fibroblast growth factor enhancement of adenovirus-mediated delivery of the herpes simplex virus thymidine kinase gene results in augmented therapeutic benefit in a murine model of ovarian cancer, *Clin. Cancer Res. 4*:2455 (1998).

24.  M. A. Printz, A. M. Gonzalez, M. Cunningham, D. L. Gu, M. Ong, G. F. Pierce, and S. L. Aukerman, Fibroblast growth factor 2-retargeted adenoviral vectors exhibit a modified biolocalization pattern and display reduced toxicity relative to native adenoviral vectors, *Hum. Gene Ther. 11*:191 (2000).

25.  S. S. Hong, A. Galaup, R. Peytavi, N. Chazal, and P. Boulanger, Enhancement of adenovirus-mediated gene delivery by use of an oligopeptide with dual binding specificity, *Hum. Gene Ther. 10*:2577 (1999).

26.  H. Romanczuk, C. E. Galer, J. Zabner, G. Barsomian, S. C. Wadsworth, and C. R. O'Riordan, Modification of an adenoviral vector with biologically selected peptides: a novel strategy for gene delivery to cells of choice [see comments], *Hum. Gene Ther. 10*:2615 (1999).

27.  E. Vigne, I. Mahfouz, J. F. Dedieu, A. Brie, M. Perricaudet, and P. Yeh, RGD inclusion in the hexon monomer provides adenovirus type 5-based vectors with a fiber knob-independent pathway for infection, *J. Virol. 73*:5156 (1999).

28.  S. C. Stevenson, M. Rollence, N. J. Marshall, and A. McClelland, Selective tar-

geting of human cells by a chimeric adenovirus vector containing a modified fiber protein, *J. Virol. 71*:4782 (1997).

29. V. N. Krasnykh, G. V. Mikheeva, J. T. Douglas, and D. T. Curiel, Generation of recombinant adenovirus vectors with modified fibers for altering viral tropism, *J. Virol. 70*:6839 (1996).

30. D. Von Seggern, S. Huang, S. K. Fleck, S. C. Stevenson, and G. R. Nemerow, Adenovirus vector pseudotyping in fiber-expressing cell lines: improved transduction of Epstein-Barr virus-transformed B cells, *J. Virol. 74*:354 (2000).

31. M. Chillon, A. Bosch, J. Zabner, L. Law, D. Armentano, M. J. Welsh, and B. L. Davidson, Group D adenoviruses infect primary central nervous system cells more efficiently than those from group C, *J. Virol. 73*:2537 (1999).

32. D. M. Shayakhmetov, T. Papayannopoulou, G. Stamatoyannopoulos, and A. Lieber, Efficient gene transfer into human CD34(+) cells by a retargeted adenovirus vector, *J. Virol. 74*:2567 (2000).

33. T. J. Wickham, E. Tzeng, L. N. Shears, P. W. Roelvink, Y. Li, G. M. Lee, D. E. Brough, A. Lizonova, and I. Kovesdi, Increased in vitro and in vivo gene transfer by adenovirus vectors containing chimeric fiber proteins, *J. Virol. 71*:8221 (1997).

34. K. Bouri, W. G. Feero, M. M. Myerburg, T. J. Wickham, I. Kovesdi, E. P. Hoffman, and P. R. Clemens, Polylysine modification of adenoviral fiber protein enhances muscle cell transduction, *Hum. Gene Ther. 10*:1633 (1999).

35. R. Gonzalez, R. Vereecque, T. J. Wickham, M. Vanrumbeke, I. Kovesdi, F. Bauters, P. Fenaux, and B. Quesnel, Increased gene transfer in acute myeloid leukemic cells by an adenovirus vector containing a modified fiber protein, *Gene Ther. 6*: 314 (1999).

36. R. Gonzalez, R. Vereecque, T. J. Wickham, T. Facon, D. Hetuin, I. Kovesdi, F. Bauters, P. Fenaux, and B. Quesnel, Transduction of bone marrow cells by the AdZ.F(pK7) modified adenovirus demonstrates preferential gene transfer in myeloma cells, *Hum. Gene. Ther. 10*:2709 (1999).

37. Y. Yoshida, A. Sadata, W. Zhang, K. Saito, N. Shinoura, and H. Hamada, Generation of fiber-mutant recombinant adenoviruses for gene therapy of malignant glioma, *Hum. Gene Ther. 9*:2503 (1998).

38. M. J. Staba, T. J. Wickham, I. Kovesdi, and D. E. Hallahan, Modifications of the fiber in adenovirus vectors increase tropism for malignant glioma models, *Cancer Gene Ther. 7*:13 (2000).

39. G. A. McDonald, G. Zhu, Y. Li, I. Kovesdi, T. J. Wickham, and V. P. Sukhatme, Efficient adenoviral gene transfer to kidney cortical vasculature utilizing a fiber modified vector, *J. Gene Med. 1*:103 (1999).

40. I. Dmitriev, V. Krasnykh, C. R. Miller, M. Wang, E. Kashentseva, G. Mikheeva, N. Belousova, and D. T. Curiel, An adenovirus vector with genetically modified fibers demonstrates expanded tropism via utilization of a coxsackievirus and adenovirus receptor-independent cell entry mechanism, *J. Virol. 72*:9706 (1998).

41. K. Kasono, J. L. Blackwell, J. T. Douglas, I. Dmitriev, T. V. Strong, P. Reynolds, D. A. Kropf, W. R. Carroll, G. E. Peters, R. P. Bucy, D. T. Curiel, and V. Krasnykh, Selective gene delivery to head and neck cancer cells via an integrin targeted adenoviral vector, *Clin. Cancer Res. 5*:2571 (1999).

42. I. Kirby, E. Davison, A. J. Beavil, C. P. Soh, T. J. Wickham, P. W. Roelvink, I. Kovesdi, B. J. Sutton, and G. Santis, Identification of contact residues and definition of the CAR-binding site of adenovirus type 5 fiber protein, *J. Virol. 74*:2804 (2000).

43. D. A. Einfeld, D. E. Brough, P. W. Roelvink, I. Kovesdi, and T. J. Wickham, Construction of a pseudoreceptor that mediates transduction by adenoviruses expressing a ligand in fiber or penton base, *J. Virol. 73*:9130 (1999).

44. J. T. Douglas, C. R. Miller, M. Kim, I. Dmitriev, G. Mikheeva, V. Krasnykh, and D. T. Curiel, A system for the propagation of adenoviral vectors with genetically modified receptor specificities, *Nat. Biotech. 17*:470 (1999).

45. J. S. Bartlett, J. Kleinschmidt, R. C. Boucher, and R. J. Samulski, Targeted adeno-associated virus vector transduction of nonpermissive cells mediated by a bispecific F(ab/gamma)2 antibody, *Nat. Biotechnol. 17*:181 (1999).

46. A. Girod, M. Ried, C. Wobus, H. Lahm, K. Leike, J. Kleinschmidt, G. Deleage, and M. Hallek, Genetic capsid modifications allow efficient re-targeting of adeno-associated virus type 2, *Nat. Med. 5*:1052 (1999).

47. Q. Yang, M. Mamounas, G. Yu, S. Kennedy, B. Leaker, J. Merson, S. F. Wong, M. Yu, and J. R. Barber, Development of novel cell surface CD34-targeted recombinant adenoassociated virus vectors for gene therapy, *Hum. Gene Ther. 9*:1929 (1998).

48. N. Kasahara, A. M. Dozy, and Y. W. Kan, Tissue-specific targeting of retroviral vectors through ligand-receptor interactions, *Science 266*:1373 (1994).

49. K. W. Peng, R. Vile, F. L. Cosset, and S. Russell, Selective transduction of protease-rich tumors by matrix-metalloproteinase-targeted retroviral vectors, *Gene Ther. 6*:1552 (1999).

50. K. W. Peng, Strategies for targeting therapeutic gene delivery, *Mol. Med. Today 5*:448 (1999).

51. A. C. Bell, A. G. West, and G. Felsenfeld, The protein CTCF is required for the enhancer blocking activity of vertebrate insulators, *Cell 98*:387 (1999).

52. J. H. Chung, A. C. Bell, and G. Felsenfeld, Characterization of the chicken beta-globin insulator, *Proc. Natl. Acad. Sci. USA 94*:575 (1997).

53. D. S. Steinwaerder, and A. Lieber, Insulation from viral transcriptional regulatory elements improves inducible transgene expression from adenovirus vectors in vitro and in vivo, *Gene Ther. 7*:556 (2000).

54. G. Vassaux, H. C. Hurst, and N. R. Lemoine, Insulation of a coditionally expressed transgene in an adenoviral vector, *Gene Ther. 6*:1192 (1999).

55. R. T. Jaggar, H. Y. Chan, A. L. Harris, and R. Bicknell, Endothelial cell-specific expression of tumor necrosis factor-alpha from the KDR or E-selectin promoters following retroviral delivery, *Hum. Gene Ther. 8*:2239 (1997).

56. T. Walton, J. L. Wang, A. Ribas, S. H. Barsky, J. Economou, and M. Nguyen, Endothelium-specific expression of an E-selectin promoter recombinant adenoviral vector, *Anticancer Res. 18*:1357 (1998).

57. W. Graulich, D. M. Nettelbeck, D. Fischer, T. Kissel, and R. Müller, Cell type specificity of the human endoglin promoter, *Gene 227*:55 (1999).

58. B. E. Huber, C. A. Richards, and T. A. Krenitsky, Retroviral-mediated gene therapy for the treatment of hepatocellular carcinoma: an innovative approach for cancer therapy, *Proc. Natl. Acad. Sci. USA 88*:8039 (1991).

59. S. Kaneko, P. Hallenbeck, T. Kotani, H. Nakabayashi, G. McGarrity, T. Tamaoki, W. F. Anderson, and Y. L. Chiang, Adenovirus-mediated gene therapy of hepatocellular carcinoma using cancer-specific gene expression, *Cancer Res. 55*:5283 (1995).

60. A. Ido, K. Nakata, Y. Kato, K. Nakao, K. Murata, M. Fujita, N. Ishii, T. Tamaoki, H. Shiku, and S. Nagataki, Gene therapy for hepatoma cells using a retrovirus vector carrying herpes simplex virus thymidine kinase gene under the control of human alpha-fetoprotein gene promoter, *Cancer Res. 55*:3105 (1995).

61. P. B. Arbuthnot, M. P. Bralet, J. C. Le, J. F. Dedieu, M. Perricaudet, C. Brechot, and N. Ferry, In vitro and in vivo hepatoma cell-specific expression of a gene transferred with an adenoviral vector, *Hum. Gene Ther. 7*:1503 (1996).

62. F. Kanai, K. H. Lan, Y. Shiratori, T. Tanaka, M. Ohashi, T. Okudaira, Y. Yoshida, H. Wakimoto, H. Hamada, H. Nakabayashi, T. Tamaoki, and M. Omata, In vivo gene therapy for alpha-fetoprotein-producing hepatocellular carcinoma by adenovirus-mediated transfer of cytosine deaminase gene, *Cancer Res. 57*:461 (1997).

63. T. Osaki, Y. Tanio, I. Tachibana, S. Hosoe, T. Kumagai, I. Kawase, S. Oikawa, and T. Kishimoto, Gene therapy for carcinoembryonic antigen-producing human lung cancer cells by cell type-specific expression of herpes simplex virus thymidine kinase gene, *Cancer Res. 54*:5258 (1994).

64. T. Tanaka, F. Kanai, S. Okabe, Y. Yoshida, H. Wakimoto, H. Hamada, Y. Shiratori, K. Lan, M. Ishitobi, and M. Omata, Adenovirus-mediated prodrug gene therapy for carcinoembryonic antigen-producing human gastric carcinoma cells in vitro, *Cancer Res. 56*:1341 (1996).

65. K. H. Lan, F. Kanai, Y. Shiratori, M. Ohashi, T. Tanaka, T. Okudaira, Y. Yoshida, H. Hamada, and M. Omata, In vivo selective gene expression and therapy mediated by adenoviral vectors for human carcinoembryonic antigen-producing gastric carcinoma, *Cancer Res. 57*:4279 (1997).

66. T. Tanaka, F. Kanai, K. H. Lan, M. Ohashi, Y. Shiratori, Y. Yoshida, H. Hamada, and M. Omata, Adenovirus-mediated gene therapy of gastric carcinoma using cancer-specific gene expression in vivo, *Biochem. Biophys. Res. Commun. 231*:775 (1997).

67. K. Brand, P. Loser, W. Arnold, T. Bartels, and M. Strauss, Tumor cell-specific transgene expression prevents liver toxicity of the adeno-HSVtk/GCV approach, *Gene Ther. 5*:1363 (1998).

68. G. Cao, S. Kuriyama, J. Gao, M. Kikukawa, L. Cui, T. Nakatani, X. Zhang, H. Tsujinoue, X. Pan, H. Fukui, and Z. Qi, Effective and safe gene therapy for colorectal carcinoma using the cytosine deaminase gene directed by the carcinoembryonic antigen promoter, *Gene Ther. 6*:83 (1999).

69. G. U. Dachs, A. V. Patterson, J. D. Firth, P. J. Ratcliffe, K. M. Townsend, I. J. Stratford, and A. L. Harris, Targeting gene expression to hypoxic tumor cells, *Nat. Med. 3*:515 (1997).

70. J. D. Harris, A. A. Gutierrez, H. C. Hurst, K. Sikora, and N. R. Lemoine, Gene therapy for cancer using tumour-specific prodrug activation, *Gene Ther. 1*:170 (1994).

71. C. J. Ring, J. D. Harris, H. C. Hurst, and N. R. Lemoine, Suicide gene expression

induced in tumour cells transduced with recombinant adenoviral, retroviral and plasmid vectors containing the ERBB2 promoter, *Gene Ther. 3*:1094 (1996).

72. M. A. Stackhouse, D. J. Buchsbaum, S. R. Kancharla, W. E. Grizzle, C. Grimes, K. Laffoon, L. C. Pederson, and D. T. Curiel, Specific membrane receptor gene expression targeted with radiolabeled peptide employing the erbB-2 and DF3 promoter elements in adenoviral vectors, *Cancer Gene Ther. 6*:209 (1999).

73. L. Chen, D. Chen, Y. Manome, Y. Dong, H. A. Fine, and D. W. Kufe, Breast cancer selective gene expression and therapy mediated by recombinant adenoviruses containing the DF3/MUC1 promoter, *J. Clin. Invest. 96*:2775 (1995).

74. Y. T. Tai, T. Strobel, D. Kufe, and S. A. Cannistra, In vivo cytotoxicity of ovarian cancer cells through tumor-selective expression of the BAX gene, *Cancer Res. 59*: 2121 (1999).

75. E. S. Massuda, E. J. Dunphy, R. A. Redman, J. J. Schreiber, L. E. Nauta, F. G. Barr, I. H. Maxwell, and T. P. Cripe, Regulated expression of the diphtheria toxin A chain by a tumor-specific chimeric transcription factor results in selective toxicity for alveolar rhabdomyosarcoma cells, *Proc. Natl. Acad. Sci. USA 94*:14701 (1997).

76. J. G. Judde, G. Spangler, I. Magrath, and K. Bhatia, Use of Epstein-Barr virus nuclear antigen-1 in targeted therapy of EBV-associated neoplasia, *Hum. Gene Ther. 7*:647 (1996).

77. M. I. Gutierrez, J. G. Judde, I. T. Magrath, and K. G. Bhatia, Switching viral latency to viral lysis: a novel therapeutic approach for Epstein-Barr virus-associated neoplasia, *Cancer Res. 56*:969 (1996).

78. M. Franken, A. Estabrooks, L. Cavacini, B. Sherburne, F. Wang, and D. T. Scadden, Epstein-Barr virus-driven gene therapy for EBV-related lymphomas, *Nat. Med. 2*:1379 (1996).

79. R. R. Weichselbaum, D. E. Hallahan, M. A. Beckett, H. J. Mauceri, H. Lee, V. P. Sukhatme, and D. W. Kufe, Gene therapy targeted by radiation preferentially radiosensitizes tumor cells, *Cancer Res. 54*:4266 (1994).

80. L. P. Seung, H. J. Mauceri, M. A. Beckett, D. E. Hallahan, S. Hellman, and R. R. Weichselbaum, Genetic radiotherapy overcomes tumor resistance to cytotoxic agents, *Cancer Res. 55*:5561 (1995).

81. D. E. Hallahan, H. J. Mauceri, L. P. Seung, E. J. Dunphy, J. D. Wayne, N. N. Hanna, A. Toledano, S. Hellman, D. W. Kufe, and R. R. Weichselbaum, Spatial and temporal control of gene therapy using ionizing radiation, *Nat. Med. 1*:786 (1995).

82. T. Joki, M. Nakamura, and T. Ohno, Activation of the radiosensitive EGR-1 promoter induces expression of the herpes simplex virus thymidine kinase gene and sensitivity of human glioma cells to ganciclovir, *Hum. Gene Ther. 6*:1507 (1995).

83. H. J. Mauceri, N. N. Hanna, J. D. Wayne, D. E. Hallahan, S. Hellman, and R. R. Weichselbaum, Tumor necrosis factor alpha (TNF-alpha) gene therapy targeted by ionizing radiation selectively damages tumor vasculature, *Cancer Res. 56*:4311 (1996).

84. Y. Manome, T. Kunieda, P. Y. Wen, T. Koga, D. W. Kufe, and T. Ohno, Transgene expression in malignant glioma using a replication-defective adenoviral vector containing the Egr-1 promoter: activation by ionizing radiation or uptake of radioactive iododeoxyuridine, *Hum. Gene Ther. 9*:1409 (1998).

85. Y. Kawashita, A. Ohtsuru, Y. Kaneda, Y. Nagayama, Y. Kawazoe, S. Eguchi, H. Kuroda, H. Fujioka, M. Ito, T. Kanematsu, and S. Yamashita, Regression of hepatocellular carcinoma in vitro and in vivo by radiosensitizing suicide gene therapy under the inducible and spatial control of radiation, *Hum. Gene Ther.* *10*:1509 (1999).

86. W. Walther, J. Wendt, and U. Stein, Employment of the mdr1 promoter for the chemotherapy-inducible expression of therapeutic genes in cancer gene therapy, *Gene Ther.* *4*:544 (1997).

87. C. J. Sherr, Cancer cell cycles, *Science* *274*:1672 (1996).

88. M. J. Parr, Y. Manome, T. Tanaka, P. Wen, D. W. Kufe, W. J. Kaelin, and H. A. Fine, Tumor-selective transgene expression in vivo mediated by an E2F-responsive adenoviral vector, *Nat. Med.* *3*:1145 (1997).

89. J. Zwicker, F. C. Lucibello, L. A. Wolfraim, C. Gross, M. Truss, K. Engeland, and R. Müller, Cell cycle regulation of the cyclin A, cdc25C and cdc2 genes is based on a common mechanism of transcriptional repression, *EMBO J.* *14*:4514 (1995).

90. D. M. Nettelbeck, J. Zwicker, F. C. Lucibello, C. Gross, N. Liu, S. Brüsselbach, and R. Müller, Cell cycle regulated promoters for the targeting of tumor endothelium, *Adv. Exp. Med. Biol.* *451*:437 (1998).

91. H. Ishikawa, K. Nakata, F. Mawatari, T. Ueki, S. Tsuruta, A. Ido, K. Nakao, Y. Kato, N. Ishii, and K. Eguchi, Utilization of variant-type of human α-fetoprotein promoter in gene therapy targeting hepatocellular carcinoma, *Gene Ther.* *6*:465 (1999).

92. U. Stein, W. Walther, and R. H. Shoemaker, Vincristine induction of mutant and wild-type human multidrug-resistance promoters is cell-type-specific and dose-dependent, *J. Cancer Res. Clin. Oncol.* *122*:275 (1996).

93. X. Li, E. M. Eastman, R. J. Schwartz, and A. R. Draghia, Synthetic muscle promoters: activities exceeding naturally occurring regulatory sequences, *Nat. Biotechnol.* *17*:241 (1999).

94. D. M. Nettelbeck, V. Jérôme, and R. Müller, A strategy for enhancing the transcriptional activity of weak cell type-specific promoters, *Gene Ther.* *5*:1656 (1998).

95. V. Jérôme, and R. Müller, Tissue-specific, cell cycle-regulated chimeric transcription factors for the targeting of gene expression to tumor cells, *Hum. Gene Ther.* *9*:2653 (1998).

96. N. Liu, F. C. Lucibello, K. Körner, L. A. Wolfraim, J. Zwicker, and R. Müller, CDF-1, a novel E2F-unrelated factor, interacts with cell cycle-regulated repressor elements in multiple promoters, *Nucl. Acids Res.* *25*:4915 (1997).

97. J. R. Bischoff, D. H. Kirn, A. Williams, C. Heise, S. Horn, M. Muna, L. Ng, J. A. Nye, J. A. Sampson, A. Fattaey, and F. McCormick, An adenovirus mutant that replicates selectively in p53-deficient human tumor cells, *Science* *274*:373 (1996).

98. A. R. Hall, B. R. Dix, S. J. O'Carroll, and A. W. Braithwaite, p53-dependent cell death/apoptosis is required for a productive adenovirus infection [see comments], *Nat. Med.* *4*:1068 (1998).

99. F. D. Goodrum, and D. A. Ornelles, p53 status does not determine outcome of E1B 55-kilodalton mutant adenovirus lytic infection, *J. Virol.* *72*:9479 (1998).

100. T. Rothmann, A. Hengstermann, N. J. Whitaker, M. Scheffner, and H. H. zur, Repli-

cation of ONYX-015, a potential anticancer adenovirus, is independent of p53 status in tumor cells, *J. Virol. 72*: 9470 (1998).

101. J. G. Hay, N. Shapiro, H. Sauthoff, S. Heitner, W. Phupakdi, and W. N. Rom, Targeting the replication of adenoviral gene therapy vectors to lung cancer cells: the importance of the adenoviral E1b-55kD gene, *Hum. Gene Ther. 10*:579 (1999).

102. D. Kirn, T. Hermiston, and F. McCormick, ONYX-015: clinical data are encouraging, *Nat. Med. 4*:1341 (1998).

103. C. Heise, J. A. Sampson, A. Williams, F. McCormick, H. D. Von, and D. H. Kirn, ONYX-015, an E1B gene-attenuated adenovirus, causes tumor-specific cytolysis and antitumoral efficacy that can be augmented by standard chemotherapeutic agents [see comments], *Nat. Med. 3*:639 (1997).

104. C. C. Heise, A. M. Williams, S. Xue, M. Propst, and D. H. Kirn, Intravenous administration of ONYX-015, a selectively replicating adenovirus, induces antitumoral efficacy, *Cancer Res. 59*:2623 (1999).

105. C. M. Vollmer, A. Ribas, L. H. Butterfield, V. B. Dissette, K. J. Andrews, F. C. Eilber, L. D. Montejo, A. Y. Chen, B. Hu, J. A. Glaspy, W. H. McBride, and J. S. Economou, p53 selective and nonselective replication of an E1B-deleted adenovirus in hepatocellular carcinoma, *Cancer Res. 59*:4369 (1999).

106. I. Ganly, D. Kirn, S. G. Eckhardt, G. I. Rodriguez, D. S. Soutar, R. Otto, A. G. Robertson, O. Park, M. L. Gulley, C. Heise, H. D. Von, and S. B. Kaye, A phase I study of Onyx-015, an E1B attenuated adenovirus, administered intratumorally to patients with recurrent head and neck cancer, *Clin. Cancer Res. 6*:798 (2000).

107. J. Fueyo, M. C. Gomez, R. Alemany, P. S. Lee, T. J. McDonnell, P. Mitlianga, Y. X. Shi, V. A. Levin, W. K. Yung, and A. P. Kyritsis, A mutant oncolytic adenovirus targeting the Rb pathway produces anti-glioma effect in vivo, *Oncogene 19*: 2 (2000).

108. S. O. Freytag, K. R. Rogulski, D. L. Paielli, J. D. Gilbert, and J. H. Kim, A novel three-pronged approach to kill cancer cells selectively: concomitant viral, double suicide gene, and radiotherapy, *Hum. Gene Ther. 9*:1323 (1998).

109. O. Wildner, R. M. Blaese, and J. C. Morris, Therapy of colon cancer with oncolytic adenovirus is enhanced by the addition of herpes simplex virus-thymidine kinase, *Cancer Res. 59*:410 (1999).

110. O. Wildner, J. C. Morris, N. N. Vahanian, H. J. Ford, W. J. Ramsey, and R. M. Blaese, Adenoviral vectors capable of replication improve the efficacy of HSVtk/GCV suicide gene therapy of cancer, *Gene Ther. 6*:57 (1999).

111. K. R. Rogulski, S. O. Freytag, K. Zhang, J. D. Gilbert, D. L. Paielli, J. H. Kim, C. C. Heise, and D. H. Kirn, In vivo antitumor activity of ONYX-015 is influenced by p53 status and is augmented by radiotherapy [in process citation], *Cancer Res. 60*:1193 (2000).

112. K. R. Rogulski, M. S. Wing, D. L. Paielli, J. D. Gilbert, J. H. Kim, and S. O. Freytag, Double suicide gene therapy augments the antitumor activity of a replication-competent lytic adenovirus through enhanced cytotoxicity and radiosensitization, *Hum. Gene Ther. 11*:67 (2000).

113. M. Aghi, T. C. Chou, K. Suling, X. O. Breakefield, and E. A. Chiocca, Multimodal cancer treatment mediated by a replicating oncolytic virus that delivers the

oxazaphosphorine/rat cytochrome P450 2B1 and ganciclovir/herpes simplex virus thymidine kinase gene therapies, *Cancer Res. 59*:3861 (1999).

114. O. Wildner, R. M. Blaese, and J. C. Morris, Synergy between the herpes simplex virus tk/ganciclovir prodrug suicide system and the topoisomerase I inhibitor topotecan, *Hum. Gene Ther. 10*:2679 (1999).

115. R. Rodriguez, E. R. Schuur, H. Y. Lim, G. A. Henderson, J. W. Simons, and D. R. Henderson, Prostate attenuated replication competent adenovirus (ARCA) CN706: a selective cytotoxic for prostate-specific antigen-positive prostate cancer cells, *Cancer Res. 57*:2559 (1997).

116. D. C. Yu, G. T. Sakamoto, and D. R. Henderson, Identification of the transcriptional regulatory sequences of human kallikrein 2 and their use in the construction of calydon virus 764, an attenuated replication competent adenovirus for prostate cancer therapy, *Cancer Res. 59*:1498 (1999).

117. D. C. Yu, Y. Chen, M. Seng, J. Dilley, and D. R. Henderson, The addition of adenovirus type 5 region E3 enables calydon virus 787 to eliminate distant prostate tumor xenografts, *Cancer Res. 59*:4200 (1999).

118. R. Alemany, S. Lai, Y. C. Lou, H. Y. Jan, X. Fang, and W. W. Zhang, Complementary adenoviral vectors for oncolysis, *Cancer Gene Ther. 6*:21 (1999).

119. P. L. Hallenbeck, Y. N. Chang, C. Hay, D. Golightly, D. Stewart, J. Lin, S. Phipps, and Y. L. Chiang, A novel tumor-specific replication-restricted adenoviral vector for gene therapy of hepatocellular carcinoma, *Hum. Gene Ther. 10*:1721 (1999).

120. S. Miyatake, A. Iyer, R. L. Martuza, and S. D. Rabkin, Transcriptional targeting of herpes simplex virus for cell-specific replication, *J. Virol. 71*:5124 (1997).

121. S. I. Miyatake, S. Tani, F. Feigenbaum, P. Sundaresan, H. Toda, O. Narumi, H. Kikuchi, N. Hashimoto, M. Hangai, R. L. Martuza, and S. D. Rabkin, Hepatoma-specific antitumor activity of an albumin enhancer/promoter regulated herpes simplex virus in vivo, *Gene Ther. 6*:564 (1999).

122. R. Y. Chung, Y. Saeki, and E. A. Chiocca, B-myb promoter retargeting of herpes simplex virus gamma34.5 gene-mediated virulence toward tumor and cycling cells, *J. Virol. 73*:7556 (1999).

123. R. G. Vile, and I. R. Hart, In vitro and in vivo targeting of gene expression to melanoma cells, *Cancer Res. 53*:962 (1993).

124. R. G. Vile, and I. R. Hart, Use of tissue-specific expression of the herpes simplex virus thymidine kinase gene to inhibit growth of established murine melanomas following direct intratumoral injection of DNA, *Cancer Res. 53*:3860 (1993).

125. R. G. Vile, J. A. Nelson, S. Castleden, H. Chong, and I. R. Hart, Systemic gene therapy of murine melanoma using tissue specific expression of the HSVtk gene involves an immune component, *Cancer Res. 54*:6228 (1994).

126. B. W. Hughes, A. H. Wells, Z. Bebok, V. K. Gadi, R. J. Garver, W. B. Parker, and E. J. Sorscher, Bystander killing of melanoma cells using the human tyrosinase promoter to express the *Escherichia coli* purine nucleoside phosphorylase gene, *Cancer Res. 55*:3339 (1995).

127. R. M. Diaz, T. Eisen, I. R. Hart, and R. G. Vile, Exchange of viral promoter/enhancer elements with heterologous regulatory sequences generates targeted hybrid long terminal repeat vectors for gene therapy of melanoma, *J. Virol. 72*:789 (1998).

128.  R. G. Vile, R. M. Diaz, N. Miller, S. Mitchell, A. Tuszyanski, and S. J. Russell, Tissue-specific gene expression from Mo-MLV retroviral vectors with hybrid LTRs containing the murine tyrosinase enhancer/promoter, *Virology 214*:307 (1995).

129.  S. Pang, S. Taneja, K. Dardashti, P. Cohan, R. Kaboo, M. Sokoloff, C. L. Tso, J. B. Dekernion, and A. S. Belldegrun, Prostate tissue specificity of the prostate-specific antigen promoter isolated from a patient with prostate cancer, *Hum. Gene Ther. 6*: 1417 (1995).

130.  C. H. Lee, M. Liu, K. L. Sie, and M. S. Lee, Prostate-specific antigen promoter driven gene therapy targeting DNA polymerase-alpha and topoisomerase II alpha in prostate cancer, *Anticancer Res. 16*:1805 (1996).

131.  W. R. Martiniello, A. J. Garcia, M. M. Daja, P. Russell, G. W. Both, P. L. Molloy, L. J. Lockett, and P. J. Russell, In vivo gene therapy for prostate cancer: preclinical evaluation of two different enzyme-directed prodrug therapy systems delivered by identical adenovirus vectors, *Hum. Gene Ther. 9*:1617 (1998).

132.  Y. Zhang, R. Stein, W. Wang, M. Steiner, and Y. Lu, Comparison of prostate specific adenoviral vectors for prostate cancer gene therapy, *Tumor Targeting 4*:158 (1999).

133.  M. S. Steiner, Y. Zhang, J. Carraher, and Y. Lu, In vivo expression of prostate-specific adenoviral vectors in a canine model, *Cancer Gene Ther. 6*:456 (1999).

134.  D. G. Hafenrichter, X. Wu, S. D. Rettinger, S. C. Kennedy, M. W. Flye, and K. P. Ponder, Quantitative evaluation of liver-specific promoters from retroviral vectors after in vivo transduction of hepatocytes, *Blood 84*:3394 (1994).

135.  S. Connelly, J. M. Gardner, A. McClelland, and M. Kaleko, High-level tissue-specific expression of functional human factor VIII in mice, *Hum. Gene Ther. 7*: 183 (1996).

136.  G. Ferrari, G. Salvatori, C. Rossi, G. Cossu, and F. Mavilio, A retroviral vector containing a muscle-specific enhancer drives gene expression only in differentiated muscle fibers, *Hum. Gene Ther. 6*:733 (1995).

137.  N. Larochelle, H. Lochmuller, J. Zhao, A. Jani, P. Hallauer, K. E. Hastings, B. Massie, S. Prescott, B. J. Petrof, G. Karpati, and J. Nalbantoglu, Efficient muscle-specific transgene expression after adenovirus-mediated gene transfer in mice using a 1.35 kb muscle creatine kinase promoter/enhancer, *Gene Ther. 4*:465 (1997).

138.  A. Fassati, A. Bardoni, M. Sironi, D. J. Wells, N. Bresolin, G. Scarlato, M. Hatanaka, S. Yamaoka, and G. Dickson, Insertion of two independent enhancers in the long terminal repeat of a self-inactivating vector results in high-titer retroviral vectors with tissue-specific expression, *Hum. Gene Ther. 9*:2459 (1998).

139.  Y. Miyao, K. Shimizu, S. Moriuchi, M. Yamada, K. Nakahira, K. Nakajima, J. Nakao, S. Kuriyama, T. Tsujii, K. Mikoshiba, et al., Selective expression of foreign genes in glioma cells: use of the mouse myelin basic protein gene promoter to direct toxic gene expression, *J. Neurosci. Res. 36*:472 (1993).

140.  M. Hashimoto, J. Aruga, Y. Hosoya, Y. Kanegae, I. Saito, and K. Mikoshiba, A neural cell-type-specific expression system using recombinant adenovirus vectors, *Hum. Gene Ther. 7*:149 (1996).

141.  E. A. McKie, D. I. Graham, and S. M. Brown, Selective astrocytic transgene expression in vitro and in vivo from the GFAP promoter in a HSV RL1 null mutant vector–potential glioblastoma targeting, *Gene Ther. 5*:440 (1998).

142. D. Vandier, O. Rixe, M. Brenner, A. Gouyette, and F. Besnard, Selective killing of glioma cell lines using an astrocyte-specific expression of the herpes simplex virus-thymidine kinase gene, *Cancer Res. 58*:4577 (1998).

143. A. E. Morelli, A. T. Larregina, A. J. Smith, R. A. Dewey, T. D. Southgate, B. Ambar, A. Fontana, M. G. Castro, and P. R. Lowenstein, Neuronal and glial cell type-specific promoters within adenovirus recombinants restrict the expression of the apoptosis-inducing molecule Fas ligand to predetermined brain cell types, and abolish peripheral liver toxicity, *J. Gen. Virol. 80*:571 (1999).

144. A. L. Peel, S. Zolotukhin, G. W. Schrimsher, N. Muzyczka, and P. J. Reier, Efficient transduction of green fluorescent protein in spinal cord neurons using adeno-associated virus vectors containing cell type-specific promoters, *Gene Ther. 4*:16 (1997).

145. G. Cao, S. Kuriyama, J. Gao, A. Mitoro, L. Cui, S. Nagao, X. Zhang, H. Tsujinoue, X. Pan, H. Fukui, and Z. Qi, In vivo gene transfer of a suicide gene under the transcriptional control of the carcinoembryonic antigen promoter results in bone marrow transduction but can avoid bone marrow suppression, *Int. J. Oncol. 15*: 107 (1999).

146. L. M. Anderson, S. Swaminothan, I. Zackon, A.-K. Tajuddin, B. Thimmapaya, and S. A. Weitzmann, Adenovirus-mediated tissue-targeted expression of the HSVtk gene for the treatment of bread cancer, *Gene Ther. 6*:854 (1999).

147. S. C. Ko, J. Cheon, C. Kao, A. Gotoh, T. Shirakawa, R. A. Sikes, G. Karsenty, and L. W. Chung, Osteocalcin promoter-based toxic gene therapy for the treatment of osteosarcoma in experimental models, *Cancer Res. 56*:4614 (1996).

148. T. Shirakawa, S. C. Ko, T. A. Gardner, J. Cheon, T. Miyamoto, A. Gotoh, L. W. Chung, and C. Kao, In vivo suppression of osteosarcoma pulmonary metastasis with intravenous osteocalcin promoter-based toxic gene therapy, *Cancer Gene Ther. 5*: 274 (1998).

149. T. A. Gardner, S. C. Ko, C. Kao, T. Shirakawa, J. Cheon, A. Gotoh, T. T. Wu, R. A. Sikes, H. E. Zhau, Q. Cui, G. Balian, and L. W. K. Chung, Exploiting stromal-epithelial interaction for model development and new strategies of gene therapy for prostate cancer and osteosarcoma matastases, *Gene Ther. Mol. Biol. 2*:41 (1998).

150. R. J. Garver, K. T. Goldsmith, B. Rodu, P. C. Hu, E. J. Sorscher, and D. T. Curiel, Strategy for achieving selective killing of carcinomas, *Gene Ther. 1*:46 (1994).

151. M. W. R. Robertson, M. Wang, G. P. Siegal, M. Rosenfeld, R. S. N. Ashford, R. D. Alvarez, R. I. Garver, and D. T. Curiel, Use of a tissue-specific promoter for targeted expression of the herpes simplex virus thymidine kinase gene in cervical carcinoma cells, *Cancer Gene Ther. 5*:331 (1998).

152. G. Gazit, S. E. Kane, P. Nichols, and A. S. Lee, Use of the stress-inducible grp78/BiP promoter in targeting high level gene expression in fibrosarcoma in vivo, *Cancer Res 55*:1660 (1995).

153. G. Gazit, G. Hung, X. Chen, W. F. Anderson, and A. S. Lee, Use of the glucose starvation-inducible glucose-regulated protein 78 promoter in suicide gene therapy of murine fibrosarcoma, *Cancer Res. 59*:3100 (1999).

154. I. Chung, P. E. Schwartz, R. G. Crystal, G. Pizzorno, J. Leavitt, and A. B. Deisseroth, Use of L-plastin promoter to develop an adenoviral system that confers

transgene expression in ovarian cancer cells but not in normal mesothelial cells, *Cancer Gene Ther.* 6:99 (1999).

155.  M. M. Katabi, H. L. Chan, S. E. Karp, and G. Batist, Hexokinase type II: a novel tumor-specific promoter for gene-targeted therapy differentially expressed and regulated in human cancer cells, *Hum. Gene Ther.* 10:155 (1999).

156.  M. J. Staba, H. J. Mauceri, D. W. Kufe, D. E. Hallahan, and R. R. Weichselbaum, Adenoviral TNF-alpha gene therapy and radiation damage tumor vasculature in a human malignant glioma xenograft, *Gene Ther.* 5:293 (1998).

157.  R. V. Blackburn, S. S. Galoforo, P. M. Corry, and Y. J. Lee, Adenoviral-mediated transfer of a heat-inducible double suicide gene into prostate carcinoma cells, *Cancer Res.* 58:1358 (1998).

158.  D. M. Nettelbeck, V. Jérôme, and R. Müller, A dual-specificity promoter system combining cell-cycle regulated and tissue-specific transcriptional control. *Gene Ther.* 6:1276 (1999).

# Index